Internet of Things from Hype to Reality

Ammar Rayes • Samer Salam

Internet of Things from Hype to Reality

The Road to Digitization

Third Edition

 Springer

Ammar Rayes
Cisco
San Jose, CA, USA

Samer Salam
Cisco
Beirut, Lebanon

ISBN 978-3-030-90160-8   ISBN 978-3-030-90158-5   (eBook)
https://doi.org/10.1007/978-3-030-90158-5

This Springer imprint is published by the registered company Springer Nature Switzerland AG
The registered company address is: Gewerbestrasse 11, 6330 Cham, Switzerland

To invent, you need a good imagination and a pile of junk.

—Thomas A. Edison

To invent, you need the Internet, communication, good imagination and a pile of things.

—Ammar Rayes

Creativity is just connecting things. When you ask creative people how they did something, they feel a little guilty because they didn't really do it, they just saw something. It seemed obvious to them after a while. That's because they were able to connect experiences they've had and synthesize new things.

—Steve Jobs

How the Internet of Things will bend and mold the IP hourglass in the decades to come will certainly be fascinating to witness. We, as engineers, developers, researchers, business leaders, consumers and human beings are in the vortex of this transformation.

—Samer Salam

Foreword I

In California, just a few months after two people stepped foot on the Moon for the first time, two computers began sending messages to each other using protocols designed to make it easy for other computers to connect and join the party [2]. On October 29, 1969, a computer in Leonard Kleinrock's lab at UCLA and a computer in Doug Engelbart's lab at SRI forged the first two nodes in what would become known as the Internet. Vint Cerf and two colleagues coined the term Internet as a shortened version of internetworking in December 1974. It did not take long for more computers and their peripherals, as well as more networks of computers, and even industrial equipment to connect and begin communicating messages, including sharing sensor data and remote control instructions. In early 1982, a soda machine at CMU became arguably the first Internet-connected appliance, announced by a broadly distributed email that shared its instrumented and interconnected story with the world. By 1991, it was clear to Mark Weiser that more and more things would someday have embedded computers, including mobile phones, cars, even door knobs, and someday even clothing [3]. Today, spacecraft are Internet-connected devices on missions exploring other planets and heading to deep space beyond our solar system. Courtesy of NASA engineers, some are even sending tweets to millions of followers here on Earth about their progress.

The Internet of Things (also known as the Internet of Everything) continues to grow rapidly today. In fact, the Internet of Things (IoT) forms the basis of what has become known as the Fourth Industrial Revolution and digital transformation of business and society [1]. The first industrial revolution was the steam engine as the focal machine, the second revolution included the machines of mass production, the third revolution was based on machines with embedded computers, and the fourth revolution (today) interconnected machines and things, including information about the materials and energy usage flowing into and out of a globally interconnected cyber physical system of systems. The level of instrumentation and interconnection is laying the infrastructure for more intelligence, including cognitive computing to be incorporated.

Why does the IoT continue to grow so rapidly? What are the business and societal drivers of its rapid growth? How does IoT relate to the Internet, what types of things make up the IoT, and what are the fundamental and new protocols being used today? How are the specific layers of the IoT protocol stack related to each other? What is the fog layer? What is the Services Platform layer? How are the security and data privacy challenges being resolved? What are the economic and business consequences of IoT, and what new ecosystems are forming? What are the most important open standards associated with IoT, and how are they evolving?

In this introductory IoT textbook, Dr. Ammar Rayes and Samer Salam guide the reader through answers to the above questions. Faculty will find well-crafted questions and answers at the end of each chapter, suitable for review and in classroom discussion topics. In addition, the material in the book can be used by engineers and technical leaders looking to gain a deep technical understanding of IoT as well as by managers and business leaders looking to gain a competitive edge and understand innovation opportunities for the future. Information systems departments based in schools of management, engineering, or computer science will find the approach used in this textbook suitable as either a primary or secondary source of course material.

In closing, and on a personal note, it has been a pleasure to call Dr. Ammar Rayes a colleague and friend for nearly a decade. He has given generously of his time as founding President of the International Society of Service Innovation Professionals (ISSIP.org), a professional association dedicated to helping multidisciplinary students, faculty, practitioners, policy-makers, and others learn about service innovation methods for business and societal applications. Ammar is one of those rare technical leaders who contributes to business, academics, and professional association contexts. My thanks to Ammar and Samer for this excellent introduction to Internet of Things, as it is one more in a line of their contributions that will help inspire the next generation of innovators to learn, develop professionally, and make their own significant contributions.

IBM, San Jose, CA, USA   Jim Spohrer

References

1. J. Lee, H.A. Kao, S. Yang, Service innovation and smart analytics for industry 4.0 and big data environment. Procedia CIRP. **16**, 3–8 (2014)
2. B.M. Leiner, V.G. Cerf, D.D. Clark, R.E. Kahn, L. Kleinrock, D.C. Lynch, J. Postel, L.G. Roberts, S. Wolff, A brief history of the internet. ACM SIGCOMM Comput. Commun. Rev. **39**(5), 22–31 (2009)
3. M. Weiser, The computer for the 21st century. Scientific American. **265**(3), 94–104 (1991)

Foreword II

The Internet of Things (IoT) has been many years in the making. Indeed, the concept of using sensor devices to collect data and then transfer it to applications across a network has been around for several decades. For example, legacy programmable logic controller (PLC) systems already provide data collection and remote actuator control using specialized networking protocols and topologies. Even though these setups have limited footprints and are rather costly, they are still widely used in many industrial settings. Meanwhile, academic researchers have also studied the use of networked sensors for various applications in recent years.

However, continuing market shifts and technology trends in the past decade have dramatically altered the value proposition of interconnected sensors and actuators. Namely, the combination of low-cost hardware and high-speed networking technologies—both wired and wireless—have enabled a new generation of compact sensor devices with ubiquitous connectivity across the wider Internet. These systems are facilitating real-time data collection/sharing and providing unprecedented visibility and control of assets, personnel, operations, and processes. The further use of cloud-based computing/storage facilities is introducing even more advanced data analysis capabilities, ushering in a new era of intelligent decision-making, control, and automation. Broadly, these new paradigms are termed as the *Internet of Things* (IoT).

Indeed there is considerable excitement, perhaps even hype, associated with the IoT. However, as technological advances and business drivers start to align here, related paradigms are clearly poised at an inflection point of growth. For example, a wide range of business and mission-critical IoT systems are already being deployed in diverse market sectors, i.e., including defense, energy, transportation, civil infrastructure, healthcare, home automation/security, and agriculture. New cloud and fog computing services are also emerging to deliver actionable insights for improving business productivity and reducing cost/risk. As these new business models start to take hold, the projected IoT market opportunity is huge, widely projected to be in the trillions of dollars in the coming decade.

In light of the above, this text presents a very timely and comprehensive look at the IoT space. The writing starts by introducing some important definitions and reviewing the key market forces driving IoT technology growth. The fundamental

IoT building blocks are then presented, including networking systems and sensor technologies. Most notably, IoT-specific networking challenges and requirements are first overviewed, including device constraints, identification, performance determinism, security, and interoperability. Emerging, streamlined IoT protocol stacks are then detailed, covering topics such as layering, routing, and addressing. The main types of sensing technologies are also discussed here along with actuator control devices. Note that the initial part of this text focuses on core IoT concepts and frameworks, leaving more industry and application-specific treatments to later.

The text then addresses broader topics relating to intelligent data management and control for IoT. Namely, the distributed fog computing platform is outlined first, including market drivers, prerequisites, and enabling technologies within the context of IoT. The crucial notion of an IoT service platform is also presented, touching upon issues such as deployment, configuration, monitoring, and troubleshooting. The writing also outlines critical security and privacy concerns relating to IoT, i.e., by categorizing a range of threat scenarios and highlighting effective countermeasures and best practices.

Finally, the latter part of the text progresses into some more business-related aspects of IoT technology. This includes a critical look at emerging vertical markets and their interconnected ecosystems and partnerships, i.e., across sectors such as energy, industrial, retail, transportation, finance, healthcare, and agriculture. Sample business cases are also presented to clearly tie in industry verticals with earlier generalized IoT concepts and frameworks. Finally, the critical role and efforts of IoT standardization organizations is reviewed along with a look at some important open source initiatives.

Overall, both authors are practicing engineers in the networking industry and actively involved in research, technology development, standards, and business marketing initiatives. As a result, they bring together wide-ranging and in-depth field experience across many diverse areas, including network management, data security, intelligent services, software systems, data analytics, and machine learning, etc. They are also widely published in the research literature and have contributed many patent inventions and standardization drafts. Hence, this team is uniquely qualified to write on this subject.

In summary, this text provides a very compelling study of the IoT space and achieves a very good balance between engineering/technology focus and business context. As such, it is highly recommended for anyone interested in this rapidly expanding field and will have broad appeal to a wide cross section of readers, i.e., including engineering professionals, business analysts, university students, and professors. Moreover, each chapter comes with a comprehensive, well-defined set of questions to allow readers to test their knowledge on the subject matter (and answer guides are also available for approved instructors). As such, this writing also provides an ideal set of materials for new IoT-focused graduate courses in engineering and business.

Department of Electrical Engineering & Florida   Nasir Ghani
Center for Cybersecurity (FC2)
University of South Florida,
Tampa, FL, USA

Preface

Technology is becoming embedded in nearly everything in our lives. Just look around you and you will see how the Internet has affected many aspects of our existence. Virtually anything you desire can be ordered instantly, at a push of a button, and delivered to your door in a matter of days if not hours. We all see the impact of smart phones, smart appliances, and smart cars to cite a few.

Today, manufacturers are installing tiny sensors in effectively every device they make and utilizing the Internet and cloud computing to connect such devices to data centers capturing critical information. By connecting things with cloud technology and leveraging mobility, desired data is captured and shared at any location and any time. The data is then analyzed to provide businesses and consumers with value that was unattainable just a decade or less ago.

Up to the minute information is provided about the states and locations of services. Further, businesses use the sensors to collect mission-critical data throughout their entire business process, allowing them to gain real-time visibility into the location, motion and state of assets, people, and transactions and enabling them to make smarter decisions.

As more objects become embedded with sensors and the ability to communicate, new business models become possible across the industry. These models offer to improve business processes, reduce costs and risks, and more importantly create huge business opportunities in a way that changes the face and the pace of business. Experts agree that the Internet of Things will revolutionize businesses beyond recognition in the decades to come.

At the core of the success of the Internet, and one of its foundational principles, is the presence of a common protocol layer, the IP layer, which provides normalization of a plethora of applications (e.g., email, web, voice, video) over numerous transport media (e.g., Ethernet, Wi-Fi, cellular). Graphically, this can be rendered as an hourglass with IP in the middle: IP being the thin waist of this proverbial hourglass. This model has served well; especially since the Internet, over the past three decades, has been primarily concerned with enabling connectivity: interconnecting networks across the globe. As the Internet evolves into the Internet of Things, the focus shifts from connectivity to data. The Internet of Things is primarily about data and gaining actionable insights from that data, as discussed above. From a technology perspective, this can be achieved with the availability of networking protocols that meet the requirements and satisfy the constraints of new Internet of Things devices, and more importantly with the availability of standard interfaces and mechanisms for application services including data access, storage, analysis, and management. How does this translate to the proverbial hourglass? At the very least, a second thin waist is required which provides a common normalization layer for application services.

The road to a standards-based Internet of Things is well underway. The industry has made significant strides toward converging on the Internet Protocol as the common basis. Multiple standards have been defined or are in the process of being defined to address the requirements of interconnecting "Things" to the Internet. However, many gaps remain especially with respect to application interoperability, common programmable interfaces, and data semantics. How the Internet of Things will bend and mold the IP hourglass in the decades to come will certainly be fascinating to witness. We, as engineers, developers, researchers, business leaders, consumers, and human beings, are in the vortex of this transformation.

In this book, we choose to introduce the Internet of Things (IoT) concepts and framework in the earlier chapters and avoid painting examples that tie the concepts to a specific industry or to a certain system. In later chapters, we provide examples and use cases that tie the IoT concepts and framework presented in the earlier chapters to industry verticals.

Therefore, we concentrate on the core concepts of IoT and try to identify the major gaps that need to be addressed to take IoT from the hype stage to concrete reality. We also focus on equipping the reader with the basic knowledge needed to comprehend the vast world of IoT and to apply that knowledge in developing verticals and solutions from the ground up, rather than providing solutions to specific problems. In addition, we present detailed examples that illustrate the implementation and practical application of abstract concepts. Finally, we provide detailed business and engineering problems with answer guides at the end of each chapter.

The following provides a chapter-by-chapter breakdown of this book's material. Chapter 1 introduces the foundation of IoT and formulates a comprehensive definition. The chapter presents a framework to monitor and control things from anywhere in the world and provides business justifications on why such monitoring and control of things is important to businesses and enterprises. It then introduces the 12 factors that make IoT a present reality.

The 12 factors consist of (1) the current convergence of IT and OT; (2) the astonishing introduction of creative Internet-based businesses with emphasis on Uber, Airbnb, Square, Amazon, Tesla, and the self-driving cars; (3) mobile device explosion; (4) social network explosion; (5) analytics at the edge; (6) cloud computing and virtualization; (7) technology explosion; (8) digital convergence/transformation; (9) enhanced user interfaces; (10) fast rate of IoT technology adoption (five times more than electricity and telephony); (11) the rise of security requirements; and (12) the nonstop Moore's law. The last section of this chapter presents a detailed history of the Internet.

Chapter 2 describes the "Internet" in the "Internet of Things." It starts with a summary of the well-known Open System Interconnection (OSI) model layers. It then describes the TCP/IP model, which is the basis for the Internet. The TCP/IP protocol has two big advantages in comparison with earlier network protocols: reliability and flexibility to expand. The TCP/IP protocol was designed for the US Army addressing the reliability requirement (resist breakdowns of communication lines in times of war). The remarkable growth of Internet applications can be attributed to this reliable expandable model.

Chapter 2 then compares IP version 4 with IP version 6 by illustrating the limitations of IPv4, especially for the expected growth to ten billions of devices with IoT. IPv4 has room for about 4.3 billion addresses, whereas IPv6, with a 128-bit address space, has room for 2^{128} or 340 trillion trillion trillion addresses. Finally, detailed description of IoT network level routing is described and compared with classical routing protocols. It is mentioned that routing tables are used in routers to send and receive packets. Another key feature of TCP/IP routing is the fact that IP packets travel through an internetwork one router hop at a time, and thus the entire route is not known at the beginning of the journey. The chapter finally discusses the IoT network level routing that includes Interior and Exterior Routing Protocols.

Chapter 3 defines the "Things" in IoT and describes the key requirements for things to be able to communicate over the Internet: sensing and addressing. Sensing is essential to identify and collect key parameters for analysis and addressing is necessary to uniquely identify things over the Internet. While sensors are very crucial in collecting key information to monitor and diagnose the "Things," they typically lack the ability to control or repair such "Things" when action is required. The chapter answers the question: why spend money to sense "Things" if they cannot be controlled? It illustrates that actuators are used to address this important question in IoT. With this in mind, the key requirements for "Things" in IoT now consist of sensing, actuating, and unique identification. Finally, the chapter identifies the main sensing technologies that include physical sensors, RFID, and video tracking and discusses the advantages and disadvantages of these solutions.

Chapter 4 discusses the requirements of IoT which impact networking protocols. It first introduces the concept of constrained devices, which are expected to comprise a significant fraction of new devices being connected to the Internet with IoT. These are devices with limited compute and power capabilities; hence, they impose special design considerations on networking protocols which were traditionally built for powerful mains-connected computers. The chapter then presents the impact of IoT's massive scalability on device addressing in light of IPv4 address exhaustion, on credentials management and how it needs to move toward a low-touch lightweight model, on network control plane which scales as a function of the number of nodes in the network, and on the wireless spectrum that the billions of wireless IoT devices will contend for.

After that, the chapter goes into the requirements for determinism in network latency and jitter as mandated by real-time control applications in IoT, such as factory automation and vehicle control systems. This is followed by an overview of the security requirements brought forward by IoT. Then, the chapter turns into the requirements for application interoperability with focus on the need for standard abstractions and application programmatic interfaces (APIs) for application, device, and data management, as well as the need for semantic interoperability to ensure that all IoT entities can interpret data unambiguously.

Chapter 5 defines the IoT protocol stack and compares it to the existing Internet Protocol stack. It provides a layer-by-layer walkthrough of that stack and, for each such layer, discusses the challenges brought forward by the IoT requirements of the previous chapter, the industry progress made to address those challenges, and the remaining gaps that require future work.

Starting with the link layer, the chapter discusses the impact of constrained device characteristics, deterministic traffic characteristics, wireless access characteristics, and massive scalability on this layer. It then covers the industry response to these challenges in the following standards: IEEE 802.15.4, TCSH, IEEE 802.11ah, LoRaWAN, and Time-Sensitive Networking (TSN). Then, shifting to the Internet layer, the chapter discusses the challenges in Low Power and Lossy Networks (LLNs) and the industry work on 6LowPAN, RPL, and 6TiSCH. After that, the chapter discusses the application protocols layer, focusing on the characteristics and attributes of the protocols in this layer as they pertain to IoT and highlighting, where applicable, the requirements and challenges that IoT applications impose on these protocols. The chapter also provides a survey and comparison of a subset of the multitude of available protocols, including CoAP, MQTT, and AMQP to name a few. Finally, in the application services layer, the chapter covers the motivation and drivers for this new layer of the protocol stack as well as the work in ETSI M2M and oneM2M on defining standard application middleware services.

Chapter 6 defines fog computing, a platform for integrated compute, storage, and network services that is highly distributed and virtualized. This platform is typically located at the network edge. The chapter discusses the main drivers for fog: data deluge, rapid mobility, reliable control, and finally data management and analytics. It describes the characteristics of fog, which uniquely distinguish it from cloud computing.

The chapter then focuses on the prerequisites and enabling technologies for fog computing: virtualization technologies such as virtual machines and containers, network mobility solutions including EVPN and LISP, fog orchestration solutions to manage topology, things connectivity and provide network performance guarantees, and last but not least data management solutions that support data in motion and distributed real-time search. The chapter concludes with the various gaps that remain to be addressed in orchestration, security, and programming models.

Chapter 7 introduces the IoT Service Platform, which is considered to be the cornerstone of successful IoT solutions. It illustrates that the Service Platform is responsible for many of the most challenging and complex tasks of the solution. It automates the ability to deploy, configure, troubleshoot, secure, manage, and monitor IoT entities, ranging from sensors to applications, in terms of firmware installation, patching, debugging, and monitoring to name just a few. The Service Platform also provides the necessary functions for data management and analytics, temporary caching, permanent storage, data normalization, policy-based access control, and exposure.

Given the complexity of the Services Platform in IoT, the chapter groups the core capabilities into 11 main areas: Platform Manager, Discovery and Registration Manager, Communication (Delivery Handling) Manager, Data Management and Repository, Firmware Manager, Topology Management, Group Management, Billing and Accounting Manager, Cloud Service Integration Function/Manager, API Manager, and finally Element Manager addressing Configuration Management, Fault Management, Performance Management, and Security Management across all IoT entities.

Chapter 8 focuses on defining the key IoT security and privacy requirements. Ignoring security and privacy will not only limit the applicability of IoT but will also have serious results on the different aspects of our lives, especially given that all the physical objects in our surroundings will be connected to the network. In this chapter, the IoT security challenges and IoT security requirements are identified. A three-domain IoT architecture is considered in the analysis where we analyze the attacks targeting the cloud domain, the fog domain, and the sensing domain. The analysis describes how the different attacks at each domain work and what defensive countermeasures can be applied to prevent, detect, or mitigate those attacks.

The chapter ends by providing some future directions for IoT security and privacy that include fog domain security, collaborative defense, lightweight cryptography, lightweight network security protocols, and digital forensics.

Chapter 9 describes IoT Vertical Markets and Connected Ecosystems. It first introduces the top IoT verticals that include agriculture and farming, energy, enterprise, finance, healthcare, industrial, retail, and transportation. Such verticals include a plethora of sensors producing a wealth of new information about device status, location, behavior, usage, service configuration, and performance. The chapter then presents a new business model driven mainly by the new information and illustrates the new business benefits to the companies that manufacture, support, and service IoT products, especially in terms of customer satisfaction. It then presents the key requirements to deliver "Anything as a Service" in IoT followed by a specific use case.

Finally, Chap. 9 combines IoT verticals with the new business model and identifies opportunities for innovative partnerships. It shows the importance of ecosystem partnerships given the fact that no single vendor would be able to address all the business requirements.

Chapter 10 discusses blockchain in IoT. It briefly introduces the birth of blockchain technology and its use in Bitcoin. In addition, it describes Bitcoin as an application of blockchain and distinguishes blockchain as a key technology, one that has various use cases outside of Bitcoin. Next, it dives into how blockchains work and outlines the features of the technology; these features include consensus algorithms, cryptography, decentralization, transparency, trust, and smart contracts. The chapter then introduces how blockchain may impact notable use cases in IoT including healthcare, energy management, and supply chain management. It reviews the advantages and disadvantages of blockchain technology and highlights security considerations within blockchain and IoT.

Chapter 11 provides an overview of the IoT standardization landscape and a glimpse into the main standards defining organizations involved in IoT as well as a snapshot of the projects that they are undertaking. It highlights the ongoing convergence toward the Internet Protocol as the normalizing layer for IoT. The chapter covers the following industry organizations: IEEE, IETF, ITU, IPSO Alliance, OCF, IIC, ETSI, oneM2M, AllSeen Alliance, Thread Group, ZigBee Alliance, TIA, Z-Wave Alliance, OASIS, and LoRa Alliance. The chapter concludes with a summary of the gaps and provides a scorecard of the industry progress to date.

Chapter 12 defines open source in the computer industry and compares the development cycles of open source and closed source projects. It discusses the drivers to open source from the perspective of the consumers of open source projects as well as contributors of these projects. The chapter then goes into discussing the interplay between open source and industry standards and stresses the tighter collaboration ensuing among them.

The chapter then provides a tour of open source activities in IoT ranging from hardware and operating systems to IoT Service Platforms.

Finally, Appendix A presents a comprehensive IoT Glossary that includes the definitions of over 1200 terms using information from various sources that include key standards and latest research. Appendixes B-F presents examples of IoT Projects.

San Jose, CA, USA   Ammar Rayes
Beirut, Lebanon   Samer Salam

Acknowledgments

We realize that the completion of this book could not have been possible without the support of many people whose names go far beyond the list that we can recognize here. Their effort is sincerely appreciated, and support is gratefully recognized. With that in heart and mind, we would like to express our appreciation and acknowledgment particularly to the following.

First, we would like to express our gratitude to members of the Cisco executive team for their support. In particular, thanks to Ghaida Nouchy, Senior Director; Grace Francisco, VP of engineering; and Pradeep Kathail, Chief Technology Officer of the Enterprise Networking Business, for their support in the planning and preparation of this book.

We also would like to express our gratefulness to Dr. Jim Spohrer of IBM Research and Professor Nasir Ghani of the University of South Florida for taking the time to write comprehensive forewords. We are very grateful to Alumni Distinguished Graduate Professor and IEEE Fellow, Harry Perros of North Carolina State University, and Dr. Alex Clemn for peer-reviewing the book proposal.

Ammar is ceaselessly thankful to his wife Rana and his children Raneem, Merna, Tina, and Sami for their love and patience during the long process of writing this book.

Samer would like to thank his parents, to whom he is eternally grateful, his wife Zeina and children Kynda, Malek, and Ziyad for their love, support, and encouragement that made this book possible. Last but not least, he would like to thank Samir, especially for his help on the use cases and his sense of humor.

Disclaimer

The recommendations and opinions expressed in this book are those of the authors and contributors and do not necessarily represent those of Cisco Systems.

Contents

About the Authors

 **Ammar Rayes** is a Distinguished Engineer at Cisco Systems. His current works include DevNet, network analytics, machine learning, and security. Previously at Cisco, he led solution teams focusing on mobile wireless, network management NMS/OSS, and Metro Ethernet. Prior to join Cisco, he was the Director of Traffic Engineering at Bellcore (formally Bell Labs).Ammar has authored 4 books, over 100 publications in refereed journals and conferences on advances in software and networking-related technologies, and over 35 patents.Ammar is the Founding President and board member of the International Society of Service Innovation Professionals (www.issip.org), Adjunct Professor at SJSU, Editor in Chief of *Advances in Internet of Things* journal, and Board Member on Transactions on Industrial Networks and Intelligent Systems. He has served as Associate Editor of *ACM Transactions on Internet Technology* and *Wireless Communications and Mobile Computing* journals, Guest Editor of multiple journals and several *IEEE Communications Magazine* issues, Co-chaired the Frontiers in Service Conference and appeared as Keynote Speaker at multiple IEEE and industry conferences.At Cisco, Ammar is the Founding Chair of Cisco Services Research board and program and the Founding Chair of Cisco Services Patent council. He received Cisco Chairman's Choice Award for IoT Excellent Innovation & Execution in 2013.He received his BS and MS degrees in EE from the University of Illinois at Urbana—Champaign, IL, USA, and his PhD degree in EE from Washington University in St. Louis, MO, USA, where he received the Outstanding Graduate Student Award in Telecommunications.

 **Samer Salam** is a Distinguished Engineer at Cisco Systems. He focuses on the system and software architecture for networking products in addition to technology incubation. His work covers the areas of machine reasoning, immersive visualization, IoT data management and analytics, machine-to-machine communication, as well as next-generation Layer 2 networking solutions and protocols.Previously at Cisco, he held multiple technical leadership and software development positions working on Layer 2 VPNs, Carrier/Metro Ethernet services, OAM, network resiliency, system scalability, software quality, multiservice edge, broadband, MPLS, and dial solutions.He holds over 90 US and international patents, is author of 13 IETF RFCs, and has authored several articles in telecommunications industry journals. He is also a speaker at CiscoLive and blogs on networking technology at http://blogs.cisco.com/author/samersalam.He holds an MS degree in Computer Engineering from the University of Southern California in Los Angeles and a BE degree in Computer and Communications Engineering, with Distinction, from the American University of Beirut, where he received the Faculty of Engineering and Architecture Dean's Award for Creative Achievement.

Chapter 1
Internet of Things (IoT) Overview

The Internet of Things (IoT) has gained significant mindshare, let alone attention, in academia and the industry especially over the past few years. The reasons behind this interest are the potential capabilities that IoT promises to offer. On the personal level, it paints a picture of a future world where all the things in our ambient environment are connected to the Internet and seamlessly communicate with each other to operate intelligently. The ultimate goal is to enable objects around us to efficiently sense our surroundings, inexpensively communicate, and ultimately create a better environment for us: one where everyday objects act based on what we need and like without explicit instructions.

IoT's promise for business is more ambitious. It includes leveraging automatic sensing and prompt analysis of thousands of service or product-related parameters and then automatically taking action before a service experience or product operation is impacted. It also includes collecting and analyzing massive amounts of structured and unstructured data from various internal and external sources, such as social media, for the purpose of gaining competitive advantage by offering better services and improving business processes. This may seem like a bold statement, but consider the impact that the Internet has already had on education, communication, business, science, government, climate control, and humanity. Many believe that IoT will create the largest technology opportunity that we have ever seen.

The term "Internet of Things" was first coined by Kevin Ashton in a presentation that he made at Procter & Gamble in 1999. Linking the new idea of RFID (radio-frequency identification) in Procter & Gamble's supply chain to the then-red-hot topic of the Internet was more than just a good way to get executive attention. He has mentioned "The Internet of Things has the potential to change the world, just as the Internet did. Maybe even more so." Afterward, the MIT Auto-ID center presented their IoT vision in 2001. Later, IoT was formally introduced by the International Telecommunication Union (ITU) *Internet Report* in 2005.

IoT is gaining momentum, especially in modern wireless telecommunications, as evidenced in the increasing presence around us of smart objects or things (e.g.,

© The Author(s), under exclusive license to Springer Nature Switzerland AG 2022   1
A. Rayes, S. Salam, *Internet of Things from Hype to Reality*,
https://doi.org/10.1007/978-3-030-90158-5_1

smartphones, smart watches, smart home automation systems, etc.), which are able to communicate with each other and collaborate with other systems to achieve certain goals.

Undeniably, the main power of IoT is the high impact it is already starting to have on business and personal lives. Companies are already employing IoT to create new business models, improve business processes, and reduce costs and risks. Personal lives are improving with advanced health monitoring, enhanced learning, and improved security just to name few examples of possible applications.

1.1 What Is the Internet of Things (IoT)?

Before defining IoT, it may be worthwhile listing the most generic enablement components. In its simple form, IoT may be considered as a network of physical elements empowered by:

- *Sensors*: to collect information.
- *Identifiers*: to identify the source of data (e.g., sensors, devices).
- *Software*: to analyze data.
- *Internet connectivity*: to communicate and notify.

Putting it all together, *IoT is the network of things, with clear element identification, embedded with software intelligence, sensors, and ubiquitous connectivity to the Internet.* IoT enables things or objects to exchange information with the manufacturer, operator, and/or other connected devices utilizing the telecommunications infrastructure of the Internet. It allows physical objects to be sensed (to provide specific information) and controlled remotely across the Internet, thereby creating opportunities for more direct integration between the physical world and computer-based systems and resulting in improved efficiency, accuracy, and economic benefit. Each thing is uniquely identifiable through its embedded computing system and is able to interoperate within the existing Internet infrastructure.

There is no disagreement between businesses and/or technical analysts that the number of things in IoT will be massive. At the time of writing this book, over 20 billion devices have been already deployed. This includes networked devices, machine-to-machine devices, phones, TVs, PCs, tablets, and other connected devices. Any object with a simple microcontroller, modest on-off switch, or even with QR (Quick Response) code[1] will be connected to the Internet in the near feature. Such a view is supported by Moore's Law, with the observation that the number of transistors in a dense integrated circuit approximately doubles every 18 months, as we will illustrate in Sect. 1.3.

---

[1] Quick Response Code is the trademark for a type of matrix barcode.

The main idea of IoT is to physically connect anything/everything (e.g., sensors, devices, machines, people, animals, trees) and processes over the Internet for monitoring and/or controlling functionality. Connections are not limited to information sites, they are actual and physical connections allowing users to reach "things" and take control when needed. Hence, connecting objects together is not an objective by itself, but gathering intelligence from such objects to enrich products and services is.

1.1.1 Background and More Complete IoT Definition

Before we give historical overview of the Internet and consequently delve into the Internet of Things, it is worthwhile providing a definition and the fundamental requirements of IoT as a basis for the inexperienced reader.

We assume that the Internet is well known and bears no further definition. The question is what do we really mean by "Things"? Well, things are actually "anything" and "everything" from appliances to buildings to cars to people to animals to trees to plants, etc. Hence, IoT in its simplest form may be considered as the intersection of the Internet, things, and data as shown in Fig. 1.1.

A more complete definition, we believe, should also include "Standards" and "Processes" allowing "Things" to be connected over the "Internet" to exchange "Data" using industry "Standards" that guarantee interoperability and enabling useful and mostly automated "Processes," as shown in Fig. 1.2.

Some companies (e.g., Cisco) refer to IoT as the IoE (Internet of Everything) with four key components: people, process, data, and Things. In this case, IoE connects:

- People: Connecting people in more relevant ways.
- Data: Converting data into intelligence to make better decisions.
- Process: Delivering the right information to the right person or machine at the right time.

Fig. 1.1 IoT definition in its simplest form

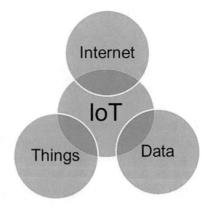

Fig. 1.2 IoT—more complete definition

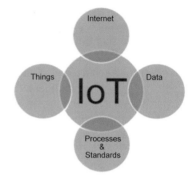

- Things: Physical devices and objects connected to the Internet and each other for intelligent decision-making, often called IoT.

They correctly believe that today's Internet is the "Internet of People," i.e., today's Internet is mainly connecting applications that are used by people. People are taking action based on notifications from connected applications. IoT is envisioned to connect "things" where "things" (not people) will be taking action, when needed, by communicating with each other intelligently. IoE is then combining the Internet of People and the Internet of Things. In this book, and in most of the recent literature, however, IoT refers to anything and everything (including people).

With this in mind, we can state a more comprehensive definition of IoT as follows: *IoT is the network of things, with device identification, embedded intelligence, and sensing and acting capabilities, connecting people and things over the Internet.*

As we already mentioned above, we will use the term "IoT" to refer to all objects/things/anything connected over the Internet including appliances, buildings, cars, people, animals, trees, plants, etc.

The basic promise of IoT is to monitor and control "things" from anywhere in the world. The first set of fundamental questions an engineer may ask are: How to monitor and control things from anywhere in the world? Why do we want to do so? Who will perform the monitoring and control? How is security guaranteed? In the remainder of this section, we will provide high-level answers to these questions. More detailed answers will be provided throughout the various chapters of this book.

1.1.2   How to Monitor and Control Things from Anywhere in the World?

Let us start with the first question. The basic requirements for IoT are the unique identity per "thing" (e.g., IP address), the ability to communicate between things (e.g., wireless communications), and the ability to sense specific information about the thing (sensors). With these three requirements, one should be able to monitor things from anywhere in the world. Another foundation requirement is a medium to

Fig. 1.3 Basic
requirements for an IoT
solution

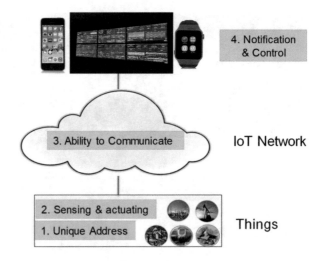

communicate. Such requirement is typically handled by a telecommunications network. Figure 1.3 presents the very basic requirements of an IoT solution.

1.1.3   Why Do We Want to Monitor and Control Things?

There are many reasons to monitor and control things remotely over the Internet: monitoring and controlling things by experts (e.g., a patient's temperature or blood pressure while the patient is at the comfort of his or her own home); learning about things by pointing a smartphone to a thing of interest, for instance; searching for things that search engines (e.g., Google) do not provide today (e.g., where are my car keys); allowing authorities to manage things in smart cities in an optimal manner (e.g., energy, driver licenses, and other documents from Department Motor Vehicle, senior citizen); and, finally, providing more affordable entertainment and games for children and adults. All of these are examples of huge business and service opportunities to boost the economic impact for consumers, businesses, governments, hospitals, and many other entities.

1.1.4   Who Will Monitor and Control?

Generally speaking, monitoring and control of IoT services may be done by any person or any machine. For example, a homeowner monitoring his own home on a mobile device based on a security system she or he has installed and configured. The homeowner may also control lights, turn on the air conditioning, shut off the heater, etc. Another example is for a service provider to monitor and control services for its customers in a network operations center (NOC) as shown in Fig. 1.4.

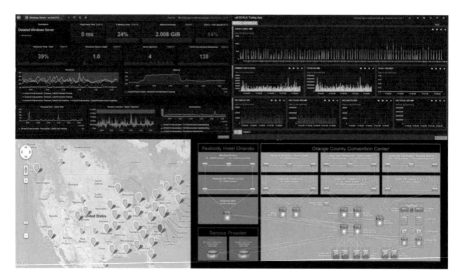

Fig. 1.4 Example of monitoring systems in a network operations center

Obviously, security is a major concern to prevent access by non-authorized people and, more importantly, prevent a malicious hacker from gaining access to the system and sending old views to the homeowner while a thief is breaking in. The areas of control are far more critical for enterprise-sensitive applications such as healthcare monitoring of patients and banking applications, as we will see in Chap. 8.

1.1.5   How Is Security Guaranteed?

Securing IoT is perhaps the biggest opportunity for technology companies and will remain so far some time in the future. Before IoT, information technology security professionals worked in a bubble as they literally owned and controlled their entire networks and secured all devices behind firewalls. With IoT, data will be collected from external, often mobile, sensors that are placed in public sites (e.g., city streets) allowing strangers to send harmful data to any network. Bring your own device (BYOD) is another example where third-party devices and hence noncorporate data sources are allowed to enter the network. IoT areas that are considered to be most vulnerable include:

- Accessing data during transport (network and transport security). Data will be transported in IoT networks at all time, for example, from sensors to gateways and from gateways to data centers in enterprises or from sensors to gateways for residential services such as video from home monitoring system to the homeowner's smartphone while he is in a coffee shop. This data may be sniffed by the man in the middle unless the transport protocols are fully secured and encrypted.

- Having control of IoT devices (control of the APIs) allows unauthorized persons to take full control of entire networks. Examples include shutting down cameras at home and shutting down patient monitoring systems.
- Having access to the IoT data itself. Is the data easily accessible? Is it stored encrypted? Shared storage in the cloud is another problem where customer may log in as customer B and look at his data. Another common problem is spoofing data via Bluetooth. Many companies are adding Bluetooth support to their devices making it more feasible for unauthorized persons to access the device's data.
- Stealing official user or network identity (stealing user or network credentials). Many websites provide default passwords for vendors. We have dedicated Chap. 8 to IoT security.

1.2 IoT Reference Framework

In this book, we will follow a reference framework that divides IoT solutions into four main levels: IoT devices (things), IoT network (infrastructure transporting the data), IoT Services Platform (software connecting the things with applications and providing overall management), and IoT applications (specialized business-based applications such as customer relation management (CRM), Accounting and Billing, and Business Intelligence (BI) applications). Control is passed down from one level to the one below, starting at the application level and proceeding to the IoT devices level and backup the hierarchy.

1. *IoT Device Level* includes all IoT sensors and actuators (i.e., the Things in IoT). The device layer will be covered in Chap. 3.
2. *IoT Network Level* includes all IoT network components including IoT gateways, routers, switches, etc. The Internet in IoT will be covered in Chap. 2.
3. *IoT Application Services Platform Level* includes the key management software functions to enable the overall management of IoT devices and network. It also includes main functions connecting the device and network levels with the application layer. It will be covered in Chap. 7.
4. *IoT Application Level* includes all applications operating in the IoT network, and this will be covered in Chap. 9.

Figure 1.5 shows an overview of the IoT levels. It describes how information is transferred from one IoT component into another. Advantages of the proposed IoT four-level model include:

- Reduced Complexity: It breaks IoT elements and communication processes into smaller and simpler components, thereby helping IoT component development, design, and troubleshooting.

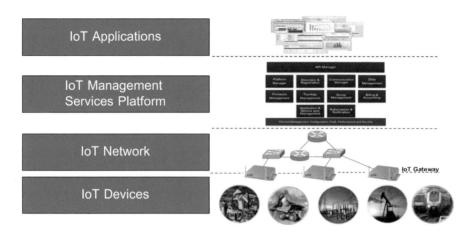

Fig. 1.5 IoT levels

- Standardized Components and Interfaces: The model standardizes the specific components within each level (e.g., what are the key components for general IoT Services Platform) as well as the interfaces between the various levels. This would allow different vendors to develop joint solutions and common support models.
- Module Engineering: It allows various types of IoT hardware and software systems to communicate with each other.
- Interoperability between vendors by ensuring the various technology building blocks can interwork and interoperate.
- Accelerate Innovation: It allows developers to focus on solving the main problem at hand without worrying about basic functions that can be implemented once across different business verticals.
- Simplified Education: It breaks down the overall complex IoT solution into smaller more manageable components to make learning easier.

1.3   Why Now? The 12 Factors for a Perfect Storm

IoT has already become a powerful force for business transformation, and its disruptive impact is already felt across all industries and all areas of society. There is a perfect storm of market disruptions happening at an unprecedented pace triggered by technology as well as new business and social requirements. This Section introduces the top 12 factors driving the explosion of IoT as shown in Fig. 1.6.

Fig. 1.6 IoT 12 driving factors

1.3.1 Convergence of IT and OT

Operation technology (OT) is the world of industrial plants and industrial control and automation equipment that include machines and systems to run the business, controllers, sensors, and actuators. Information technology (IT) is the world of end-to-end information systems focusing on compute, data storage, and networking to support business operation in some context such as business process automation systems, customer relation management (CRM) systems, supply chain management systems, logistics systems, and human resources systems.

Historically, IT and OT were always managed by two separate organizations with different cultures, philosophies, and set of technologies. IT departments were originally created by companies to create efficient and effective forms of telephony communication among various departments. Then they were extended to provide video and web conferences and network internal communications and secure external electronic communications such as emails. Often the final decision with the selection of communication systems, website hosting, and backup servers was the responsibility of the IT department.

OT relies on real-time data that drives safety, security, and control. It depends on very well-defined, tested, and trusted processes. Many plants need to run 24 × 7 with zero downtime (e.g., City Water Filtration System), and thus industrial processes cannot tolerate shutdown for software updates. IT is more lenient with software updates, introduction of new technologies, etc.

"When you take people with an IT background and bring them into an industrial control system environment, there's a lack of understanding from operations why they're there and there is a lack of understanding of the specific controls environment needs from IT," says Tim Conway, technical director, ICS and SCADA for the SANS Institute. He points out that typically IT professionals are trained and driven to perform a task: "They work on a box, a VM (virtual machine), a storage area

Fig. 1.7 The merger of IT
and OT

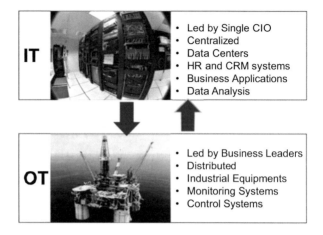

network, or a firewall. They don't realize that they're a part of a larger control system operation, and how the things that they do can impact others."

IoT is having a major impact on OT and the traditional IT operational model. With the fast introduction of business-specific technologies (e.g., Internet-based oil rig monitoring systems), IT operations can no longer scale, keep up with the fast-evolving requirements, nor provide the required expertise. Traditional IT departments simply lack the required resources to introduce IoT solutions in a timely fashion, effectively operate and monitor such solutions, or react to the massive amount of monitoring data that is generated by IoT devices (Fig. 1.7).

The bottom line is that IT is moving fast into plant floors. With the pressure of IoT technology adoption by cutting-edge businesses, OT is forced to accept a greater level of integration. Hence, traditional IT and OT functions are expected to merge or quickly risk the loss of the business to cutting-edge competitors (why? See problem 11). IT operations leaders must move closer to the business and adapt their employee skill sets, their processes, and their tools to monitor IoT availability and performance in order to support business initiatives as shown in Fig. 1.7.

1.3.2   The Astonishing Introduction of Creative Internet-Based Businesses

1.3.2.1   Uber

Many are familiar with Uber's story where the co-founders were attending a conference in Paris in 2008. Travis Kalanick and Garrett Camp were complaining about finding a cab especially while carrying luggage and under the rain. When they started to brainstorm the next day, they came up with three main requirements: the solution had to be Internet-based (i.e., request and track service from mobile device), it had to provide the service fast, and the rides had to be picked up from any location.

The key component of Uber's solution is the Internet-based platform connecting customers (passengers) with the service providers (car drivers). Because the consumers are not Uber's employees and because there is practically an infinite number of cars that could potentially join Uber, Uber has the requirement to scale at an incredibly fast rate at zero marginal cost.

Uber uses sensor technologies in driver's smartphones to track their behaviors. If you ride with Uber and your driver speeds, breaks too hard, or takes you on a wildly lengthy route to your destination, it is no longer your word against theirs. Uber is using Gyrometer and GPS data to track the behavior of their drivers. Gyrometers in smartphones measure small movements, while GPS combined with accelerometers shows how often a vehicle starts and stops and the overall speed.

The idea is to gradually improve safety and customer satisfaction, though there is no word on whether or not you might be able to actively seek out a faster driver if that is what you are after.

Today Uber is one of the leading transportation services in the world with a market value over 20 billion dollars.

1.3.2.2   Airbnb

Airbnb is an Internet-based service for people to list, find, and rent lodging. It was founded in 2008 in San Francisco, California, by Brian Check and Joe Gebbia shortly after creating AirBed and Breakfast during a conference. The original site offered rooms, breakfast, and business networking opportunity for the conference attendees who were unable to find a hotel. In February 2008, technical architect Nathan Blecharczyk joined Airbnb as the third co-founder. Shortly thereafter, the newly created company focused on high-profile events where alternative lodging was very limited.

Incredibly similar to the Uber model, Airbnb utilizes a platform business model. This means they facilitate the exchange between consumers (travelers) and service providers (homeowners). Airbnb also required a scalable Internet-based platform supporting from a few customers to hundreds of thousands during major events. More importantly, Airbnb is partnering with Internet companies (e.g., Nest of Google) to deliver remote keyless solutions to customers by unlocking doors (with IoT digital keys) over the Internet.

Just like Uber, Airbnb found a multibillion dollar business based on an Internet platform connecting people and places together that competently disrupted the traditional hotel business model. These linear businesses have to invest millions into building new hotels, while Airbnb does not have to deal with that.

Just like Uber, today Airbnb is one of the leading hotel services in the world (Fig. 1.8).

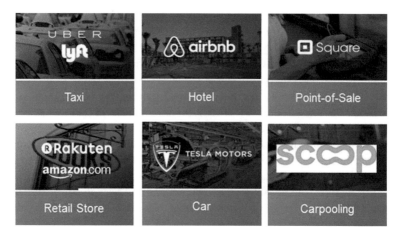

Fig. 1.8 Examples of Internet-based businesses

Fig. 1.9 Square credit
card reader. (Source:
Square Inc.)

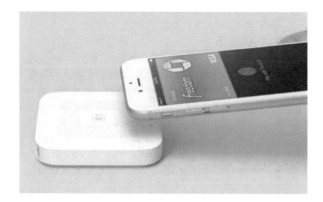

1.3.2.3 Square

Square Inc., also San Francisco based, was inspired by Jack Dorsey in 2008 when his friend, Jim McKelvey, in St. Louis at the time, was unable to complete a $2000 sale of his glass faucets and fittings because he could not accept credit cards. Jack and Jim started the point-of-sale software financial services company in 2010. The company allows small business mobile individuals and merchants to make secure payments using applications like Square Capital and Square Payroll. The Internet-based software solution allows customers and small business owners to enter credit card information manually or to swipe the card via the Square Reader (see Fig. 1.9), a small plastic device that plugs into the audio jack of supported smart mobile devices with an interface resembling a traditional cash register.

Square has introduced an application that integrates its reader with a smart-phone's motion sensor. The application can determine that the card reader is failing by analyzing the motion sensor data to detect movements indicating multiple card

Fig. 1.10   Intuit GoPayment Reader. (Source: Intuit)

swipes. If the card reader did not read any data during the card swipes, the application can deduce that the card reader is broken. This solution allows Square to send a replacement card reader to swap the broken card in a timely fashion.

Square also launched Square Cash applications allowing individuals and businesses to transfer money with a unique username. In 2015, Square introduced Customer Engagement, a suite of CRM tools which includes email marketing services. These tools allow businesses to target specific customer segments with customized promotions based on actual purchase history. Square also introduced Square Payroll tool for small business owners to process payroll for their employees.

Other financial companies have also introduced Internet-based mobile payment solutions including Intuit GoPayment Reader, which is integrated with Intuit's host of products and software (Fig. 1.10), PayPal Here Reader, and others.

Just like Uber and Airbnb, Square found a novel business based on Internet platform connecting small business owners and customers together that competently disrupted the traditional small business payment models.

1.3.2.4   Amazon

Amazon.com is the largest Internet retailer company in the worldIt started, in 1994, as an Internet-based book seller and swiftly expanded into music, movies, electronics, and household goods; Amazon utilized the Internet to break the traditional retailer model. It did not need to stock many of the merchandises it was selling on its website. Instead, it identified matching partner companies and issued customer orders over a secure Internet-based platform.

Amazon also offers businesses the capability to sell online via Amazon Services. Another part of its retail strategy is to serve as the channel for other retailers to sell their products and take a percentage of every purchase.

Retail is only part of Amazon.com business. It also offers cloud-based services known as Amazon Web Services or AWS with Software as a Services (SaaS), Platform as a Services (PaaS), and Infrastructure as a Services (IaaS) as well as other types of businesses. Amazon is perhaps one of the first companies to develop a set of businesses based on an Internet platform connecting end customers (e.g., retail customer, businesses) to products and services (e.g., merchandise, cloud services) thereby disrupting traditional retail models.

1.3.2.5 Tesla

Tesla Motors was founded in 2003 by a group of engineers in Silicon Valley with a mission to develop a successful luxurious electrical car and then invest the resulting profits to make a less expensive electric car. With instant torque, incredible power, and zero emissions, Tesla's products would be cars without compromise.

Tesla's engineers first designed a power train for a sports car built around an AC induction motor, patented in 1888 by Nikola Tesla, the inventor who inspired the company's name. The resulting Tesla Roadster was launched in 2008 with an incredible range of 245 miles per charge of its lithium ion battery. The Roadster was able to set a new standard for electric mobility. In 2012, Tesla launched Model S, the world's first premium electric sedan.

Tesla is considered as the best example yet of IoT. It did not only bend the traditional industry manufacturing model to Internet-based model with thousands of sensors (Fig. 1.11), but it also demonstrated the tremendous value of IoT with the 2014 recalls. In early 2014, Traffic Safety Administration published two recall announcements, one for Tesla Motors and one for GM. Both were related to problems that

Fig. 1.11 Tesla Factory in Fremont, California. (Source: Tesla Motors Inc.)

could cause fires. Tesla's fix was conducted for 29,222 cars as an "over-the-air" software update without requiring owners to bring their cars to the dealer.

1.3.2.6   Self-Driving Cars

Self-driving cars are no longer a fantasy. There are already thousands of self-driving cars with features that allow them to brake, speed, and steer with limited or no driver interaction.

Self-driving cars can be divided into two main categories: semiautonomous and fully autonomous. A semiautonomous car performs certain self-driving tasks (e.g., fully brakes when it gets too close to an object, drives itself on the freeway), while a fully autonomous car drives itself from origin to destination without any driver interaction. Fully autonomous cars are further divided into user-operated and driverless.

Safety is considered one of the biggest advantages of self-driving cars. In general, self-driving cars are equipped with a large number of sensors including laser range finders (to measure a subject's distance and take photos that are in sharp focus), radars, and video cameras collecting information from the road. They are also equipped with actuators to control steering and braking. The collected data (from sensors, radars, and video) is promptly processed with the positional information from the car's GPS unit and the navigation system to determine its position and to build a three-dimensional model of its surroundings.

The resulting model is then processed by the car's control system to make navigation decisions. Self-driving car control systems typically use stored maps to find optimal path to destination, avoid obstacles, and send decisions to the car's actuators. IoT applies to interactions and communications between self-driving care components, between the car and roadside infrastructure, as well as among self-driving cars (Fig. 1.12).

Fig. 1.12 Google self-driving car. (Source: Google)

Finally, it is worth noting that there are various other examples of companies that have used the Internet for new and creative business models, with various levels of success, including Scoop Inc. for carpooling and Pandora in the music industry.

1.3.3 Mobile Device Explosion

There is an unprecedented explosion in the number of new things being connected to the Internet every day, where it is not just sheer volume of mobile devices and sensors, but things that normally have not been connected to the network, such as those found in manufacturing, utilities, and transportation, are all becoming networked devices. Because of the mobile explosion that has touched our home and work lives, we have already seen over 5 million mobile applications[2] developed in the past several years resulting in billions of downloads.

Mobile data traffic has grown 18-fold in the last few years. According to Cisco's Visual Networking Index, smartphone traffic grew from 1.74 exabyte per month in 2014 to more than 18 exabyte per months in 2019 as shown in Fig. 1.10 and such growth rate has even accelerated in recent years.

The increase in mobile data traffic is driven by two factors: the increase in the number of users and the data consumption per user. The average smartphone generated 4 GB of traffic per month in 2019, as shown in Fig. 1.13. This growth is fueled by IoT connecting things with people and more importantly allowing people to monitor and control things from anywhere in the world in real time.

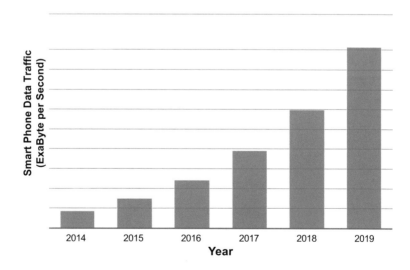

Fig. 1.13 Smartphone traffic over time

---

[2] According to Buildfire.com, the number of applications developed by Apple Store and Google Play alone were over 4.8 million in 2021.

1.3.4 Social Network Explosion

Social networks, such as Facebook, Instagram, Twitter, and YouTube, and the adoption of cloud-based services, such as Amazon's AWS and Salesforce.com, are all examples of the large-scale migration to the cloud across virtually every industry. In fact, two-thirds of all data center traffic will be from the cloud in 3 years. All of this leads to data explosion, where, already, the data being created on the Internet each day is equal to half of all the data that has been accumulated since the dawn of humanity (Fig. 1.14).

1.3.5 Analytics at the Edge

Before introducing the different versions of analytics, it is important to define the terms: big data, structured data, and unstructured data. Big data refers to the extremely large amount of data being generated and accumulated by IT systems as the result of the operation of an associated system. The latter could be a product, process, service, etc. This massive amount of data can be analyzed to identify patterns and gain insights into the operation of the associated system. The analysis often involves applying statistical techniques since human processing is not viable due to the sheer volume of the data.

Structured data refers to organized data that can fit in rows and columns. Examples of such data include customer data, sales data, and stock records. Structured data is often high value, cleansed, and indexed. Unstructured data, on the other hand, is difficult to organize or bring together. Examples of unstructured data include images, X-rays, video, social media data, and some machine outputs mixed with text.

Analytics 1.0 refers to the process of collecting structured data from various sources and sending the collected data to a *centralized* location to be correlated and analyzed using predefined queries and descriptive/historic views. Businesses and enterprises have been collecting structured data from internal systems (e.g., CRM,

Fig. 1.14 Examples of social network explosion

Sale Records, RMA Records, and Case Records), sending such data to a centralized data center to be stored in traditional tables and databases. The data is then parsed and often correlated with other types of data to produce business intelligence (e.g., offer discounts for customers in a certain location due to large unused inventory). The process of collecting, transferring, correlating, and analyzing the structured data can take hours or days.

Analytics 1.0 then evolved to Analytics 2.0 or big data and analytics with action-able insight. Analytics 2.0 basically collects structured and unstructured data from various sources but still sends the collected data to a *centralized* location to be cor-related and analyzed using complex queries along with forward-looking and pred-icative views this time. Examples of unstructured data for enterprises include call center logs, mobility data, and social media data where users are conversing and providing feedback about an enterprise's service, product, or solutions.

With the deployment of complex systems to capture and analyze big data in a data center, the overall process of collecting, transferring, correlating, and analyzing the structured and structured data is reduced to minutes or seconds.

Today, massive amounts of data are being created at the edge of the network, and the traditional ways of performing analytics over that data are no longer viable. Minutes or even seconds of delay in data processing are no longer effective for many businesses. Take, for example, a sensor in an oil rig. If the pressure was to drop substantially, the rig needs to be shut off instantaneously and before the system breaks and causes a major disaster.

Companies are realizing that they just cannot keep moving massive amounts of data to centralized data stores. The data is too big, is changing too fast, and is too geographically distributed. Certain analysis must be performed in real time and can-not withstand the delays of sending the raw data to a centralized data center to be analyzed and then send back the result to the source. In addition, certain industries (e.g., Healthcare, Defense) have the requirement to analyze the data close to the source due to data privacy or security.

Analytics 3.0 allows companies to collect, parse, analyze, and correlate (with stored data) structured as well as unstructured data at or close to the edge (the source of the data). To support this, companies have introduced massive solutions (hard-ware and software) that allow enterprises to capture, process, and analyze data at the edge. Can you think of examples of such companies (see problem 15)?

Analytics 4.0 is expected to be around application development and automated network services where businesses develop and deploy integrated application, sen-sors, networks with APIs.

Analytics 1.0, 2.0, and 3.0 are compared in Table 1.1 and in Fig. 1.15. Table 1.1 shows a comparison of key factors, while Fig. 1.15 displays a process summary.

Table 1.1 Comparison of key factors for Analytics 1.0, 2.0, and 3.0

	Analytics 1.0	Analytics 2.0	Analytics 3.0
Collected data type	Structured	Structured and unstructured	Structured and unstructured
Data analysis location	Centralized data center	Centralized data center	At edge and in data center
Time to analyze data	Days–hours	Hours–minutes	Seconds–microseconds
Data volume	Small data	Big data	Big data

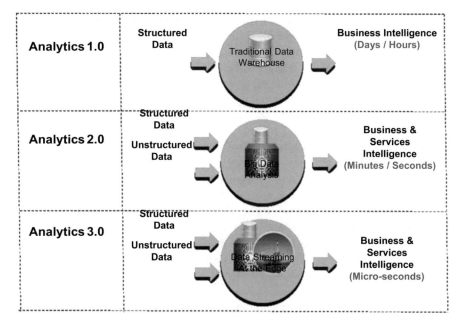

Fig. 1.15 Analytics 1.0, 2.0, and 3.0

1.3.6   Cloud Computing and Virtualization

In the past, enterprises (companies or businesses) were forced to deploy and manage their own computing infrastructures. Cloud computing, which was introduced in 2008, allows enterprises to outsource their computing infrastructure fully or partially to public cloud provides (e.g., Amazon AWS, Microsoft Azure, Google Compute Engine). Data showed that the average network computing and storage infrastructure for a start-up in year 2000 was $5 million. The cost in year 2016 has dropped to $5 thousand. This enormous 99% decline in cost was made possible by cloud computing and vitalization.

Public Cloud providers deliver cloud services, on demand, over the Internet. Enterprises pay only for the CPU cycles, storage, or bandwidth they consume.

Enterprises also have the choice to deploy *Private Cloud* solutions in their own data centers and deliver computing services to their internal sub-businesses/users. Such model offers flexibility and convenience while preserving management, control, and security to their IT departments.

Cloud computing may be also offered in a *Hybrid Cloud* model that consists of a combination of public and private clouds allowing enterprises to create a scalable solution by utilizing the public cloud infrastructure while still preserving full control over critical data.

Cloud computing is attractive to many enterprises allowing them to self-provision their own services for any type of workload on demand. They can start small and then scale up almost instantly with minimum expertise and pre-planning, while they pay only for what they use, typically, in addition to a basic subscription charge.

Cloud computing has been classified into three main service categories: Infrastructure as a Service (IaaS), Platform as a Service (PaaS), and Software as Service (SaaS). PaaS, for instance, allow enterprises to utilize a third-party platform and permit them to focus on developing and managing their own software applications without the complexity of building and maintaining the required infrastructure.

Cloud computing has been made possible by the advent of *virtualization* technologies. Rather than dedicating distinct IT infrastructure (e.g., servers, storage nodes, networking nodes) to a single business entity (e.g., customer or enterprise), virtualization allows cloud providers to divide a physical machine (e.g., server) into multiple virtual entities thereby creating an isolated virtual server, a virtual storage device, and virtual network resources for each enterprise, all running over the same shared physical IT infrastructure. Virtual machines are one form of virtualization that allows running multiple operating systems over the same physical server hardware.

Containers are another form of virtualization. In containers, the virtualization layer runs as a service on top of a common operating system kernel. The operating system's kernel runs on the hardware node with several isolated guest process groups installed on top of it. The isolated guest process groups are called containers. They share the same operating system kernel but are completely isolated at the application level.

Containers are intended to run separate applications. Examples of containers include Linux containers (LXC) and open-source Docker.

As with Analytics (Sect. 1.3.5), Cloud may be divided into Cloud 1.0 and Cloud 2.0. Cloud 1.0 is SaaS, PaaS, and IaaS. Cloud 2.0 is Cloud 1.0 with machine learning to extract business intelligence from the data using algorithms that learn from data pattern. It should be noted that traditional techniques and machine learning programs work without specific instructions on where to look for data pattern.

1.3.7 Technology Explosion

IoT hardware (e.g., sensors, inexpensive computers such as Raspberry Pi, open-source microcontrollers such as Arduino) and software technologies are not only being developed faster than ever before but with much lower prices. Such devices are already transforming user behaviors and creating new business opportunities. Business leaders are realizing that unless their organizations quickly adapt to such changes, their businesses will soon become irrelevant or inefficient to survive in an increasingly competitive marketplace.

1.3.8 Digital Convergence/Transformation

Digital convergence has initially started with a limited scope: move to "paperless" operation and save trees. Now, it is transforming the future in profound ways. Digital convergence is being adopted by key industries with extended goals to move to digital operation, extract data from various sources including the devices and processes that are enabled by digitization, and then analyze the extracted data and correlate it with other data sources to extract intelligence that improves products, customer experience, security, sales, etc. Many healthcare organizations (e.g., Kaiser Permanente) have been using digital convergence with extended goals of improving the patient experience, improving population health, and reducing healthcare costs.

With the connection of billions smart objects to the Internet, companies are realizing the upcoming challenges and are adding to their executive boards the role of a Chief Digital Officer (CDO) who can oversee the full range of digital strategies and drive change across the organization. CDOs are expected to significantly impact existing systems, solutions, and business processes and more importantly intrinsically enable new types of innovation and creativity.

1.3.9 Enhanced User Interfaces

User experience (UX) or human to machine interaction, where applicable, is very essential for the success of IoT. A core IoT UX principle is meeting the basic needs for the usage of a product or a service without aggravation or difficulty. Overengineering or including too much intelligence into products can backfire and be counterproductive. User interfaces that are frustrating to use and slow to extract relevant information can lead to customer desertion. A toaster, for example, ultimately exists to make toast. But if we overengineer with too much information, switches, and options, we risk building products that are so annoying that our customers will not want to use them.

There is now a wealth of technology and markup languages (e.g., HTML 5) that allow software engineers to adapt key UX principles and meet the so-called KISS (keep it short and simple) principle. KISS states that most systems work best if they are kept simple. Top UX principles include:

- *Simple and Easy Principle*: Best UX system is a system without UI. Simplicity should be a key goal in design, and unnecessary complexity should be avoided. Make sure you reduce the user's cognitive workload whenever possible. Make sure the UI is consistent/stable, intuitive, and establish a clear visual hierarchy.
- *Contextual Principle*: Make sure that users are contextually aware of where they are within a system.
- *Human Principle*: Make sure the UI provides human interactions above the machine-like interactions.
- *Engagement Principle*: Make sure that the UI fully engages the user, delivers value, and provides a strong information sense.
- *Beauty and Delight Principle*: Make sure the UX is enjoyable and make the user wants to use the system or service.

1.3.10 Fast Rate of IoT Technology Adoption (Five Times More than Electricity and Telephony)

Many of us are changing our mobile devices and tablets at faster rate than ever before. Experts believe that there was a point of inflexion sometime between 2009 and 2010, where the number of connected devices began outnumbering the planet's human population. And these are not just laptops, mobile phones, and tablets—they also include sensors and everyday objects that were previously unconnected. Surveys and detailed analysis indicated that the adoption rate of such technology is five times faster than that of electricity and telephony growth. Traditionally the adoption of technology was always proportional to population growth. Hence, IoT adoption gap has already widened exponentially over the last several years, with the number of sensors, objects, and other "things." This is best illustrated by global IP traffic growth, as shown in Fig. 1.16. According to June 2016 Cisco Visual Networking Index (VNI) forecast, global IP traffic in 2015 stands at 72.5 exabytes (EB, 10^{18} byte) per month and nearly tripled by 2020, to reach 194.4 EB per month. Consumer IP traffic reached 162.2 EB per month, and business IP traffic surpassed 32.2 EB per month in 2020.

Adding all these physical objects to IP networks imposes new and novel requirements on existing networking models. ITC will need to deal with those requirements in a relatively short order.

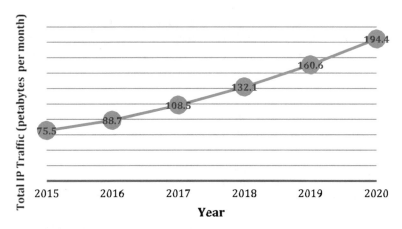

Fig. 1.16   Global IP traffic growth, 2015–2020. (Source: 2016 Cisco VIN)

1.3.11   The Rise of Security Requirements

Protection of business and personal data and systems has been an issue since the inception of data networks. With the commercialization of the Internet, security concerns expanded to cover personal privacy, financial transactions, and the threat of cyber robbery. Today, security of the network is being expanded to include safety or physical security.

Many of us are buying and deploying smart gadgets all over our homes. Examples include smart cameras that notify our smartphones during business hours when movement is detected, smart doors that open remotely, and the smart fridges that notify us when we are short of milk. Imagine now the level of control that an attacker can gain by hacking those smart gadgets if the security of those devices was to be overlooked. In fact, the damage caused by cyberattacks in the IoT era will have a direct impact on all the physical objects that you use in your daily life. The same applies to smart cars as the number of integrated sensors continues to grow rapidly and as the wireless control capabilities increase significantly over time, giving an attacker who hacks a car the ability to control the windshield wipers, the radio, the door lock, and even the brakes and the steering wheel of the vehicle. Our bodies will not also be safe from cyberattacks. In fact, researchers have shown that an attacker can control remotely implantable and wearable health devices (e.g., insulin pumps and heart pacemakers) by hacking the communication link that connects them to the control and monitoring system.

1.3.12   The Nonstop Moore's Law

It is possible to summarize Moore's Law impact with three key observations:

1. Over the history of computing hardware, computer power has been doubling approximately every 18 months. This relates to the fact that the number of transistors in a dense integrated circuit has been growing by twofold every 18 months since the transistor was invented in 1947 by John Bardeen, Walter Brattain, and William Shockley in Bell Labs, as shown in Fig. 1.17.

   Now, the largest existing networks contain millions of nodes and billions of connections. Human brains, on the other hand, are about a hundred thou sand times more powerful. A human brain has one hundred thousand billion nodes and a hundred trillion connections. Hence, with Moore's Law, a computer should be as powerful as the human brain in about 25 years!

2. Silicon transistor storage technology size has continued to shrink over the years and is approaching atomic level. For years now, we have been putting more power and more storage on the same size device. To illustrate this idea, the number of all transistors in all PCs in 1995, a peak year for Microsoft, was about

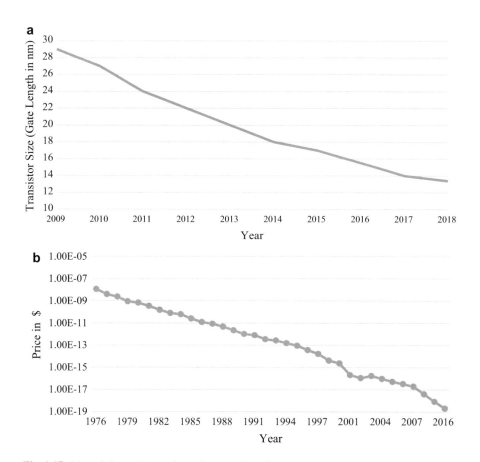

Fig. 1.17 Moore's Law: (**a**) transistor size over time, (**b**) transistor price over time

800 trillion transistors. Today, 800 trillion transistors are included in one week-end's sales of Apple's iPhones.
3. The price of the transistor is being reduced by more than 50% every year. In 1958 Fairchild Semiconductor procured its first order for 100 transistors at $150 apiece from IBM's Federal Systems Division. Today, you can buy over one million transistors for 8 cents. Figure 1.17 shows such trend over time.

There is no exact number for the estimated IoT revenue for the next 10 years, but all industry leaders have agreed that the opportunity is indeed huge.

A study by General Electric, which likened the IoT trend to the industrial revolution of the eighteenth and nineteenth centuries, concluded that the IoT over the next two decades could add as much as US $15 trillion to the global gross domestic product (GDP)—which is roughly the size of today's US economy.

As we mentioned before, Gartner says 64 billion devices will be in use in 2025. That translates to eight devices for every person of the eight billion people that are expected to be around in a few years.

Gartner also published the number of "things" connected over the Internet as shown in Table 1.2. Without automotive, the total number of IoT installed based devices was close to 21 billion in 2020. This includes 4.9 billion in 2015 and 6.4 billion connected things in use in 2016 (about 7% from 2015). These numbers are fueled by major digital shifts by the forces of mobile, cloud computing, and social media combined with IoT. Many businesses feel that they are at a competitive disadvantage unless they pursue IoT. Gartner believes consumer applications will drive the number of connected things, while enterprises will account for most of the revenue.

A separate analysis from Morgan Stanley believes that the number can actually be as high as 75 billion and also claims that there are unique consumer devices or equipment that could be connected to the Internet.

Regardless of which study to agree with, the bottom line is that the stakes are high, and people will be the beneficiaries of this new IoT economy. Using IoT-developed innovations, for example, we can reduce waste, protect our environment, boost farm production, get early warnings of structural weaknesses in bridges and dams, and enable remotely controlled lights, sprinkler systems, washing machines, sensors, actuators, and gadgets.

This revolution is based on the transformational role of digital technologies, in particular Internet-based cloud, mobility, and application technologies. But the real power of IoT is moving from an "open-loop" world characterized by people in the

Table 1.2 IoT units installed base by category, excluding automotive

Category	2014	2015	2016	2020
Consumer	2277	3023	4024	13,509
Generic business	623	815	1092	4408
Vertical business	898	1065	1276	2880
Grand Total	3807	4902	6392	20,797

Source: Gartner

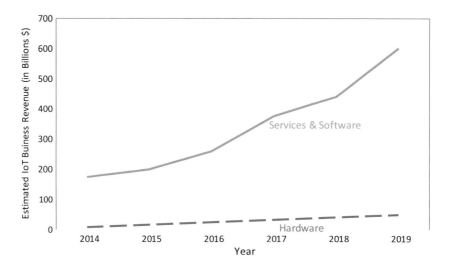

Fig. 1.18 IoT business revenue from enterprise

process to one that will be an automated "closed loop." In this model, humans will only intervene in the process as an exception, for example, if a robot, jet engine, driverless truck, or gas turbine requires a part within itself to be changed (in some cases, even these will be automated!).

There is no reason to doubt that devices connected to IoT will soon be flooding the mass market. We will see compact, connected sensors and actuators make their way onto everyday consumer electronics and household appliances and on general infrastructure.

Networks and semiconductor manufacturers no doubt will benefit from this movement, but big data vendors should also be cheering, with all things connected to the Internet that opens up more real-time data inventory to sell (Fig. 1.18).

1.4 History of the Internet

Before the advent of the Internet, the world's main communication networks were based on circuit-switching technology: the traditional telephone circuit, wherein each telephone call is allocated a dedicated, end-to-end, electronic connection between the two communicating stations (stations might be telephones or computers). Circuit-switching technology was not suitable for computer networking.

The history of the Internet begins with the development of electronic computers in the 1950s where the initial concepts of packet switching were introduced in several computer science laboratories. Various versions of packet switching were later announced in the 1960s. In the early 1980s, the TCP/IP (Transmission Control Protocol/Internet Protocol) stack was introduced. Then, the commercial use of the

Internet started in the late 1980s. Later, the World Wide Web (WWW) became available in 1991, which made the Internet more popular and stimulated the rapid growth. The Web of Things (WoT), which based on WWW, is considered a part of IoT.

To illustrate the importance of packet-switching technologies, consider computer A (in Los Angeles) wants to communicate with Computer B (in New York) in a circuit-switched network. One common way is to select a path in the network connecting computers A and B. In this case, the selected path would be dedicated to A and B for the duration of their message exchange. The problem with circuit switching is that the line is tied up regardless of how much information is exchanged (i.e., no other computers are allowed to utilize the line between A and B even with free bandwidth). Unlike voice traffic, circuit switching is a problem for computers because their information exchange is typically "bursty" rather than smooth or constant. Two computers might want to exchange a file, but after that file is exchanged, the computers may not engage in communication again for quite some time.

Packet switching was introduced as the alternative technology to circuit switching for computer communications. It has been reported that packet-switching work was done during the time of the Cold War, and a key part of motivation for developing packet switching was the design of a network that could withstand a nuclear attack. Such theory was denied by the Advanced Research Projects Agency Network (ARPANET), an early packet-switching network adopter and the first network to implement the Internet protocol suite TCP/IP. However, the later work on internetworking emphasized robustness and survivability, including the capability to withstand losses of large portions of the underlying networks.

To understand the fundamental of packet switching, consider sending a container of goods from Los Angeles to New York City. Rather than sending the entire container over a particular route, it is divided into packages (called packets). Packets are assembled, addressed, and sent in a particular way such that:

- The packets are numbered so they can be reassembled in the correct sequence at the destination.
- Each packet contains destination and return addresses.
- The packets are transmitted over the network of routes as capacity becomes available.
- The packets are forwarded across the network separately and do not necessarily follow the same route; if a particular link of a given path is busy, some packets might take an alternate route.

Packet switching is a generic philosophy of network communication, not a specific protocol. The protocol used by the Internet is called TCP/IP. The TCP/IP protocol was invented by Robert Kahn and Vint Cerf. The IP in TCP/IP stands for Internet protocol: the protocol used by computers to communicate with each other on the Internet. TCP is responsible for the data delivery of a packet, and IP is responsible for the logical addressing. In other words, IP obtains the address, and TCP guarantees delivery of data to that address. Both technologies became the technical foundation of the Internet.

The earliest ideas for a computer network, intended to allow general communications among computer users, were formulated by computer scientist J. C. R. Licklider of Bolt, Beranek, and Newman (BBN), in April 1963, in memoranda discussing the concept of the "Intergalactic Computer Network." Those ideas encompassed many of the features of the contemporary Internet. In October 1963, Licklider was appointed head of the Behavioral Sciences and Command and Control programs at the Defense Department's Advanced Research Projects Agency (ARPA). He convinced Ivan Sutherland and Bob Taylor that this network concept was very important and merited development although Licklider left ARPA before any contracts were assigned for development [5].

Devices using the Internet must implement the IP stack. Packets that follow the IP specification are called IP datagrams. These datagrams have two parts: header information and data. To continue with the letter analogy, think of the header as the information that would go on an envelope and the data as the letter that goes inside the envelope. The header information includes such things as total length of the packet, destination IP address, source IP address, time to live (the time to live is decremented by routers as the packet passes through them; when it hits zero, the packet is discarded; this prevents packets from getting into an "infinite loop" and tying up the network), and error checking information.

- The IP packets are independent of the underlying hardware structure. In order to travel across different types of networks, the packets are encapsulated into frames. The underlying hardware understands the particular frame format and can deliver the encapsulated packet.
- The TCP in TCP/IP stands for Transmission Control Protocol. This is a protocol that, as the name implies, is responsible for assembling the packets in the correct order and checking for missing packets. If packets are lost, the TCP endpoint requests new ones. It also checks for duplicate packets. The TCP endpoint is responsible for establishing the session between two computers on a network. The TCP and IP protocols work together.
- An important aspect of packet switching is that the packets have forwarding and return addresses. What should an address for a computer look like? Since it is a computer and computers only understand binary information, the most sensible addressing scheme is one based on binary numbers. Indeed, this is the case, and the addressing system used by IP version 4 software is based on a 32-bit IP address, and IP version 6 is based on 128-bit IP address as will be explained in Chap. 2 (Fig. 1.19).

1.5   Summary

We would like to conclude this chapter by restating our definition of IoT as the network of things, with clear element identification, embedded with software intelligence, sensors, and ubiquitous connectivity to the Internet. IoT is empowered by

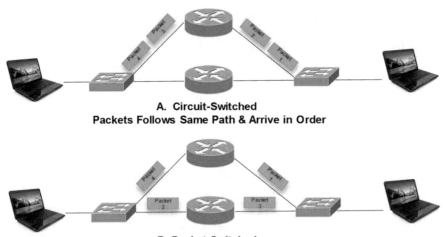

A. Circuit-Switched
Packets Follows Same Path & Arrive in Order

B. Packet-Switched
Packets May Follow Different Paths & Arrive Out-of-Order

Fig. 1.19 Circuit switched vs. packet switched

four main elements: sensors to collect information, identifiers to identify the source of data, software to analyze the data, and Internet connectivity to communicate and enable notifications. Sensors may be physical (e.g., sensors capturing the temperature) or logical (e.g., embedded software measurements such as CPU utilization). IoT's ultimate goal is to create a better environment for humanity, where objects around us know what we like, what we want, and what we need and act accordingly without explicit instructions.

IoT is fueled by explosion in technologies including the IT and OT convergence; the introduction of Internet-based business at a fast rate; the explosion in smart mobile devices; the explosion in social networking applications; the overall technology explosion; the massive digital transformation; the enhanced user interfaces allowing people to communicate by a simple touch, voice command, or even an observing command; the faster than ever technology adoption; the increased demand for security applications and solutions; and of course Moore's Law. Securing IoT is viewed as a challenge and colossal business opportunity at the same time with areas that embrace securing the data at rest, securing the transport of the data, securing APIs/interfaces among systems and various sources of data, and of course controlling sensors and applications.

Problems and Exercises

1. What is the simple definition of IoT? What is the "more complete definition"? What is the main difference?
2. IoT components were listed for the simple definition to include the intersection of the Internet, Things, and data. Process and standards were added to the complete definition. Why are process and standards important for the success of IoT?

3. What are the main four components that empower IoT? List the main function of each component.
4. What is IoT's promise? What is IoT's ultimate goal?
5. Cisco estimated that the IoT will consist of almost 30 billion objects by 2023. Others have higher estimates. What was their logic?
6. What is Moore's Law? When was it first observed? Why is it relevant to IoT?
7. In a table, list the 12 factors that are fueling IoT with a brief summary of each factor.
8. What are the top three challenges for IoT? Why are those challenges also considered as opportunities?
9. What is BYOD? Why is it considered a security threat for the network?
10. How do companies deal with BYOD today? List an example of BYOD system.
11. Why is operation technology (OT) under pressure to integrate with information technology (IT)?
12. Uber is using smartphone Gyrometer data to monitor speeding drivers. What is "Gyrometer"? How does it work? Where was it first used?
13. What is KISS? What are the top five principles for KISS user experience?
14. Section 1.3.10 stated the following three facts: (a) over the history of computing hardware, computer power has been doubling every 18 months, (b) biggest networks we have today have millions of nodes and billions of connection, and (c) a human brain has a hundred thousand billion nodes and a hundred trillion connections. It then stated that using (a)–(b), in year 2015, a computer should be as powerful as a human brain in about 25 years! How did the author arrive at 25? How long would it take if the computer power was doubling every 2 years instead of 18 months and why?
15. What are the key four differences between Analytics 1.0, 2.0, and 3.0?
16. List examples of solutions that offer Analytics 3.0.
17. What are the top three benefits of cloud computing? What do they mean?
18. In a table format, compare IaaS, PaaS, and SaaS. List an example for each.
19. What are the main differences between virtual machines and containers in virtualization? Provide an example of container technology. Which approach do you prefer and why?
20. List two main functions that TCP/IP protocol, the bread and butter of today's Internet.
21. Why do we need both TCP and IP protocols?
22. It is often said by User Experience Experts that the "Best Interface for a system is no User Interface." What does such statement mean? When does it typically apply? Provide an example in networking technologies.
23. This question has four parts:

    (a) What is circuit-switched technology? What is packet-switched technology?
    (b) What are circuit-switched networks and packet-switched networks used for? List an example of each use.
    (c) Why did we need packet-switched technology?
    (d) In a table, list three main differences between packet switching and circuit switching?
    (e) Which approach is better for the Internet and why?

24. What is a connection-oriented protocol? What is a connectionless protocol? Provide an example of each.
25. Some companies use the term IoE instead of IoT. What is their logic?
26. What is Cloud 1.0 and Cloud 2.0? What is the main difference between cloud 1.0 and cloud 2.0? How does machine learning differ from traditional approaches to extract business intelligence form the data?
27. Circuit-switched networks are designed with either frequency-division multiplexing (FDM) or time-division multiplexing (TDM). For TDM link, time is divided into frames of fixed duration,num and each frame is divided onto a fixed number of time slots as shown below (for a network link supporting up to three connections/circuits).

1	2	3	1	2	3	1	2	3	1	2	3
Slot	Slot	Slot	Slot	Slot	Slot	Slot	Slot	Slot	Slot	Slot	Slot

← Frame → ← Frame → ← Frame → ← Frame →

--→ Time

When the network establishes a connection across a link, the network dedicates one time slot in every frame to this connection. These slots are dedicated for the sole use of that connection, with one time slot available for use (in every frame) to transmits the connection's data.

(a) How does FDM work in circuit-switched networks?
(b) What is the typical frequency band in tradition circuit-switched-based telephone networks/public-switched telephone network (PSTN)?
(c) Compare FDM with TDM.
(d) Draw FDM and TDM for a tradition circuit-switched network link supporting up to five connections/circuits.

28. Refer again to problem 27 above. Let us assume that all links in the circuit-switched network are T1 (i.e., have a bit rate of 1.536 Mbps with 24 slots) and use TDM

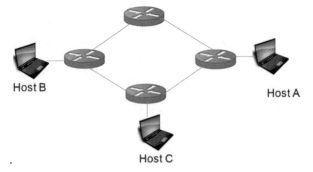

(a) Assuming setup and propagation delays are zero, how long does it take to send a file of 1.280 M bits from Host A to Host B? How about from Host A to Host C? Do you expect the answer to be the same or different and why?

(b) Let us also assume that it takes 500 ms to establish an end-to-end circuit before Host A can begin to transmit the file and 250 ms for a propagation delay between any two adjacent routers. How long does it take to send a file form Host A to Host B?

(c) What is the difference between transmission delay and prorogation delay? Which delay is a function of the distance between the routers?

References

1. A. Dohr, R. Modre-Opsrian, M. Drobics, D. Hayn, G. Schreier, The internet of things for ambient assisted living, in *Information Technology: New Generations (ITNG), 2010 Seventh International Conference on, 2010*, pp. 804–809 Online: http://ieeexplore.ieee.org/stamp/stamp.jsp?tp=&arnumber=5501633

2. M Drobics, E Fugger, B Prazak-Aram, G Schreier, *Evaluation of a Personal Drug Reminder*. (unpublished, 2009)

3. S. Haller, S. Karnouskos, C. Schroth, *The Internet of Things in an Enterprise Context* (Springer (Berlin-Heidelberg), Vienna, 2008)

4. International Telecommunication Union. ITU Internet Reports 2005 and 2015: The Internet of Things. Geneva, s.n., 2005. http://www.itu.int/internetofthings/

5. K. Ashton, "That 'Internet of Things' Thing, In the real world, things matter more than ideas", June 22, 2009, Online: http://www.rfidjournal.com/articles/view?4986

6. Top 3 Security Issues in Consumer Internet of Things (IoT) and Industrial IoT Youtube John Barrett at TEDxCIT: https://www.youtube.com/watch?v=QaTIt1C5R-M

7. Wikipedia, ARPANET, Online: http://en.wikipedia.org/wiki/ARPANET

8. Gail Honda, Kipp Martin, Essential Guide to Internet Business Technology Book, Feb 19, 2002 by Prentice Hall, Online: http://www.informit.com/articles/article.aspx?p=27569&seqNum=4

9. http://www.businessinsider.com/75-billion-devices-will-be-connected-to-the-internet-by-2020-2013-10#ixzz3YAtxfDCp

10. IoT Definitions, Online: http://gblogs.cisco.com/asiapacific/the-internet-of-everything-opportunity-for-anz-agribusiness/#more-120

11. Gartner News View: http://www.gartner.com/newsroom/id/2905717

12. Information Week IoE, Peter Waterhouse, December 2013, Online: http://www.informationweek.com/strategic-cio/executive-insights-and-innovation/internet-of-everything-connecting-things-is-just-step-one/d/d-id/1112958

13. *LG* Answers to IoT, the Latest Trend in IT-Talk Service-Oriented IoT, Online: http://www.lgcnsblog.com/features/answers-to-iot-the-latest-trend-in-it-talk-service-oriented-iot-1/

14. Driving Moore's Law with Python-Powered Machine Learning: An Insider's Perspective by Trent McConaghy PyData Berlin 2014, OnLine: http://www.slideshare.net/PyData/py-data-berlin-trent-mcconaghy-moores-law

15. Clock speed: Data from 1976–1999: E. R. Berndt, E. R. Dulberger, and N. J. Rappaport, "Price and Quality of Desktop and Mobile Personal Computers: A Quarter Century of History," July 17, 2000, http://www.nber.org/~confer/2000/si2000/berndt.pdf

16. Data from 2001–2016: ITRS, 2002 Update, On-Chip Local Clock in Table 4c: Performance and Package Chips: Frequency On-Chip Wiring Levels—Near-Term Years, p. 167. OnLine: http://www.singularity.com/charts/page62.html

17. Average transistor price: Intel and Dataquest reports (December 2002), see Gordon E. Moore, "Our Revolution," http://www.sia-online.org/downloads/Moore.pdf

18. The Internet of Things, Online: https://en.wikipedia.org/wiki/Internet_of_Things

19. L. David Roper, Silicon Intelligence Evolution: Online http://arts.bev.net/roperldavid, October 23, 2010, http://www.roperld.com/science/SiliconIntelligenceEvolution.htm

20. The Silicon Engine: A Timeline of Semiconductor in Computer, Online: http://www.computerhistory.org/semiconductor/timeline/1958-Mesa.html

21. T.E. Kurt, Disrupting and enhancing Healthcare with IoT, Health, Technology & engineering Program at USC, Arch 2, 2013, online: http://www.slideshare.net/todbotdotcom/disrupting-and-enhancing-healthcare-with-the-internet-of-things

22. Insight's The Semiconductor Laser's Cost Curve, Online: http://sweptlaser.com/semiconductor-laser-cost-curve

23. P. Welander "IT vs. OT: Bridging the divide - Traditional IT is moving more onto the plant floor. OT will have to accept a greater level of integration. Is that a problem or an opportunity?", Control engineering, 08/16/2013, Online: http://www.controleng.com/single-article/it-vs-ot-bridging-the-divide/db503d6cb9af3014f532cf19b5bf75e8.html

24. Airbnb Business Model, Online: https://www.quora.com/What-is-Airbnbs-business-model

25. Five Things You Can Learn From One of Airbnb's Earliest Hustles, Online: http://www.inc.com/alex-moazed/cereal-obama-denver-the-recipe-these-airbnb-hustlers-used-to-launch-a-unicorn.html

26. S. Ganguli, The Impact of the IoT on Infrastructure Monitoring, October 2015, Online: https://www.gartner.com/doc/3147818?srcId=1-2819006590&pcp=itg

27. Sqaure Inc, Online: https://en.wikipedia.org/wiki/Square,_Inc

28. G. Sterling, Greg, "Expanding Its Services, Square Launches Email Marketing With A Twist", April 2015, Online: http://marketingland.com/expanding-its-services-square-launches-email-marketing-with-a-twist-2-124282

29. Analysis of the Amazon Business Model, July 2015, Online: http://www.digitalbusinessmodelguru.com/

30. About Tesla, Online: http://www.teslamotors.com/about

31. A. Brisbourne, Tesla's Over The Air Fix: Best Example yet of the Internet of Things, February 2014, Online: http://www.wired.com/insights/2014/02/teslas-air-fix-best-example-yet-internet-things/

32. U. Wang, A Manufacturing Lesson From Tesla Motors, Forbes, August 2013, Online: http://www.forbes.com/sites/uciliawang/2013/08/08/a-manufacturing-lesson-from-tesla-motors/

33. How PayPal Here Stacks Up Against Other Mobile Payment Options, Online: http://mashable.com/2012/03/16/paypal-here-competitors/#cSQKd8eMwPqa

34. F. Richter, "Global Smartphone Traffic to Increase Tenfold by 2019", February 2015, Online: http://www.statista.com/chart/3227/global-smartphone-traffic-to-increase-tenfold-by-2019/

35. Security of IoT: Lessons from the Past for the Connected Future, Aa, Online: http://www.windriver.com/whitepapers/security-in-the-internet-of-things/wr_security-in-the-internet-of-things.pdf

36. Curb Your Enthusiasm, Uber Newsroom, Joe Sullivan, Chief Security Officer, January 26, 2016

37. Fundamental Principles of Great UX Design | How to Deliver Great UX Design, Janet M. Six, Nov 17, 2014, Online: http://www.uxmatters.com/mt/archives/2014/11/fundamental-principles-of-great-ux-design-how-to-deliver-great-ux-design.php#sthash.oEzaPFAH.dpuf

38. Three Social Media Marketing Options to Consider in 2016, Hiral Rana, Jan 31, 2016, Online: https://www.google.com/search?q=social+media&rls=com.microsoft:en-US:IE-Address&source=lnms&tbm=isch&sa=X&ved=0ahUKEwiU4_y107nMAhXFMKYKHXldAEAQ_AUICCgC&biw=1577&bih=912#imgrc=ZH-8cjbgp-pIBM%3A

39. Detecting a malfunctioning device using sensors, United States Patent 8777104, Online: http://www.freepatentsonline.com/8777104.html

40. Virtual Machines Vs. Containers: A Matter Of Scope, Information Week Network Computing, May 28, 2014, Online: http://www.networkcomputing.com/cloud-infrastructure/virtual-machines-vs-containers-matter-scope/2039932943
41. "Google's Self-Driving Car Hit a Bus", American Safety Council February 29, 2016, Online: http://blog.americansafetycouncil.com/googles-self-driving-car-hit-a-bus/
42. "10 Million Self-Driving Cars will be on the Road by 2020", Business Insider, Juley 29, 2015, Online: http://www.businessinsider.com/report-10-million-self-driving-cars-will-be-on-the-road-by-2020-2015-5-6
43. "How Self-driving Cars work", Shima Rayej, Robohub Automotive, June 3, 2014, Online: http://robohub.org/how-do-self-driving-cars-work/
44. Alternative To, NetCrunch, Online: http://alternativeto.net/software/netcrunch/comments/
45. Amazon Web Services is Approaching a $10 billion-a-year business, Recorde, April 28 2016, Online: http://www.recode.net/2016/4/28/11586526/aws-cloud-revenue-growth
46. Google says welcome to the Cloud 2.0, ComuterWold, May 24, 2016 issue, Online: http://www.computerworld.com/article/3074998/cloud-computing/google-says-welcome-to-the-cloud-20.html?token=%23tk.CTWNLE_nlt_computerworld_enterprise_apps_2016-05-27&idg_eid=28bc8cb86c8c36cb5f0c09ae2e86ba26&utm_source=Sailthru&utm_medium=email&utm_campaign=Computerworld%20Enterprise%20Apps%202016-05-27&utm_term=computerworld_enterprise_apps#tk.CW_nlt_computerworld_enterprise_apps_2016-05-27
47. "Gartner Says 6.4 Billion Connected "Things" Will Be in Use in 2016, Up 30 Percent From 2015", November 10, 2015, online: http://www.gartner.com/newsroom/id/3165317
48. Cisco Visual Networking Index: Forecast and Methodology, 2015–2020, June 6, 2016, Online: http://www.cisco.com/c/en/us/solutions/collateral/service-provider/visual-networking-index-vni/complete-white-paper-c11-481360.html
49. 2021 Cisco Global Networking Trends Report: https://www.lazorpoint.com/hubfs/eBooks/2021-networking%20report.pdf

Chapter 2
The Internet in IoT

Reliable and efficient communication is considered one of the most complex tasks in large-scale networks. Nearly all data networks in use today are based on the Open Systems Interconnection (OSI) standard. The OSI model was introduced by the International Organization for Standardization (ISO), in 1984, to address this complex problem. ISO is a global federation of national standards organizations representing over 100 countries. The model is intended to describe and standardize the main communication functions of any telecommunication or computing system without regard to their underlying internal structure and technology. Its goal is the interoperability of diverse communication systems with standard protocols. The OSI is a conceptual model of how various components communicate in data-based networks. It uses "divide and conquer" concept to virtually break down network communication responsibilities into smaller functions, called layers, so they are easier to learn and develop. With well-defined standard interfaces between layers, OSI model supports modular engineering and multi-vendor interoperability.

2.1 The Open System Interconnection Model

The OSI model consists of seven layers as shown in Fig. 2.1: Physical (layer 1), Data Link (layer 2), Network (layer 3), Transport (layer 4), Session (layer 5), Presentation (layer 6), and Application (layer 7). Each layer provides some well-defined services to the adjacent layer further up or down the stack, although the distinction can become a bit less defined in layers 6 and 7 with some services overlapping the two layers.

- *OSI Layer 7—Application Layer*: Starting from the top, the Application Layer is an abstraction layer that specifies the shared protocols and interface methods used by hosts in a communications network. It is where users interact with the

© The Author(s), under exclusive license to Springer Nature Switzerland AG 2022   35
A. Rayes, S. Salam, *Internet of Things from Hype to Reality*,
https://doi.org/10.1007/978-3-030-90158-5_2

Fig. 2.1 OSI layers and
data formats

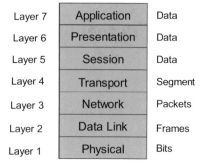

network using higher-level protocols such as DNS (Domain Naming System),
HTTP (Hypertext Transfer Protocol), Telnet, SSH, FTP (File Transfer Protocol),
TFTP (Trivial File Transfer Protocol), SNMP (Simple Network Management
Protocol), SMTP (Simple Mail Transfer Protocol), X Windows, RDP (Remote
Desktop Protocol), etc.

- *OSI Layer 6—Presentation Layer*: Underneath the Application Layer is the
  Presentation Layer. This is where operating system services (e.g., Linux, Unix,
  Windows, MacOS) reside. The Presentation Layer is responsible for the delivery
  and formatting of information to the Application Layer for additional processing
  if required. It ensures that the data can be understood between the sender and
  receiver. Thus it is tasked with taking care of any issues that might arise where
  data sent from one system needs to be viewed in a different way by the other
  system. The Presentation Layer releases the Application Layer of concerns
  regarding syntactical differences in data representation within the end-user systems. Example of a presentation service would be the conversion of an EBCDIC-
  coded text computer file to an ASCII-coded file and certain types of encryption
  such as Secure Sockets Layer (SSL) protocol.
- *OSI Layer 5—Session Layer*: Below the Presentation Layer is the Session Layer.
  The Session Layer deals with the communication to create and manage a session
  (or multiple sessions) between two network elements (e.g., a session between
  your computer and the server that your computer is getting information from).
- *OSI Layer 4—Transport Layer*: The Transport Layer establishes and manages
  the end-to-end communication between two end points. The Transport Layer
  breaks the data, it receives from the Session Layer, into smaller units called
  Segments. It also ensures reliable data delivery (e.g., error detection and retransmission where applicable). It uses the concept of windowing to decide how much
  information should be sent at a time between end points. Layer 4 main protocols
  include Transmission Control Protocol (TCP) and User Datagram Protocol
  (UDP). TCP is used for guarantee delivery applications such as FTP and web
  browsing applications, whereas UDP is used for best effort applications such as
  IP telephony and video over IP.
- *OSI Layer 3—Network Layer*: The Network Layer provides connectivity and
  path selection (i.e., IP routing) based on logical addresses (i.e., IP addresses).

Hence, routers operate at the Network Layer. The Network Layer breaks up the data it receives from the Transport Layer into packets, which are also known as IP datagrams, which contain source and destination IP address information that is used to forward the datagrams between hosts and across networks.[1] The Network Layer is also responsible for routing of IP datagrams using IP addresses. A routing protocol specifies how routers communicate with each other, exchanging information that enables them to select routes between any two nodes on a computer network. Routing algorithms determine the specific choice of routes. Each router has a priori knowledge only of networks attached to it directly. A routing protocol shares this information first among immediate neighbors and then throughout the network. This way, routers gain knowledge of the topology of the network. The major routing protocol classes in IP networks will be covered in Sect. 2.5. They include interior gateway protocol type 1, interior gateway protocol type 2, and exterior gateway protocols. The latter are routing protocols used on the Internet for exchanging routing information between autonomous systems.

- It must be noted that while layers 3 and 4 (Network and Transport Layers) are theoretically separated, they are typically closely related to each other in practice. The well-known Internet Protocol name "TCP/IP" comes from the Transport Layer protocol (TCP) and Network Layer protocol (IP).
- Packet switching networks depend upon a connectionless internetwork layer in which a host can send a message without establishing a physical connection with the recipient. In this case, the host simply puts the message onto the network with the destination address and hopes that it arrives. The message data packets may appear in a different order than they were sent in connectionless networks. It is the job of the higher layers, at the destination side, to rearrange out of order packets and deliver them to proper network applications operating at the Application Layer.
- *OSI Layer 2—Data Link Layer:* The Data Link Layer defines data formats for final transmission. The Data Link Layer breaks up the data it receives into frames. It deals with delivery of frames between devices on the same LAN using Media Access Control (MAC) Addresses. Frames do not cross the boundaries of a local network. Internetwork routing is addressed by layer 3, allowing data link protocols to focus on local delivery, physical addressing, and media arbitration. In this way, the Data Link Layer is analogous to a neighborhood traffic cop; it endeavors to arbitrate between parties contending for access to a medium, without concern for their ultimate destination. The Data Link Layer typically has error detection (e.g., Cyclical Redundancy Check (CRC)). Typical Data Link Layer devices include switches, bridges, and wireless access points (APs). Examples of data link protocols are *Ethernet* for local area networks (multi-node) and the *Point-to-Point Protocol (PPP)*.

---

[1] IP packets are referred to as IP datagrams by many experts. However, some experts used the phrase "stream" to refer to packets that are assembled for TCP and the phrase "datagram" to packets that are assembled for UDP.

- *OSI Layer 1—Physical Layer*: The Physical Layer describes the physical media access and properties. It breaks up the data it receives from the Data Link Layer into bits of zeros and ones (or "off" and "on" signals). The Physical Layer basically defines the electrical or mechanical interface to the physical medium. It consists of the basic networking hardware transmission technologies. It principally deals with wiring and caballing. The Physical Layer defines the ways of transmitting raw bits over a physical link connecting network nodes including copper wires, fiber-optic cables, optical wavelength, and wireless frequencies. The Physical Layer determines how to put a stream of bits from the Data Link Layer on to the pins for a USB printer interface, an optical fiber transmitter, or a radio carrier. The bit stream may be grouped into code words or symbols and converted to a physical signal that is transmitted over a hardware transmission medium. For instance, it uses +5 volts for sending a bit of 1 and 0 volts for a bit of 0 (Table 2.1).

2.2   End-to-End View of the OSI Model

Figure 2.2 provides an overview of how devices theoretically communicate in the OSI mode. An application (e.g., Microsoft Outlook on a User A's computer) produces data targeted to another device on the network (e.g., User B's computer or a server that User A is getting information from). Each layer in the OSI model adds its own information (i.e., headers, trailers) to the front (or both the front and the end) of the data it receives from the layer above it. Such process is called *Encapsulation*.

Table 2.1 Summary of key functions, devices, and protocols of the OSI layers

OSI layer	Main function	Examples of main devices	Examples of main protocol
Application	Provides network services to the end-host's applications	Server, laptops, PCs	HTTPS, FTP, Telnet, SSH
Presentation	Ensures the data can be understood between two end hosts	N/A	Data encoding, data formatting, and serialization
Session	Manages multiple sessions between end hosts	N/A	Connection management, error recovery
Transport	Establishes end-to-end connectivity and ensures reliable data delivery	Firewalls	TCP, UDP
Network	Connectivity and path selection based on logical addresses	Routers, firewalls	IPv4, IPv6
Data link	Defines data format for transmission	Switches, APs	IEEE 802.1 (Ethernet), PPP
Physical	Defines physical media access and properties	Fiber optics, category 5 cables, coaxial cables	IEEE 802.3

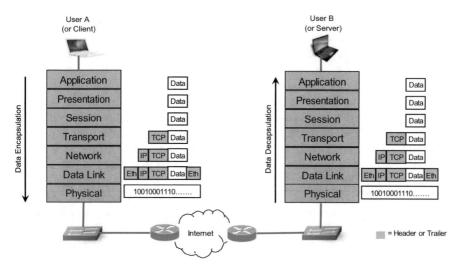

Fig. 2.2 Illustration of OSI model

For instance, the Transport Layer adds a TCP header, the Network Layer adds an IP header, and the Data Link Layer adds Ethernet header and trailer.

Encapsulated data is transmitted in protocol data units (PDUs): Segments on the Transport Layer, Packets on the Network Layer, and Frames on the Data Link Layer and Bits on the Physical Layer, as was illustrated in Fig. 2.2. PDUs are passed down through the stack of layers until they can be transmitted over the Physical Layer. The Physical Layer then slices the PDUs into bits and transmits the bits over the physical connection that may be wireless/radio link, fiber-optic, or copper cable. +5 volts are often used to transmit 1 s and 0 volts are used to transmit 0 s on copper cables. The Physical Layer provides the physical connectivity between hosts over which all communication occurs. The Physical Layer is the wire connecting both computers on the network. The OSI model ensures that both users speak the same language on the same layer allowing sending and receiving layers (e.g., networking layers) to virtually communicate. Data passed upward is decapsulated before being passed further up. Such process is called *decapsulation*. Thus, the Physical Layer chops up the PDUs and transmits the PDUs over the physical connection.

2.3 Transmission Control Protocol/Internet Protocol (TCP/IP)

TCP/IP (Transmission Control Protocol/Internet Protocol) is a connection-oriented transport protocol suite that sends data as an unstructured stream of bytes. By using sequence numbers and acknowledgment messages, TCP can provide a sending node with delivery information about packets transmitted to a destination node. Where data has been lost in transit from source to destination, TCP can retransmit the data until either a timeout condition is reached or until successful delivery has been achieved. TCP can also recognize duplicate messages and will discard them appropriately. If the sending computer is transmitting too fast for the receiving computer, TCP can employ flow control mechanisms to slow data transfer. TCP can also communicate delivery information to the upper-layer protocols and applications it supports. All these characteristics make TCP an end-to-end reliable transport protocol.

TCP/IP was in the process of development when the OSI standard was published in 1984. The TCP/IP model is not exactly the same as OSI model. OSI is a seven-layered standard, but TCP/IP is a four-layered standard. The OSI model has been very influential in the growth and development of TCP/IP standard, and that is why much of the OSI terminology is applied to TCP/IP.

The TCP/IP Layers along with the relationship to OSI layers are shown in Fig. 2.3. TCP/IP has four main layers: Application Layer, Transport Layer, Internet Layer, and Network Access Layer. Some researchers believe TCP/IP has five layers: Application Layer, Transport Layer, Network Layer, Data Link Layer, and Physical Layer. Conceptually both views are the same with Network Access being equivalent to Data Link Layer and Physical Layer combined.

2.3.1 TCP/IP Layer 4: Application Layer

As with the OSI model, the Application Layer is the topmost layer of TCP/IP model. It combines the Application, Presentation, and Session Layers of the OSI model. The Application Layer defines TCP/IP application protocols and how host programs interface with Transport Layer services to use the network.

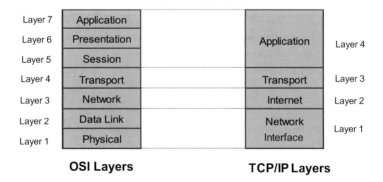

Fig. 2.3 Relationship between OSI reference model and TCP/IP

2.3.2   TCP/IP Layer 3: Transport Layer

The Transport Layer is the third layer of the four-layer TCP/IP model. Its main tenacity is to permit devices on the source and destination hosts to carry on a conversation. The Transport Layer defines the level of service and status of the connection used when transporting data. The main protocols included at the Transport Layer are TCP (Transmission Control Protocol) and UDP (User Datagram Protocol).

2.3.3   TCP/IP Layer 2: Internet Layer

The Internet Layer of the TCP/IP stack packs data into data packets known as IP datagrams, which contain source and destination address information that is used to forward the datagrams between hosts and across networks. The Internet Layer is also responsible for routing of IP datagrams.

The main protocols included at the Internet Layer are IP (Internet Protocol), ICMP (Internet Control Message Protocol), ARP (Address Resolution Protocol), RARP (Reverse Address Resolution Protocol), and IGMP (Internet Group Management Protocol).

The main TCP/IP Internet Layer (or Networking Layer in OSI) devices are routers. Routers are similar to personal computers with hardware and software components that include CPU, RAM, ROM, flash memory, NVRAM, and interfaces. Given the importance of the router's role in IoT, we will use the next section to describe its main functions.

2.3.3.1   Router Main Components

There are quite a few types and models of routers. Generally speaking, every router has the same common hardware components as shown in Fig. 2.4. Depending on the model, router's components may be located in different places inside the router.

1. *CPU (Central Processing Unit)*: CPU is an older term for microprocessor, the central unit containing the logic circuitry that preforms the instruction of a router's program. It is considered as the brain of the router or a computer. CPU is responsible for executing operating system commands including initialization, routing, and switching functions.
2. *RAM (Random Access Memory)*: As with PCs, RAM is a type of computer memory that can be accessed randomly; that is, any byte of memory can be accessed without touching the preceding bytes. RAM is responsible for storing the instructions and data that CPU needs to execute. This read/write memory contains the software and data structures that allow the router to function. RAM is volatile memory, so it loses its content when the router is powered down or restarted.

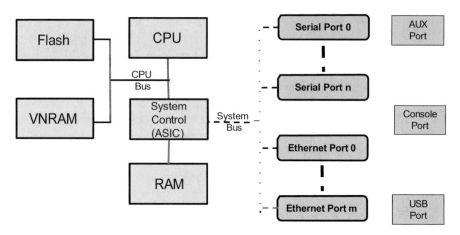

Fig. 2.4 Router main components

However, the router also contains permanent storage areas such as ROM, flash memory, and NVRAM. RAM is used to store the following:

(a) *Operating system*: The software image (e.g., Cisco's IOS) is copied into RAM during the boot process.

(b) *"Running Config" file*: This file stores the configuration commands that cisco IOS software is currently using on the router.

(c) *IP routing tables*: Routing tables are used to determine the best path to route packets to destination devices. It will be covered in Sect. 2.5.3.

(d) *ARP cache*: ARP cache contains the mapping between IP and MAC addresses. It is used on routers that have LAN interfaces such as Ethernet.

(e) *Buffer*: Packets are temporary stored in a buffer when they are received on congested interface or before they exit an interface.

3. *ROM (Read-Only Memory)*: As the name indicates, read-only memory typically refers to hardwired memory where data (stored in ROM) cannot be changed/ modified except with a slow and difficult process. Hence, ROM is a form of permanent storage used by the router. It contains code for basic functions to start and maintain the router. ROM contains the ROM monitor, which is used for router disaster recovery functions such as password recovery. ROM is nonvolatile; it maintains the memory contents even when the power is turned off.

4. *Flash Memory*: Flash memory is nonvolatile computer memory that can be electrically stored and erased. Flash is used as permanent storage for the operating system. In most models of Cisco router, Cisco IOS software is permanently stored in flash memory.

5. *NVRAM (Nonvolatile RAM)*: NVRAM is used to store the startup configuration file "startup config," which is used during system startup to configure the software. This is due to the fact that NVRAM does not lose its content when the power is turned off. In other words, the router's configuration is not erased when the router is reloaded.

Recall that all configuration changes are stored in the "running config" file in RAM. Hence, to save the changes in the configuration in case the router is restarted or loses power, the "running config" must be copied to NVRAM, where it is stored as the "startup configuration" file.

Finally, NVRAM contains the software *Configuration Register*, a configurable setting in Cisco IOS software that determines which image to use when booting the router.

6. *Interfaces*: Routers are accessed and connected to the external world via the interfaces. There are several types of interfaces. The most common interfaces include:

(a) *Console (Management) Interface*: Console port or interface is the management port which is used by administrators to log on to a router directly (i.e., without using a network connection) via a computer with an RJ-45 or mini-USB connector. This is needed since there is no display device for a router. The console port is typically used for initial setup given the lack of initial network connections such as SSH or HTTPS. A terminal emulator application (e.g., HyperTerminal or PuTTy) is required to be installed on the PC to connect to router. Console port connection is a way to connect to the router when a router cannot be accessed over the network.

(b) *Auxiliary Interface*: Auxiliary port or interface allows a direct, non-network connection to the router, from a remote location. It uses a connector type to which modems can plug into, which allows an administrator from a remote location to access the router like a console port. Auxiliary port is used as a way to dial in to the router for troubleshooting purposes should regular connectivity fail. Unlike the console port, the auxiliary port supports hardware flow control, which ensures that the receiving device receives all data before the sending device transmits more. In cases where the receiving device's buffers become full, it can pass a message to the sender asking it to temporarily suspend transmission. This makes the auxiliary port capable of handling the higher transmission speeds of a modem.

Much like the console port, the auxiliary port is also an asynchronous serial port with an RJ-45 interface. Similarly, a rollover cable is also used for connections, using a DB-25 adapter that connects to the modem. Typically, this adapter is labeled "MODEM."

(c) *USB Interface*: It is used to add a USB flash drive to a router.

(d) *Serial Interfaces (Asynchronous and Synchronous)*: Configuring the serial interface allows administrators to enable applications such as wide area network (WAN) access, legacy protocol transport, console server, and remote network management.

(e) *Ethernet Interface*: Ethernet is the most common type of connection computers use in a local area network (LAN). Some vendors categorize Ethernet ports into three areas:

• *Standard/Classical Ethernet (or just Ethernet)*: Usual speed of Ethernet is 10 Mbps.

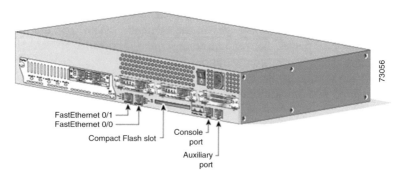

Fig. 2.5 Example of a router's rear panel. (Source: Cisco)

- *Fast Ethernet*: Fast Ethernet was introduced in 1995 with a speed of 100 Mbps (10× faster than standard Ethernet). It was upgraded by improving the speed and reducing the bit transmission time. In standard Ethernet, a bit is transmitted in 1 s, and in Fast Ethernet it takes 0.01 μs for 1 bit to be transmitted. So, 100 Mbps means transferring speed of 100 Mbits per second.
- *Gigabit Ethernet*: Gigabit Ethernet was introduced in 1999 with a speed of 1000 Mbps (10× faster than Fast Ethernet and 100× faster than classical Ethernet) and became very popular in 2010. Gigabit Ethernet maximum network limit is 70 km if single-mode fiber is used as a medium. Gigabit Ethernet is deployed in high-capacity backbone network links. In 2000, Apple's Power Mac G4 and PowerBook G4 were the first mass-produced personal computers featuring the 1000BASE-T connection [2]. It quickly became a built-in feature in many other computers.
  Faster Gigabit Ethernet speeds have been introduced by vendors including 10 Gbps and 100 Gbps, which is supported, for example, by the Cisco Nexus 7700 F3-Series 12-Port 100 Gigabit Ethernet module (Fig. 2.5).

Table 2.2 outlines the main functions of each of the router's components.

2.3.4   TCP/IP Layer 1: Network Access Layer

The Network Access Layer is the first layer of the four-layer TCP/IP model. It combines the Data Link and the Physical Layers of the OSI model. The Network Access Layer defines details of how data is physically sent through the network. This includes how bits are electrically or optically signaled by hardware devices that interface directly with a network medium, such as coaxial cable, optical fiber, radio links, or twisted pair copper wire. The most common protocol included in the Network Access Layer is Ethernet. Ethernet uses Carrier Sense Multiple Access/

Table 2.2 Main functions of the router's component

Router component	Main function	Volatile/ nonvolatile
CPU	Executes operating system commands: initialization, routing, and switching functions	Nonvolatile
RAM	Stores the instruction and data that CPU needs to execute (considered the working area of memory storage used by the CPU) Stores: "running config" file, routing tables, ARP cache, and buffer	Volatile
ROM	Contains *code for basic functions* to start and maintain the router	Nonvolatile
Flash	Permanently stores the *operating system* (e.g., where a router finds and boots its IOS image)	Nonvolatile
NVRAM	Stores the "startup config" file, holds configuration register software	Nonvolatile
Interfaces/ ports	Routers are accessed and connected to the external world via the interfaces	N/A

Collision Detection (CSMA/CD) method to access the media, when Ethernet operates in a shared media. Such Access Method determines how a host will place data on the medium.

2.4   IoT Network Level: Key Performance Characteristics

As we illustrated in Chap. 1, the IoT reference framework consists of four main levels: IoT Device Level (e.g., sensors and actuators), IoT Network Level (e.g., IoT gateways, routers, switches), IoT Application Services Platform Level (the IoT Platform, Chap. 7), and IoT Application Level.

The IoT Network Level is in fact the TCP/IP Layers as shown in Fig. 2.6. It should be noted that we have removed TCP/IP's Application Layer to prevent overlap with the IoT Application Level.

In this section we will discuss the most important performance characteristics of IoT network elements. Such features are essential in evaluating and selecting IoT network devices especially IoT gateways, routers, and switches.

IoT Network Level key characteristics may be grouped into three main areas: end-to-end delay, packet loss, and network element throughput. Ideally, engineers want the IoT network to move data between any end points (or source and destination) instantaneously, without any delay or packet loss. However, the physical laws in the Internet constrain the amount of packets that can be transferred between end points per second (known as throughput), present various types of delays to transfer packets from source to destination, and can indeed lose packets.

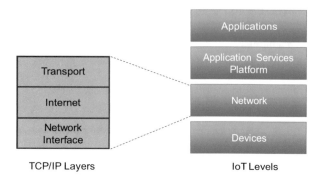

Fig. 2.6 Mapping of IoT reference framework to TCP/IP Layers

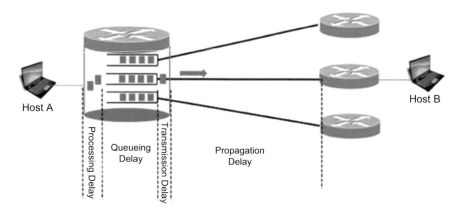

Fig. 2.7 End-to-end delay from host A to host B with illustration at router A

2.4.1   End-to-End Delay

End-to-end delay across the IoT network is perhaps the most essential performance characteristic for real-time applications especially in wide area networks (WAN) that connect multiple geographies. It may be defined as the amount of time (typically in fractions of seconds) for a packet to travel across the network from source to destination (e.g., from host A to host B as shown in Fig. 2.7). Measuring the end-to-end delay is not a trivial task as it typically varies from one instance to another. Engineers, therefore, are required to measure the delay over a specific period of time and report the average delay, the maximum delay, and the delay variation during such period (known as jitter). Hence, jitter is defined as the variation in the delay of received packets between a pair of end points.

In general, there are several contributors to delay across the network (as shown in Fig. 2.7). The main ones are the following:

- *Processing delay*: which is defined as the time a router takes to process the packet header and determine where to forward the packet. It may also include the time needed to check for bit-level errors in the packet (typically in the order of microseconds).

- *Queuing delay*: which is defined as the time the packet spends in router queues as it awaits to be transmitted onto the outgoing link. Clearly Queuing delay depends on the number of earlier-arriving packets in the same queue (typically in the order of microseconds to milliseconds).
- *Transmission delay*: which is defined as the time it takes to push the packet's bits onto the link. Transmission delay of packet of length L bits is defined L/R where R is the transmission rate of a link between two devices. For example, for a packet of length 1000 bits and a link of speed of 100 Mbps, the delay is 0.01 ms. (Transmission delay is typically in the order of microseconds to milliseconds.)
- *Propagation delay*: which is defined as the time for a bit (of the packet) to propagate from the beginning of a link (once it leaves the source router) to reach its destination router. Hence, Propagation delay on a given link depends on the physical medium of the link itself (e.g., twisted pair copper, fiber, coaxial cable) and is equal to the distance of the link (between two routers) divided by the propagation speed (e.g., speed of light). (Propagation delay is typically in the order of milliseconds). It should be noted that unlike Transmission delay (i.e., the amount of time required to push a packet out), Propagation delay is independent of the packet length.

Hence, the total delay (d_{Total}), between two end points, is the sum of the Processing delay ($d_{Process}$), the Queuing delay (d_{Queue}), the Transmission delay (d_{Trans}), and the Propagation delay (d_{Prop}) across utilized network elements in the path, i.e.,

$$D_{Total} = d_{Process} + d_{Queue} + d_{Trans} + d_{Prop}$$

End-to-end delay is typically measured using *Traceroute* utility (available on many modern operating systems) as well as vendor-specific tools (e.g., Cisco's IP SLA (service-level agreement) that continuously collects data about delay, jitter, response time, and packet loss). What is the other utility/command that returns only the final roundtrip times from the destination point (see Problem 27)?

A Traceroute utility's output displays the route taken between two end systems, listing all the intermediate routers across the network. For each intermediate router, the utility also shows the roundtrip delay (from source to the intermediate router) and time to live (a mechanism that limits the lifetime of the traceroutes packet). Other advantageous of Traceroute utility includes troubleshooting (showing the network administrator bottlenecks and why connections to a destination server are poor) and connectivity (showing how systems are connected to each other and how a service provider connects to the Internet).

Figure 2.8 shows a simple example of Traceroute utility to trace a path from a client (connected to router A) to the server. In this case, the client enters the command "traceroute 10.1.3.2." Traceroute utility will display four roundtrip delays, based on three different test packets, sent from the client's computer to router A, client's computer to router B, client's computer to router C, and finally client's computer to the server.

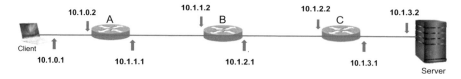

Fig. 2.8 Traceroute example

The output shows that the roundtrip delay from the client's computer to router A (ingress port) is 1 ms for the first test packet, 2 ms for the second test packet, and 1 ms for the third and final test. It should be noted that "three test packets" is a typical default value in Traceroute tool and can be adjusted as needed. Also, other parameters may be reported by the tool (e.g., time to live (TTL)) depending on the user's tool configurations.

A# trceroute 10.1.3.2

Type escape sequence to abort. Tracing the route to 10.1.3.2

1 10.1.0.2 1 ms, 2 ms, 1 ms
2 10.1.1.2 13 ms, 14 ms, 15 ms
3 10.1.2.2 26 ms, 31 ms, 29 ms
4 10.1.3.2 41 ms, 43 ms, 44 ms

2.4.2   Packet Loss

Packet loss occurs when at least one packet of data traveling across a network fails to reach its destination. In general, packets are dropped and consequently lost when the network is congested (i.e., one of the network elements is already operating at full capacity and cannot keep up with arriving packets). This is due to the fact that both queues and links have finite capacities. Hence, a main reason for packet loss is link or queue congestion (i.e., a link between two devices, and its associated queues, is fully occupied when data arrives). Another reason for packet loss is router performance (i.e., links and queues have adequate capacity, but the device's CPU or memory is fully utilized and not able to process additional traffic). Less common reasons include faulty software deployed on the network device itself or faulty cables.

It should be noted that packet loss may not be as bad as it first seems. Many applications are able to gracefully handle it without impacting the end user, i.e., the application realizes that a packet was lost, adjusts the transfer speed, and requests data retransmission. This works well for file transfer and emails. However, it does not work well for real-time applications such as video conferencing and voice over IP.

2.4.3   Throughput

Throughput may be defined as the maximum amount of data moved successfully between two end points in a given amount of time. Related measures include the link and device speed (how fast a link or a device can process the information) and response time (the amount of time to receive a response once the request is sent).

Throughput is one of the key performance measures for network and computing devices and is typically measured in bits per second (bps) or gigabits per second (Gbps) at least for larger network devices. The system throughput is typically calculated by aggregating all throughputs across end points in a network (i.e., sum of successful data delivered to all destination terminals in a given amount of time).

The simplest way to show how throughput is calculated is through examples. Assume host A is sending a data file to host B through three routers and the speed (e.g., maximum bandwidth) of link i is R_i as shown in Fig. 2.9. Also assume that each router speed (processing power) is higher than the speed of any link and no other host is sending data. In this example, the throughput is

$$\min\left(R_1, R_2, R_3 \text{ and } R_4\right).$$

Thus if $R_1 = R_2 = R_3 = 10$ Mbps and $R_4 = 1$ Mbps, the throughput is 1 Mbps.

Estimating the throughput is more complicated when multiple paths are allowed in the network. Figure 2.10, for instance, shows that data from host A to host B may take path R_1, R_2, R_3, and R_4 or R_1, R_5, R_6, and R_4.

Using the pervious example assumptions (i.e., the speed of each router is higher than the speed of any link and no other host is sending data) and the following new assumptions:

- $R_2 = R_3 = R_5 = R_6 = 10$ Mbps.
- $R_1 = R_4 = 1$ Mbps.
- Data is equally divided between the two paths.

The throughput for this example is still 1 Mbps.
Now, if links R_1 and R_4 are upgraded to 100 Mbps, i.e.,

- $R_2 = R_3 = R_5 = R_6 = 10$ Mbps.
- $R_1 = R_4 = 100$ Mbps.
- Data is equally divided between the two paths.

Then, the throughput will be 20 Mbps (see Problem 25).

Fig. 2.9 Throughput for a file transfer from host A to host B with a single route

Fig. 2.10 Throughput for a file transfer from host A to host B with multiple routes

Table 2.3 Examples of Internet protocol suite (partial list)

TCP/IP layer	Top protocols
Application layer	BGP, DHCP, DNS, HTTP, IMAP, LDAP, MGCP, POP, ONC/RPC, RTP, RTSP, RIP, SIP, SNMP, SSH, Telnet, SSL, SMTP (Email), XMPP
Transport layer	TCP, UDP, DCCP, SCTP, RSVP
Internet layer	IPv4, IPv6, ICMP, ICMPv6, IGMP, IPSec, OSPF, EIGRP
Network Interface layer	ARP, PPP, MAC

2.5   Internet Protocol Suite

As we mentioned earlier, TCP/IP provides end-to-end connectivity specifying how data should be packetized, addressed, transmitted, routed, and received at the destination. Table 2.3 lists top (partial list) protocols at each layer.

The objective of this chapter is not to provide an exhaustive list of the TCP/IP protocols but rather to provide a summary of the key protocols that are essential for IoT.

The remainder of this chapter focuses on the main Internet Layer address protocols, namely, IP version 4 and IP version 6. It then describes the main Internet routing protocols, namely, OSPF, EIRGP, and BGP.

2.5.1   IoT Network Level: Addressing

As we mentioned earlier in this chapter, Internet Protocol (IP) provides the main internetwork routing as well as error reporting and fragmentation and reassembly of information units called datagrams for transmission over networks with different maximum data unit sizes. IP addresses are globally unique numbers assigned by the Network Information Center. Globally unique addresses permit IP networks anywhere in the world to communicate with each other. Most of existing networks today use IP version 4 (IPv4). Advanced networks use IP version 6 (IPv6).

2.5.1.1   IP Version 4

IPv4 addresses are normally expressed in dotted-decimal format, with four numbers separated by periods, such as 192.168.10.10. It consists of 4-octets (32-bit) number that uniquely identifies a specific TCP/IP (or IoT) network and a host (computer, printer, router, IP-enabled sensor, any device requiring a network interface card) within the identified network. Hence, an IPv4 address consists of two main parts: the network address part and the host address part. A subnet mask is used to divide an IP address into these two parts. It is used by the TCP/IP protocol to determine whether a host is on the local subnet or on a remote network.

IPv4 Subnet Mask

It is important to recall that in TCP/IP (or IoT) networks, the routers that pass packets of data between networks do not know the exact location of a host for which a packet of information is destined. Routers only know what network the host is a member of and use information stored in their route table to determine how to get the packet to the destination host's network. After the packet is delivered to the destination's network, the packet is delivered to the appropriate host. For this process to work, an IP address is divided into two parts: network address and host address.

To better understand how IP addresses and subnet masks work, IP addresses should be examined in binary notation. For example, the dotted-decimal IP address 192.168.10.8 is (in binary notation) the 32 bit number 11000000.10101000.000010 10.00001000. The decimal numbers separated by periods are the octets converted from binary to decimal notation.

The first part of an IP address is used as a network address and the last part as a host address. If you take the example 192.168.10.8 and divide it into these two parts, you get the following: 192.168.10.0 network address and .8 host address or 192.168.10.0 network address and 0.0.0.8 host address.

In TCP/IP, the parts of the IP address that are used as the network and host addresses are not fixed, so the network and host addresses above cannot be determined unless you have more information. This information is supplied in another 32-bit number called a subnet mask. In the above example, the subnet mask is 255.255.255.0. It is not obvious what this number means unless you know that 255 in binary notation equals 11111111; so, the subnet mask is

$$11111111.11111111.11111111.0000000$$

Lining up the IP address and the subnet mask together, the network and host portions of the address can be separated:

$$11000000.10101000.00001010.10001000 --IP\,address\,(192.168.10.8)$$

$$11111111.11111111.11111111.00000000 --Subnet\,mask\,(255.255.255.0)$$

The first 24 bits (the number of ones in the subnet mask) are identified as the network address, with the last 8 bits (the number of remaining zeros in the subnet mask) identified as the host address. This gives you the following:

$$11000000.10101000.00001010.00000000 --\text{Network address}(192.168.10.0)$$

$$00000000.00000000.00000000.00001000 --\text{Host address}(000.000.000.8)$$

IPv4 Classes

Five classes (A, B, C, D, and E) have been established to identify the network and host parts. All the five classes are identified by the first octet of IP address. Classes A, B, and C are used in actual networks. Class D is reserved for multicasting (data is not destined for a particular host; hence there is no need to extract host address from the IP address). Class E is reserved for experimental purposes.

Figure 2.11 shows IPv4 address formats for classes A, B, and C. Class A networks provide only 8 bits for the network address field and 24 bits for host address. It is intended mainly for use with very large networks with large number of hosts. The first bit of the first octet is always set to 0 (zero). Thus the first octet ranges from 1 to 127, i.e., 00000001–011111111. Class A addresses only include IP starting from 1.x.x.x to 126.x.x.x only. The IP range 127.x.x.x is reserved for loopback IP addresses. The default subnet mask for class A IP address is 255.0.0.0 which implies that class A addressing can have 126 networks (2^7–2) and 16,777,214 hosts (2^{24}–2).

Class B networks allocate 16 bits for the network address field and 16 bits for the host address filed. An IP address which belongs to class B has the first two bits in the first octet set to 10, i.e., 10000000–10111111 or 128–191 in decimal. Class B IP

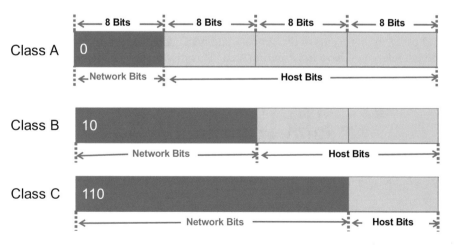

Fig. 2.11 IPv4 address formats for classes A, B, and C

addresses range from 128.0.x.x to 191.255.x.x. The default subnet mask for class B is 255.255.x.x. Class B has 16,384 (2^{14}) network addresses and 65,534 (2^{16}–2) host addresses.

Class C networks allocate 24 bits for the network address field only 8 bits for the host field. Hence, the number of hosts per network may be a limiting factor. The first octet of Class C IP address has its first 3 bits set to 110, that is: 1110 0000–1110 1111 or 224–239 in decimal.

Class C IP addresses range from 192.0.0.x to 223.255.255.x. The default subnet mask for Class C is 255.255.255.x. Class C gives 2,097,152 (2^{21}) Network addresses and 254 (2^8–2) Host addresses.

Finally, IP networks may also be divided into smaller units called subnetworks or subnets for short. Subnets provide great flexibility for network administrators. For instance, assume that a network has been assigned a Class A address and all the nodes on the network use a Class A address. Further assume that the dotted-decimal representation of this network's address is 28.0.0.0. The network administrator can subdivide the network using sub-netting by "borrowing" bits from the host portion of the address and using them as a subnet field.

2.5.1.2 IP Version 6

IPv4 has room for about 4.3 billion addresses, which is not nearly enough for the world's people, let alone IoT with a forecast of 20 billion devices by 2020. In 1998, the Internet Engineering Task Force (IETF) had formalized the successor protocol: IPv6. IPv6 uses a 128-bit address, allowing 2^{128} or 340 trillion trillion trillion (3.4×10^{38}) addresses. This translates to about 667×10^{21} (667 sextillion) addresses per square meter in earth. Version 4 and version 6 protocols are not designed to be interoperable, complicating the transition to IPv6. However, several IPv6 transition mechanisms have been devised to permit communication between IPv4 and IPv6 hosts.

IPv6 delivers other benefits in addition to a larger addressing space. For example, permitting hierarchical address allocation techniques that limit the expansion of routing tables simplified and expanded multicast addressing and service delivery optimization. Device mobility, security, and configuration aspects have been considered in the design of IPv6.

1. *IPv6 Addresses Are Broadly Classified Into Three Categories:*

    (a) Unicast addresses: A unicast address acts as an identifier for a single interface.
    An IPv6 packet sent to a unicast address is delivered to the interface identified by that address.
    (b) Multicast addresses: A multicast address acts as an identifier for a group/set of interfaces that may belong to different nodes. An IPv6 packet delivered to a multicast address is delivered to the multiple interfaces.
    (c) Anycast addresses: Anycast addresses act as identifiers for a set of interfaces that may belong to different nodes. An IPv6 packet destined for an anycast address is delivered to one of the interfaces identified by the address.

2.5.2   IPv6 Address Notation

The IPv6 address is 128 bits long. It is divided into blocks of 16 bits. Each 16-bit block is then converted to a 4-digit hexadecimal number, separated by colons. The resulting representation is called colon-hexadecimal. This is in contrast to the 32-bit IPv4 address represented in dotted-decimal format, divided along 8-bit boundaries, and then converted to its decimal equivalent, separated by periods.

1. *IPV6 Example*

    (a) *Binary Form*

    - 0111000111011010000000001101001100000000000000000101
      11100111011
    - 0000001010101010000000001111111111111111000101000100
      1110001011011

    (b) *16-Bit Boundaries Form*

    - 0111000111011010   0000000011010011   0000000000000000
      0010111100111011
    - 0000001010101010   0000000011111111   1111111000101000
      1001110001011011

    (c) 16-Bit Block Hexadecimal *and Delimited with Colons Form*

    - 71DA:00D3:0000:2F3B:02AA:00FF:FE28:9C5B.
    - i.e., $(0111000111011010)_2 = (71DA)_{16}$, $(0000000011010011)_2 = (D3)_{16}$, and so on.

    (d) *Final Form (16-Bit Block Hexadecimal and Delimited with Colons Form, Simplified by Removing the Leading Zeros).*

    - 71DA:D3:0:2F3B:2AA:FF:FE28:9C5B

2.5.3   IoT Network Level: Routing

Routers use routing tables to communicate: send and receive packets among themselves. TCP/IP routing specifies that IP packets travel through an internetwork one router hop at a time. Hence, the entire route is not known at the beginning of the journey. Instead, at each stop, the next router hop is determined by matching the destination address within the packet with an entry in the current router's routing table using internal information.

Before describing the main routing protocols in the Internet today, it is important to introduce a few fundamental definitions.

- *Static Routes*: Static routes define specific paths that are manually configured between two routers. Static routes must be manually updated when network changes occur. Static routes use should be limited to simple networks with predicted traffic behavior.
- *Dynamic Routes*: Dynamic routing requires the software in the routing devices to calculate routes. Dynamic routing algorithms adjust to changes in the network and repeatedly select best routes. Internet-based routing protocols are dynamic in nature. Routing tables should be updated automatically to capture changes in the network (e.g., link just went down, link that was down is no up, link speed update).
- *Autonomous System (AS)*: It is a network or a collection of networks that are managed by a single entity or organization (e.g., Department Network). An AS may have multiple subnetworks with combined routing logic and common routing policies. Routers used for information exchange within AS are called interior routers. They use a variety of interior routing protocols such as OSPF and EIGRP. Routers that move information between autonomous systems are called exterior routers, and they use the exterior gateway protocol such as Border Gateway Protocol (BGP). Interior routing protocols are used to update the routing tables of routers within an AS. In contrast, exterior routing protocols are used to update the routing tables of routers that belong to different AS. Figure 2.12 shows an illustration of two autonomous systems connected by BGP external routing protocol.
- *Routing Table*: Routing tables basically consist of destination address and next hop pairs. Figure 2.13 shows an example of a typical Cisco router routing table using the command "show ip route." It lists the set of comprehensive codes including various routing schemes. Figure 2.13 also shows that the first entry is interpreted as meaning "to get to network 29.1.0.0 (subnet 1 on network 24), the next stop is the node at address 51.29.23.12." We will refer to this figure as we introduce various routing schemes.
- *Distance Vector Routing*: A vector in distance vector routing contains both distance and direction to determine the path to remote networks using hop count as the metric. A hop count is defined as the number of hops to destination router or network (e.g., if there are two routers between a source router and destination router, the number of hops will be three). All neighbor routers will send information about their connectivity to their neighbors indicating how far other routers are from them. Hence, in distance vector routing, all routers exchange information only with their neighbors (not with all routers). One of the weaknesses of distance vector protocols is convergence time, which is the time it takes for routing information changes to propagate through all the topology.
- *Link-State Routing*: Contrast to distance vector, link-state routing requires all routers to know about the paths reachable by all other routers in the network. In this case, link-state data is flooded to the entire router in AS. Link-state routing requires more memory and processor power than distance vector routing. Also, link-state routing can degrade the network performance during the initial discov-

ery process, as it requires flooding the entire network with link-state advertisements (LSAs).

2.5.3.1   Interior Routing Protocols

Interior gateway protocols (IGPs) operate within the confines of autonomous systems. We will next describe only the key protocols that are currently popular in TCP/IP networks. For additional information, the reader is encouraged to peruse the references at the end of the chapter.

1. *Routing Information Protocol (RIP)*: RIP is perhaps the oldest interior distance vector protocol. It was developed by Xerox Corporation in the early 1980s. It uses hop count (maximum is 15) and maintains times to detect failed links. RIP has a few serious shortcomings: it ignores differences in line speed, line utilization, and other metrics. More significantly, RIP is very slow to converge for larger networks, consumes too much bandwidth to update the routing tables, and can take a long time to detect routing loops.
2. Enhanced Interior Gateway Routing Protocol (EIGRP): Cisco was the first company to solve RIP's limitations by introducing the interior gateway routing protocol (IGRP) first in the mid-1980s. IGRP allows the use of bandwidth and delay metrics to determine the best path. It also converges faster than RIP by preventing sharing hop counts and avoiding potential routing loops caused by disagreement over the next routing hop to be taken.

Cisco then enhanced IGRP to handle larger networks. The enhanced IGRP (EIGRP) combines the ease of use of traditional distance vector routing protocols with the fast rerouting capabilities of the newer link-state routing protocols. It consumes significantly less bandwidth than IGRP because it is able to limit the exchange of routing information to include only the changed information.

3. *Open Shortest Path First (OSPF)*: Open Shortest Path First (OSPF) was developed by the Internet Engineering Task Force (IETF) in RFC-2328 as a replacement for RIP. OSPF is based on work started by John McQuillan in the late 1970s and continued by Radia Perlman and Digital Equipment Corporation in the mid-1980s. OSPF is widely used as the Interior Router protocol in TCP/IP networks. OSPF is a link-state protocol, so routers inside an AS only broadcast their link-states to all the other routers. It uses configurable least cost parameters including delay, data rate/link speed, cost, and other parameters. Each router maintains a database topology of the AS to which it belongs. In OSPF every router calculates the least cost path to all destination networks using Dijkstra's algorithm. Only the next hop to the destination is stored in the routing table.

OSPF maintains three separate tables: neighbor table, link-state database table, and routing table.

(a) *Neighbor Table*: Neighbor table uses the so-called Hello Protocol to build neighbor relationship. The relationship is used to exchange information with all neighbors for the purpose of building the link-state DB table. When a new router joins the network, it sends a "Hello" message periodically to all neighbors (typically every few seconds). All neighbors will also send Hello messages. The messages maintain the state of the neighbor tables.

(b) *Link-State DB Table*: Once the neighbor tables are built, link-state advertisements (LSAs) will be sent out to all neighbors. LSAs are packets that contain information about networks that are directly connected to the router that is advertising. Neighboring routers will receive the LSAs and add the information to the link-state DB. They then increment the sequence number and forward LSAs to their neighbors. Hence, LSAs are prorogated from routers to all the neighbors with advertised information about all networks connected to them. This is considered the key to dynamical routing.

(c) *Routing Table*: Once the link-state DB tables are built, Dijkstra's algorithm (sometimes called the Shortest Path First Algorithm) is used to build the routing tables.

4. *Integrated Intermediate System to Intermediate System (IS-IS)*: Integrated IS-IS is similar in many ways to OSPF. It can operate over a variety of subnetworks, including broadcast LANs, WANs, and point-to-point links. IS-IS was also developed by IETF as an Internet Standard in RFC 1142.

2.5.3.2 Exterior Routing Protocols

Exterior Routing Protocols provide routing between autonomous systems. The two most popular Exterior Routing Protocols in the TCP/IP are EGP and BGP.

1. *Exterior Gateway Protocol (EGP)*: EGP was the first exterior routing protocol that provided dynamic connectivity between autonomous systems. It assumes that all autonomous systems are connected in a tree topology. This assumption is no longer true and made EGP obsolete.

2. *Border Gateway Protocol (BGP)*: BGP is considered the most important and widespread exterior routing protocol. Like EGP, BGP provides dynamic connectivity between autonomous systems acting as the Internet core routers. BGP was designed to prevent routing loops in arbitrary topologies by preventing routers from importing any routes that contain themselves in the autonomous system's path. BGP also allows policy-based route selection based on weight (set locally on the router), local preference (indicates which route has local preference and BGP selects the one with the highest preference), network or aggregate (chooses the path that was originated locally via an aggregate or a network), and shortest AS Path (used by BGP only in case it detects two similar paths with nearly the same local preference, weight and locally originated or aggregate addresses) just to name a few.

BGP's routing table contains a list of known routers, the addresses they can reach, and a cost metric associated with the path to each router so that the best available route is chosen. BGP is a layer 4 protocol that sits on top of TCP. It is simpler than OSPF, because it does not have to worry about functions that TCP addresses.

The latest revision of BGP, BGP4 (based on RFC4271), was designed to handle the scaling problems of the growing Internet.

2.6   Summary

This chapter focused on the "Internet" in the "Internet of Things." It started with an overview of the well-known Open System Interconnection Model Seven Layers along with the top devices and protocols. It showed how each layer divides the data it receives from end-user applications or from layer above it into protocol data units (PDUs) and then adds additional information to each PDU for tracking. This process is called the Encapsulation. Examples of PDUs include Segments on the Transport Layer, Packets on the Network Layer, and Frames on the Data Link Layer. PDUs are passed down through the stack of layers until they can be transmitted over the Physical Layer. The OSI model ensures that both users speak the same language on the same layer allowing sending and receiving layers to virtually communicate. Data passed upward is decapsulated, with the decapsulation process, before being passed further up to the destination server, user, or application.

Next, it described the TCP/IP model which is the basis for the Internet. The TCP/IP protocol has two big advantages in comparison with earlier network protocols: reliability and flexibility to expand. In fact, the TCP/IP protocol was designed for the US Army addressing the reliability requirement (resist breakdowns of communication lines in times of war). The remarkable growth of Internet applications can be attributed to its fixable expandability model.

The chapter then introduced the key IoT Network Level characteristics that included end-to-end delay, packet loss, and network element throughput. Such characteristics are vital for network design and vendor selection. The chapter next compared IP version 4 with IP version 6. It showed the limitation of IPv4, especially for the expected 50 billion devices for IoT. IPv4 has room for about 4.3 billion addresses, whereas IPv6, with a 128-bit address, has room for 2^{128} or 340 trillion trillion trillion (3.4×10^{38}) addresses. Finally detailed description of IoT Network Level routing was described and compared with classical routing protocols. It was mentioned that routing tables are used in routers to send and receive packets. Another key feature of TCP/IP routing is the fact that that IP packets travel through an internetwork one router hop at a time thus the entire route is not known at the beginning of the journey.

Problems and Exercises

1. Ethernet and Point-to-Point Protocol (PPP) are two examples of data link protocols listed in this chapter. Name two other data link protocols.
2. Provide an example of Session Layer protocol.
3. In a table format, compare the bandwidth, distance, interference rating, cost, and security of (a) twisted pair, (b) coaxial cabling, and (c) fiber optical cabling.
4. (a) What are the main components of a router? (b) Which element is considered the most essential? (c) Why?

5. What is the main function of NVRAM? Why is such function important to operate a router?
6. How do network administrators guarantee that changes in the configuration are not lost in case the router is restarted or loses power?
7. What is a disaster recovery function in a router? Which router's sub-component contains such function?
8. Many argue that routers are special computers but built to handle internetwork traffic. List three main differences between routers and personal computers.
9. There are no input devices for router like a monitor, a keyboard, or a mouse. How does a network administrator communicate with the router? List all possible scenarios. What are the main differences between such interfaces?
10. How many IPv4 addresses are available? Justify your answer.
11. What is the ratio of the number of addresses in IPv6 compared to IPv4?
12. IPv6 uses a 128-bit address, allowing 2^{128} addresses. In decimal, how many IPv6 addresses exist? How many IPv6 addresses will each human have? Why do we need billions of addresses for each human being?
13. How many IPv6 address will be available on each square meter of earth?
14. What are the major differences between interior and exterior routing protocols?
15. What is distance vector protocol? Why is it called a vector? Where is it used?
16. When would you use static routing and when would use dynamic routing? Why?
17. Most IP networks use dynamic routing to communicate between routers but may have one or two static routes. Why would you use static routes?
18. We have mentioned that in TCP/IP networks, the entire route is not known at the beginning of the journey. Instead, at each stop, the next router hop is determined by matching the destination address within the packet with an entry in the current router's routing table using internal information. IP does not provide for error reporting back to the source when routing anomalies occur.

    (a) Which Internet Protocol provides error reporting?
    (b) List two other tasks that this protocol provides?

19. Why is EGP considered to be obsolete for the current Internet?
20. In a table, compare the speed and distance Standard Ethernet, Fast Ethernet, and Giga Ethernet. Why is Ethernet connection limited to 100 m?
21. Why the Internet does require both TCP and IP protocols?
22. Are IPv4 and IPv6 protocols designed to be interoperable? How would an enterprise transition from IPv4 to IPv6?
23. What are the four different reasons for packet loss? List remediation for each reason.
24. List two factors that can affect throughput of a communication system.
25. Figure 2.10 (in Sect. 2.4.3) stated the throughput between host A and host B is 20 Mbps with the assumptions:

    • $R_2 = R_3 = R_5 = R_6 = 10$ Mbps.
    • $R_1 = R_4 = 100$ Mbps.
    • Data is equally divided between the two paths.
      How did the authors arrive at 20 Mbps?

26. Assuming host A is transferring a large file to host B. What is the throughput between host A and host B for the network shown below?

    (a) Assumptions:

        • The speed of each router is higher than the speed of any link in the network.
        • No other host is sending data.
        • $R_2 = R_3 = R_5 = R_6 = R_7 = R_8 = 10$ Mbps.
        • $R_1 = R_4 = 1$ Mbps.
        • Data is equally divided between the three paths.

    (b) Assumptions:

        • The speed of each router is higher than the speed of any link in the network.
        • No other host is sending data.
        • $R_2 = R_3 = R_5 = R_6 = R_7 = R_8 = 10$ Mbps.
        • $R_1 = R_4 = 100$ Mbps.
        • Data is equally divided between the three paths.

    (c) Assumptions:

        • The speed of each router is 1 Mbps.
        • No other host is sending data.
        • $R_2 = R_3 = R_5 = R_6 = R_7 = R_8 = 10$ Mbps.
        • $R_1 = R_4 = 100$ Mbps.
        • Data is equally divided between the three paths.

27. What is Traceroute? What does it typically report? What are the main advantageous of trace route? What is the main difference between Traceroute and Ping?
28. For the network shown below, assume the network administer is interested in measuring the end-to-end delay from router A to the server.

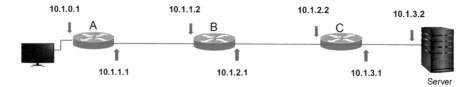

    (a) What is the Traceroute command? Hence, Traceroute command is sent from router A directly (i.e., via the shown connected terminal).

Fig. 2.12 Example of
autonomous systems

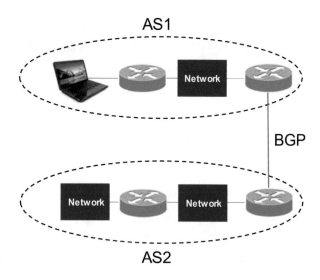

Codes: C - connected,
            S - static,
            I - IGRP,
            R - RIP,
            M - mobile,
            B - BGP
            D - EIGRP,
            EX - EIGRP external,
            O - OSPF,
            IA - OSPF inter area
            N1 - OSPF NSSA external type 1,
            N2 - OSPF NSSA external type 2
            E1 - OSPF external type 1,
            E2 - OSPF external type 2,
            E - EGP,
            i - IS-IS,
            su - IS-IS summary,
            L1 - IS-IS level-1,
            L2 - IS-IS level-2
            ia - IS-IS inter area,
            * - candidate default,
            U - per-user static route,
            o - ODR,
            P - periodic downloaded static route

Gateway of last resort is not set

24.0.0.0/16 is subnetted, 1 subnets
29.1.0.0 [110/65] via 51.29.23.12, 08:01:39, FastEthernet0/1
51.0.0.0/24 is subnetted, 1 subnets C
51.34.23.0 is directly connected, FastEthernet0/1

Fig. 2.13 Example of a routing table

(b) Which device will send their delays?

29. What is time to live command? Why is it needed?

References

1. W. Odom, CCNA Routing and Switching 200–120 Official Cert Guide Library Book, ISBN: 978–1587143878, May 2013
2. P. Browning, F. Tafa, D. Gheorghe, D. Barinic, Cisco CCNA in 60 Days, ISBN: 0956989292, March 2014
3. G. Heap, L. Maynes, *CCNA Piratical Studies Book* (Cisco Press, April 2002)
4. Information IT Online Library.: http://www.informit.com/library/content. aspx?b=CCNA_Practical_Studies&seqNum=12
5. Inter NIC (InterNIC is a registered service mark of the US Department of Commerce. It is licensed to the Internet Corporation for Assigned Names and Numbers, which operates this website)—Public Information Regarding Internet Domain Name Registration Services, Inter NIC, Online: https://lookup.icann.org/
6. Understanding TCP/IP addressing and subnetting basics, Online: https://support.microsoft. com/en-us/kb/164015
7. Tutorials Point, "IPv4 – Address Classes", Online: http://www.tutorialspoint.com/ipv4/ipv4_ address_classes.htm
8. Google IPv6, "What if the Internet ran out of room? In fact, it's already happening", Online: http://www.google.com/intl/en/ipv6/
9. Wikipedia, "Internet Protocol version 6 (IPv6):, Online: https://en.wikipedia.org/wiki/IPv6
10. IPv6 Addresses, Microsoft Windows Mobile 6.5, April 8, 2010, Online: https://msdn.micro-soft.com/en-us/library/aa921042.aspx
11. Binary to Hexadecimal Convert, Online: http://www.binaryhexconverter.com/ binary-to-hex-converter
12. Technology White Paper, Cisco Systems online: http://www.cisco.com/c/en/us/tech/ip/ip-routing/tech-white-papers-list.html
13. M. Caeser, J. Rexford, "BGP routing policies in ISP networks", Online: https://www. cs.princeton.edu/~jrex/papers/policies.pdf
14. A. Shaikh, A.M. Goyal, A. Greenberg, R. Rajan, An OSPF topology server: Design and evaluation. IEEE J. Sel. Areas Commun **20**(4) (2002)
15. Y. Yang, H. Xie, H. Wang, A. Silberschatz, Y. Liu, L. Li, A. Krishnamurthy, On route selection for interdomain traffic engineering. IEEE Netw. Mag. Spec. Issue Interdomain Rout (2005)
16. N. Feamster, J. Winick, J. Rexford, "A model of BGP routing for network engineering," in Proc. ACM SIGMETRICS, June 2004
17. N. Feamster, H. Balakrishnan, Detecting BGP configuration faults with static analysis, in *Proc. Networked Systems Design and Implementation*, (2005)
18. Apple History/Power Macintosh Gigabit Ethernet, Online: http://www.apple-history.com/ g4giga. Retrieved November 5, 2007

Chapter 3
The Things in IoT: Sensors and Actuators

3.1 Introduction

The Internet of Things (IoT) was defined in Chap. 1 as the intersection of the Internet, Things, and Data. Processes and standards were also added for a more comprehensive IoT definition. Things were defined as anything and everything stretching from appliances to buildings to cars to people to animals, to trees, to plants, etc.

Chapter 1 further categorized IoT into four main levels: IoT devices, IoT network, IoT services platform, and IoT applications. Each level has its own medium and protocols.

This chapter first defines the "Things" in IoT and then describes the key requirements for things to be able communicate over the Internet. The two main requirements for "Things" in IoT are sensing and addressing. Sensing is essential to identify and collect key parameters for analysis, and addressing is necessary to uniquely identify things over the Internet. While sensors are very crucial in collecting key information to monitor and diagnose the "Things," they typically lack the ability to control or repair such "Things" when overhaul is needed. This raise the question: why spend money to sense "Things" if they cannot be controlled? Actuators have been introduced to address this important question in IoT. With this in mind, the key requirements for "Things" in IoT now consist of sensing, actuating, and unique identification as shown in Figs. 3.1 and 3.2. It should be noted that sensing and actuating capabilities may be supported on the same device.

© The Author(s), under exclusive license to Springer Nature Switzerland AG 2022   63
A. Rayes, S. Salam, *Internet of Things from Hype to Reality*,
https://doi.org/10.1007/978-3-030-90158-5_3

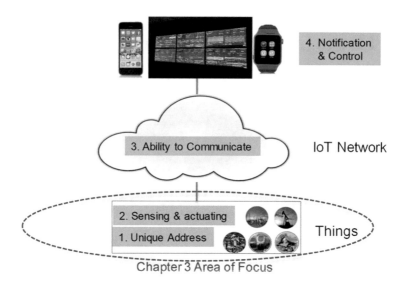

Fig. 3.1 "Thing" in IoT: definition view

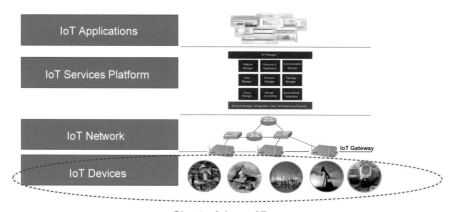

Chapter 3 Area of Focus

Fig. 3.2 "Things" in IoT: IoT level view

3.2 IoT Sensors

3.2.1 Definition

A sensor is a device (typically electronic) that detects events or changes in its physical environment (e.g., temperature, sound, heat, pressure, flow, magnetism, motion, chemical and biochemical parameters) and provides a corresponding output. Most sensors take analog inputs and deliver digital, often electrical, outputs. Because the

Fig. 3.3 Components of
smart sensors

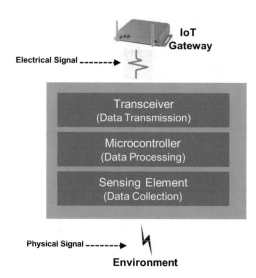

sensing element, on its own, typically produces analog output, an analog-to-digital converter is often required.

Sensors are comparable to the human five senses. They form the front end of the IoT devices, i.e., "Things." Sensors are very crucial in every IoT vertical (e.g., smart cities, smart grid, healthcare, agriculture, security and environment monitoring, and smart parking) as they bridge the world's physical objects with the Internet.

Sensors may be very simple with a core function to collect and transmit data or smart by providing additional functionality to filter duplicate data and only notify the IoT gateway when very specific conditions are met. This requires some programing logic to be present on the sensor itself. In this case, an IoT sensing device requires at least three elements—sensor(s), microcontrollers, and connectivity to send filtered data to IoT gateway or other systems. Figure 3.3 shows the components for smart sensor.

Sensors may collect large amounts of data at any time and from any location and transmit it over an IoT network in real time. The data is then analyzed and possibly correlated with other business intelligence databases to provide business insight or enhanced awareness of the environment, bringing onward opportunities and/or gains in efficiency and productivity.

3.2.2 Why Sensors

As we mentioned above, a sensor's main purpose is collecting data from its surrounding environment and providing output to its adjoining devices (e.g., gateways, actuators) or applications. Sensors typically collect data using physical interfaces (inputs) that sense the environment and then convert input signals into electrical signals (outputs) that are understood by the communication and computing devices. Output signals are then processed by the gateways and/or by applications of the IoT Platform. In some instances, sensors' outputs are processed directly by a light-weight application.

3.2.3   Sensor Types

There are many types of proprietary and nonproprietary sensors. The current IoT trend is to move away from proprietary and closed systems and embrace IP-based sensor networks. This allows native connectivity between wireless sensor networks and the Internet, enabling smart objects to participate in IoT. IP-based sensor networks require each device to be uniquely identifiable with a unique IP address so that it can be easily identifiable over a large network. Building an all-IP infrastructure from scratch, however, would be difficult because many different sensor and actuator technologies (both wired and wireless) have already been deployed over the years.

There are many different types of sensors across various technologies. The most common of which include:

1. *Temperature Sensors*: Temperature is perhaps the most commonly measured conservational quantity. This is anticipated since most physical, electronic, chemical, mechanical, and biological systems are affected by temperature. There are four types of temperature sensors:

   (a) Thermocouple Sensors: A thermocouple is a device consisting of two different and dissimilar conductors in contact. It produces a voltage as a result of the thermoelectric effect. Thermocouple sensor is made by joining two dissimilar metals at one end.
   (b) Resistance Temperature Detector (RTD) Sensors: RTDs are temperature sensing devices whose resistance changes with temperature. They have been used for many years to measure temperature in laboratory and industrial processes and have developed a reputation for accuracy, repeatability, and stability.
   (c) Thermistors: Similar to the RTD, the thermistor is a temperature sensing device whose resistance changes with temperature. Thermistors, however,

Fig. 3.4 Examples of temperature sensors and applications

are made from semiconductor materials. Resistance is determined in the same manner as the RTD, but thermistors exhibit a highly nonlinear resistance vs. temperature curve.

(d) Semiconductor Sensors: They are classified into different types like voltage output, current output, digital output, resistance output silicon, and diode temperature sensors. Modern semiconductor temperature sensors offer high accuracy and high linearity over an operating range of about 55 °C to +150 °C (−58 to 302 °F). They can also include signal processing circuitry within the same package as the sensor, thereby avoiding the need to add compensation circuits. Figure 3.4 shows examples of temperature sensor.

2. *Pressure Sensors*: Pressure sensors are used to measure the pressure of gases or liquids including water level, flow, speed, and altitude. Practical examples include sensors for pumps and compressors, hydraulic systems, and refrigerators. A pressure sensor typically acts as a transducer where it generates a signal as a function of the pressure imposed. Hence, pressure sensors are also called pressure transducers, pressure transmitters, and pressure senders, among other names.

Touchscreen smartphones, tablets, and computers come with various pressure sensors. Whenever slight pressure is applied on the touch screen through a finger, tiny pressure sensors (typically multiple sensors located at the corners of the screen; see Fig. 3.5) determine where exactly pressure is applied and consequently generate an output signal that informs the processor. Pressure sensors have also been widely used in automotive applications to measure fluid level, airbag, and antilock braking system, in biomedical applications to sense blood pressure, in aviation to maintain a balance between the atmospheric pressure and the control systems of the airplanes, and in submarines to estimate depth and ensure proper operation of electronic systems and other components. Figure 3.5 shows examples of pressure sensors.

3. *Flow Sensors*: Flow sensors are used to detect and record the rate of fluid flow in a pipe or a system. They are also used to measure the flow/transfer of heat caused by the moving medium. Sensing and measuring the flow are critical for many applications ranging from bereave machine to more serious applications such as flow monitoring for high-purity acids.

Fig. 3.5 Examples of pressure sensors. (Source: Force Sensing & Fitbit)

Fig. 3.6 Examples of flow sensor

Fig. 3.7 Examples of level sensors with Wi-Fi propane remote monitoring. (Source: Tank Utility)

A good example about the importance of flow sensing and monitoring is the water crisis in Flint, Michigan, USA, which started in April 2014 and resulted in criminal charges filed against three people in regard to the crisis by Michigan Attorney General in April 2016.

Flint basically changed its water source from treated Detroit Water that was sourced from the great lakes and the Detroit River to the Flint River. Officials basically had failed to detect a very high lead contamination creating a serious public health danger. The acidic Flint River water caused lead from aging pipes to leak into the water supply, causing extremely elevated levels of the heavy metal. Thousands of children were exposed to drinking water with very high levels of lead, and many experienced health problems (Fig. 3.6).

4. *Level Sensors*: Level sensors are used to measure the level of fluids continuously or at point values. The element to be measured can be inside a container (Fig. 3.7) or can be in its natural form such as a well in an oil rig.

   There are many uses for level sensors. Ultrasonic level sensors, for instance, are used for non-contact level sensing of highly viscous liquids and even bulk solids. They are also widely used in water treatment applications for pump control and open-channel flow measurement. Another example is the capacitance level sensors to measure the presence of a variety of solids and liquids using radio-frequency signals in the capacitance circuit.

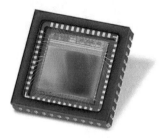 

Fig. 3.8 Examples of imaging sensors. (Source: e2v & DGDL)

5. *Imaging Sensors*: Imaging sensors are sophisticated sensors used in digital cameras, medical imaging machines, and night vision equipment. They are utilized to measure image information by capturing and then converting variable attenuation of waves into signals (Fig. 3.8).
6. *Noise Sensors*: High noise can have damaging effects on humans (e.g., cardiovascular) as well as animals (e.g., hearing loss). Such noise is often caused by machines, airplanes, trains, construction, and loud music especially in closed spaces.

    Many government agencies have started installing noise sensors to measure noise pollutions or the so-called noise disturbance (excessive noise that may harm humans or animals).

    Ambient noise sensors continuously monitor noise levels in surrounding environments. When the noise level changes, they send electronic signal to an overall ambient noise system to take action. Such action may be an automatic action (e.g., adjust music level) or a simple notification to authorities.
7. *Air Pollution Sensors*: Many governments have established agencies to monitor and control the air quality in major cities. For instance, the USA has established the EPA (Environmental Protection Agency), in 1970, with a mission to protect Americans from significant health risks by providing accurate environmental information to its citizens.

    Air pollution sensors detect and monitor the presence of air pollution in the surrounding environment. They focus on five main components: ozone, particulate matter, carbon monoxide, sulfur dioxide, and nitrous oxide.
8. *Proximity and Displacement Sensors*: Proximity sensors detect the presence or absence of objects using electromagnetic fields, light, or sound. There are many types, each suited to specific applications and environments:

    (a) Inductive Sensors: Used for close-range detection of ferrous material.
    (b) Capacitive Sensors: Used for close-range detection of nonferrous material.
    (c) Photoelectric Sensors: Used for long-range target detection.
    (d) Ultrasonic Sensors: Used for long-range detection of targets with difficult surface (Table 3.1).

9. *Infrared Sensors*: Infrared sensors are used to track an object's movement. They produce and receive infrared waves in the form of heat.

Table 3.1 Examples of proximity sensor types

Sensor technology	Sensing range (mm)	Main use
Inductive	4–40	Ferrous metal (e.g., iron, aluminum, copper) close-range detection
Capacitive	3–60	Nonferrous material (e.g., wood, plastic liquid) close-range detection
Photoelectric	1–60	Material long-range target detection
Ultrasonic	3–30	Material long-range target detection with challenges (e.g., rough service, multiple colors)

10. *Moisture and Humidity Sensors*: Moisture and humidity sensors (sometimes referred to as hygrometer sensors) are used to measure and report the relative humidity in the air. They use capacitive measurement by relying on electrical capacitance.
11. *Speed Sensors*: Speed sensors are commonly used to detect the speed of transport vehicles. Examples include wheel speed sensors, speedometers, Doppler radar, and laser surface velocimeter.

There are so many other types of sensors. Examples include acceleration sensors, biosensors, gas and chemical sensors, mass sensor, tilt sensors, and force sensors.

3.2.4 Sensor Characteristics

Most IoT applications require smaller and smarter sensors with advanced functionality to collect more data, low-power processors, longer battery life, faster response time, and shorter time to market. Sensors are expected to be dynamic in their natural surroundings with embedded ability to collect real-time data.

In general, sensors can be either self-directed (autonomous) where they work on their own once they are installed or user-controlled where collection conditions are preprogrammed by the user depending on their needs. Finally, sensors should also have the capability to send the collected data (or a subset of it) to the appropriate system via the IoT gateway as we illustrated in Fig. 3.2.

IoT sensors are expected to have the following characteristics:

1. *Data Filtering*: A sensor's core function is the ability to collect and send data to the IoT gateway or other appropriate systems. Sensors are not expected to perform deep analytical functions. However, simple filtering techniques may be required. Onboard data (or signal) processing microcontroller (as shown in Fig. 3.3) makes a smart sensor smarter. The microcontroller filters the data/signals before transmission to the IoT gateway or control network. It basically removes duplicate or unwanted data or noise before transferring the data.

   As we mentioned in Sect. 3.2.3, non-autonomous sensors are custom-programmed to produce alerts automatically when certain conditions are met (e.g., temperature is above 70 °F in a data center). They often integrate VLSI technology and MEMS devices to reduce cost and optimize integration.

2. *Minimum Power Consumption*: Several factors are driving the requirements for low-power consumptions in IoT. Sensors for multiple IoT verticals (e.g., smart grid, railways, and roadsides) will be installed in locations that are difficult to reach to replace batteries.

3. *Compact*: Space will also be limited for most IoT verticals. As such, sensors need to fit in small spaces.

4. *Smart Detection*: An important sensing category for the IoT is remote sensing, which consists of acquiring information about an object without making physical contact with it; the object can be nearby or several hundred meters away. Multiple technology options are available for remote sensing, and they can be divided into three broad functions:

    (a) Presence or proximity detection—when just determining the absence or presence of an object is sufficient (e.g., security applications). This is the simplest form of remote sensing.
    (b) Speed measurement—when the exact position of an object is not required, but accurate speed is (e.g., traffic monitoring applications).
    (c) Detection and ranging—when the position of an object relative to the sensor must be determined precisely and accurately (e.g., vehicle collision avoidance).

5. *High Sensitivity*: Sensitivity is generally the ratio between a small change in electrical output signal and a small change in physical signal. It may be expressed as the derivative of the transfer function (the functional relationship between input signal and output signal) with respect to physical signal. Sensitivity indicates how much the output of the device changes with unit change in input (quantity to be measured). For example, if the voltage of a temperature sensor changes by 1 mV for every 1 °C change in temperature, then the sensitivity of the sensor is said to be 1 mV/°C.

6. *Linearity*: Linearity is the measure of the extent to which the output is linearly proportional to the output. Nonlinearity is the maximum deviation from a linear transfer function over the specified dynamic range.

7. *Dynamic Range*: The range of input signals which may be converted to electrical signals by the sensor. Outside of this range signals cause unsatisfactory accuracy.

8. *Accuracy*: The maximum expected error between measured (actual) and ideal output signals. Manufacturers often provide the accuracy in the datasheet, e.g., high-quality thermometers may list accuracy to within 0.01% of full-scale output.

9. *Hysteresis*: When a sensor does not return the same output value when the input stimulus is driven up or down. The width of the expected error in terms of the measured quantity is defined as the hysteresis.

10. *Limited Noise*: All sensors produce some level of noise traffic with their output signals. Sensor noise is only an issue if it impacts the performance of the IoT

system. Smart sensors must filter out unwanted noise and be programmed to produce alerts on their own when critical limits are reached. Noise is generally distributed across the frequency spectrum. Many common noise sources produce a white noise distribution, which is to say that the spectral noise density is the same at all frequencies.

11. *Wide Bandwidth*: Sensors have finite response times to instantaneous changes in physical signal. Also, many sensors have decay times, which represent the time after a step change in input signal for the sensor output to decay to its original value. The bandwidth of a sensor is the frequency range between these two frequencies. When a sensor is utilized to collect measurements, it is recommended to use sensors with the widest possible bandwidth. This ensures that the basic measurement system is capable of responding linearly over the full range of interest. The disadvantage, however, is that wider bandwidth may result in sensor response to unwanted frequency.

12. *High Resolution*: The resolution of a sensor is defined as the smallest detectable signal fluctuation. It is the smallest change in the input that the device can detect. The definition of resolution must include some information about the nature of the measurement being carried out.

13. *Minimum Interruption*: Sensors must operate normally at all time with zero or near-zero interruption and be programmed to produce instant alerts on their own when their normal operation is interrupted.

14. *Higher Reliability*: Higher reliability sensor provides the assurance to rely on the accuracy of the output measurements.

15. *Ease of Use*: Ease of use is considered the top requirement for any electronic system nowadays. Clear examples we have all experienced are Apple's iPhone vs. competitor devices with the same functionality. Users are willing to pay more for easy-to-use devices, and sensors are no exceptions. The best user interface is "no user interface" where sensors are expected to work by themselves once they are connected.

Other characteristics include some data storage and self-warning of anomalous symptoms.

3.3   RFID

Another way of capturing information from "Things" is through the use of RFID (radio-frequency identification). RFID is not a sensor but a mechanism to capture information pre-embedded into the so-called Tag of a thing or an object using radio waves.

RFID consists of two parts: a tag and a reader. Further, the tag has two parts: a microchip that stores and processes information and an antenna to receive and transmit a signal. The tag contains the specific serial number for one specific object. The reader reads the information encoded on a tag, using a two-way radio transmitter-receiver, by emitting a signal to the tag using an antenna. The tag responds with the information written in its memory. The reader will then transmit the read results to an RFID computer program.

An RFID-based solution has some advantages over older reader-tag-based solutions, such as barcode, including:

- RFID tag does not need to be within direct line of sight of the reader and can be read from a distance up to 12 m for passive ultrahigh frequency (UHF) system. Battery-powered tags typically have a reading range of 100 m.
- RFID data on the tag can be modified based on business needs. The barcode data is very difficult to change once deployed.
- RFID tags are durable. Barcodes, in comparison, are printed on a product for everyone to see. They can be damaged or changed. RFID tags are hidden and may be reused across multiple products. Also RFID tags are capable of storing much more data.
- RFID data may be encrypted on the tag, thereby preventing unauthorized users from changing the data or counterfeiting.
- RFID systems can read hundreds of tags simultaneously. This is significant in a retail store as it saves the staff valuable time that they can spend on higher-value tasks.

Figure 3.9 shows the RFID main components: a programmable RFID tag for storing data, a reader with an antenna to read the tags, and an application software hosted on a computer to analyze the data.

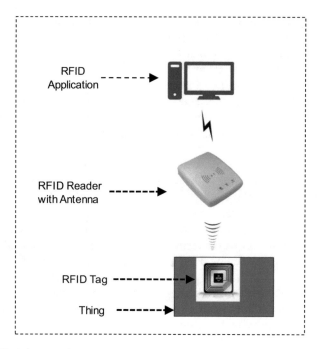

Fig. 3.9 RFID main components

Like any other technology, RFID has a number of disadvantages, but they are relatively minor. A top disadvantage is the susceptibility of the tags to jamming by blocking the RFID radio waves, for instance, by wrapping the tags with metallic material such as aluminum foil. Metallic ink on book covers can also affect the transmission of the radio waves.

Another potential disadvantage is the interference between multiple readers and tags if the overall system is not set up appropriately. Each RFID reader basically scans all the tags it picks up in its range. This may create a mix-up between tag information (e.g., charging a customer for items in someone else's shopping carts within the same range).

3.3.1 RFID Main Usage and Applications

RFID is already used by a large number of applications. Top examples include:

- *Access Control and Management*: Many companies and government agencies are using RFID tags in identification badges, replacing earlier magnetic stripe cards. With RFID, employees as well as authorized guest may be greeted by their name on a screen or by a voice message upon entering a building. Companies are currently using data collected from the information associated with each employee's badge to plan for workplace optimization.
- RFID tags are also widely used for electronic toll collections (e.g., California's E-ZPass) eliminating major delay on toll roads. Electronic toll collection system determines if the passing vehicle is enrolled in the program, automatically issues traffic citations for those that are not, and automatically withdraws the toll charges from the accounts of registered car owners.
- *Passport*: Many departments of state around the world (e.g., the USA, Canada, Norway, Malaysia, Japan, and many EU countries) are using RFID passports that can be read from a reader located up to 10 m away. In this case, passports are designed with an electronic tag that contains main information with a digital picture of the passport holder. Most solutions are also adding a thin metal lining to make it more difficult for unauthorized readers to scan information when the passport is closed. Standards for RFID passports have been established by the International Civil Aviation Organization, and are contained in ICAO Document 9303 (6th edition, 2006).
- *Healthcare*: With 2014 veteran complaints including allegations that 40 veterans may have died waiting for care at a Phoenix VA hospital, many hospitals or agencies, including the US Department of Veterans Affairs, have already started or announced plans to deploy RFID in hospitals across the USA to improve healthcare.
- RFID-based solutions in healthcare have started in private and public hospitals across the world, at least several years before the veteran's complaints, to track and manage expensive mobile medical equipment thereby allowing hospital staff

to track in real-time data relevant to healthcare equipment or personnel, monitor environment conditions, and more importantly protect healthcare workers from infections and other hazards.

- *Logistics and Supply Chain Tracking*: Major retailers in the world (e.g., Walmart), as well as the US Department of Defense, have published requirements that their vendors place RFID tags on all shipments to improve supply chain management. Such requirements allow retailers to manage their merchandise without manual data entry. RFID can also help with automatic electronic surveillance and self-checkout process for consumers. Finally, many factories are tracking their products throughout the manufacturing process using RFIDs to better estimate delivery dates for customers.
- *Athletic and Sport Event Timing*: Tracking the exact timing of runners in marathons or races is crucial, and often a portion of a second makes a difference. Athletic Timing is one of the most widespread use cases of RFID. Many runners are not even aware that they are being timed with RFID technology. Experts use such fact as an evidence of RFID's seamless ability to enhance consumer experience.
- *Animal Tracking*: Since the outbreak of mad cow disease, RFID has become critical in animal identification management, although RFID animal tagging started at least a decade before the disease. Some governments (i.e., Australia) are now requiring all cattle, sheep, and goats sold to be RFID tagged.
- *Other Applications*: RFID is also used for airport baggage tracking logistics, interactive marketing, laundry management for employers with huge number of uniforms (e.g., casinos), item level inventory tracking, conference attendee tracking, material management, IT asset management, library system, and real-time location system.

3.4   Video Tracking

Video tracking is the process of capturing and analyzing the video feeds, frame by frame, of a particular object or person over a short time interval. It is used to measure and analyze movements, visual attention, as well as behavior. Video tracking is used for customer identification, surveillance, augmented reality, traffic control, and medical imaging.

It is yet another mechanism to identify and monitor "things" when sensors or RFID tags are not available. Video tracking may also be used in conjunction with sensors and/or RFID to provide a more comprehensive solution.

Unlike preinstalled sensors and RFID tags in "things," video tracking can be turned on instantly. However, video tracking does have a major weakness, with today's technology. Video tracking is often time-consuming. It requires analyzing large amounts of video traffic and, in many cases, correlation with historical data, to arrive at accurate conclusions. Another challenge to video tracking is the complex object/image recognition techniques. This is a huge area of research in machine learning today.

3.4.1   Video Tracking Applications

- *Retailers*: Many retailers have started using video tracking solutions, often in conjunction with Wi-Fi access point data (how?; see problem 22), to increase sales and provide a better customer experience. Video traffic is analyzed using complex algorithms that track eye movements and identify fixation (e.g., desirability, obsession, and attraction to a product) and glissades (e.g., wobbling movements). The collected data is then filtered against well-established business rules to determine an internal action (e.g., change location of merchandize, add more checkout lines) or external action (e.g., offer the customer a certain discount).
- Determining the business rules is a very challenging problem. Many companies use advanced systems and techniques (e.g., machine learning, analysis of social media data, artificial intelligence) or hire a marketing firm to survey a large number of customers to arrive at such rules. Example of new rules is the fact that the faster a shopper finds the first item she/he needs, the more she/he purchases in such category. This dispels the pervious myth that the more time a shopper spends in a particular area, the more she/he buys.
- Video tracking is also used to improve the overall shopping experience in the store as a service differentiation especially if the store is a bit more expensive than similar stores in the area. The analysis of multiple grocery store traffic indicated that customers did not mind paying a bit more for faster checkout lines with friendly cashiers, bright lights, and clean belts. Analyzed data also indicated that the vast majority of customers do not pay attention to internal signs inside the store.
- *Banking*: Similar to retailers, banks have also started using video tracking solutions, often combined with Wi-Fi data. Access to Wi-Fi data in banks is easier given that most of the customers download the bank's mobile app on their smartphones. With the right setting, mobile apps often allow the bank to track the whereabouts of the customer.
- Banks use the data to quickly identify the customer (often before he lines in the queue). Depending on the priority of such customer (e.g., has large sums of money deposited at the bank), special greeting may be zero-wait private service if offered by the bank manager.
- *Other Uses*: The applications of video tracking with advanced backend analytics are unlimited, ranging from physical monitoring and security to traffic management and control and to augmented reality where an actual view is augmented by a computer-generated sensual input such as video.

3.4.2 Video Tracking Algorithms

To perform video tracking, an algorithm analyzes sequential video frames and out-puts the movement of targets between the frames. There is a variety of algorithms, each having its own strengths and weaknesses. Considering the intended application is important when choosing which algorithm to use. There are two major compo-nents of a visual tracking system: target representation and localization and filtering and data association.

Target representation and localization are mostly a bottom-up process. These methods give a variety of tools for identifying the moving object. Locating and tracking the target object successfully are dependent on the algorithm. For example, using blob tracking is useful for identifying human movement because a person's profile changes dynamically [6]. Typically, the computational complexity for these algorithms is low. The following are some common target representation and local-ization algorithms:

Kernel-based tracking (mean-shift tracking [7]): an iterative localization procedure based on the maximization of a similarity measure (Bhattacharyya coefficient).
Contour tracking: detection of object boundary (e.g., active contours or Condensation algorithm). Contour tracking methods iteratively evolve an initial contour initial-ized from the previous frame to its new position in the current frame. This approach to contour tracking directly evolves the contour by minimizing the con-tour energy using gradient descent.

Filtering and data association is mostly a top-down process, which involves incorporating prior information about the scene or object, dealing with object dynamics, and evaluation of different hypotheses. These methods allow the tracking of complex objects along with more complex object interaction like tracking objects moving behind obstructions [8]. Additionally, the complexity is increased if the video tracker (also named TV tracker or target tracker) is not mounted on rigid foundation (onshore) but on a moving ship (offshore), where typically an inertial measurement system is used to pre-stabilize the video tracker to reduce the required dynamics and bandwidth of the camera system [9]. The computational complexity for these algorithms is usually much higher. The following are some common filter-ing algorithms:

Kalman filter: an optimal recursive Bayesian filter for linear functions subjected to Gaussian noise. It is an algorithm that uses a series of measurements observed over time, containing noise (random variations) and other inaccuracies, and pro-duces estimates of unknown variables that tend to be more precise than those based on a single measurement alone [10].
Particle filter: useful for sampling the underlying state-space distribution of non-linear and non-Gaussian processes.

3.5   IoT Actuators

3.5.1   Definition

An actuator is a type of motor that is responsible for controlling or taking action in a system. It takes a source of data or energy (e.g., hydraulic fluid pressure, other sources of power) and converts the data/energy to motion to control a system.

3.5.2   Why Actuators?

As mentioned in Sect. 3.2, sensors are responsible to sense changes in their surroundings, collect relevant data, and make such data available to monitoring systems. Collecting and displaying data by a monitoring system are useless unless such data is translated into intelligence that can be used to control or govern an environment before a service is impacted. Actuators use sensor-collected and analyzed data as well as other types of data intelligence (see problem 11) to control IoT systems, for example, shutting down gas flow when the measured pressure is below a certain threshold.

3.5.3   Actuator Types

- *Electrical Actuators*: Electric actuators are devices driven by small motors that convert energy to mechanical torque. The created torque is used to control certain equipment that requires multi-turn valves or to control gates. Electric actuators are also used in engines to control different valves.
- *Mechanical Linear Actuators*: Mechanical actuators convert rotary motion to linear motion. Devices such as screws and chains are utilized in this conversion. The simplest example of mechanical liner actuators is referred to as the "screw" where leadscrew, screw jack, ball screw, and roller screw actuators all operate on the same principle by rotating the actuator's nut, the screw shaft moves in a line.
- *Hydraulic Actuators*: Hydraulic actuators are simple devices with mechanical parts that are used on linear or quarter-turn valves. They are designed based on Pascal's Law: when there is an increase in pressure at any point in a confined incompressible fluid, then there is an equal increase at every point in the container. Hydraulic actuators comprised of a cylinder or fluid motor that utilizes hydraulic power to enable a mechanical process. The mechanical motion gives an output in terms of linear, rotary, or oscillatory motion. Hydraulic actuators can be operated manually, such as a hydraulic car jack, or they can be operated through a hydraulic pump, which can be seen in construction equipment such as cranes or excavators.

- *Pneumatic Actuators*: Pneumatic actuators work on the same concept as hydraulic actuators except compressed gas is used instead of liquid.
- *Manual Actuators*: Manual actuator employs levers, gears, or wheels to enable movement, while an automatic actuator has an external power source to provide motion to operate a valve automatically. Power actuators are a necessity on valves in pipelines located in remote areas.

3.5.4 Controlling IoT Devices

There are two main philosophies to monitor and control IoT devices: local control and global control. The first approach requires an intelligent local controller (e.g., home's thermostat to control furnace and air conditioning system). The second approach is to move the control onto the cloud and simply embed inexpensive sensors everywhere (e.g., in this case, thermostat is eliminated altogether), and instead put temperature sensors around the house. An extension of this would be to pull the controller boards out of the furnace and air conditioner—connect their inputs and outputs to the Internet as well, so a cloud application can directly read their states and control their subsystems.

Clearly this approach requires many more, much finer-grained connected devices. And it offers the possibility of control strategies that would not be possible for an isolated thermostat. You could use ambient weather conditions, forecasts, and the current locations of the residents as inputs, for example, to determine an optimum strategy for making life comfortable while saving energy.

We believe the right approach is a combination of the two approaches depending on the specific IoT vertical and environment. This area will be covered in more details in Chap. 9.

3.6 How Things Are Identified in IoT?

As we mentioned in Chap. 2, the most convenient way to identify every IoT devices is to assign unique IP address to each sensor and actuator. However, IPv4 addresses are expensive and limited. IPv6 addresses are not widely deployed yet. In addition, many sensors and actuators are not IP enabled. IoT gateways, however, do have unique IP addresses. Hence, non-IP-enabled sensors and actuators may be identified by their associated gateways.

Chapter 5 will provide compressive details of various sensing protocols and illustrate how IoT sensors and actuator will be tracked and identified.

3.7   Summary

This chapter defined the "Things" in IoT. Three main techniques to identify things were discussed in details: embedded hardware sensors that sense the thing or surrounding environment and notify a client application, RFIDs with a tag to store information on a thing and a reader to read such information and pass them to an application to analyze, and finally video tracking. The advantages and disadvantages of these solutions were discussed. Once the data is analyzed (from sensors or other sources), IoT actuators are responsible for controlling or taking action if required. Finally, we have discussed the procedure to identify various devices in IoT networks.

Problems and Exercises

1. List the top three requirements for "Things" in IoT? What is the purpose behind these requirements?
2. Why are actuators required in IoT networks?
3. What is the definition of a sensor in IoT? Why is there a need for A/D converters in most sensors?
4. Why are sensors required to convert physical signals into electrical signal?
5. In a table, list and compare the various types of actuators. Which actuator type is considered to be environmentally friendly and why?
6. What are the key differences between sensors and actuators?
7. Chapter 1 (Sect. 1.2) mentioned that connecting objects together is not an objective by itself. Sections 3.1 and 3.5.2 mentioned that collecting data from sensors is not an objective by itself either. What is the business objective for connecting things and collecting data? How to achieve such objective?
8. What are the two main uses of flow sensors?
9. In a table format, list the key functionality of all sensors (A through I) listed in Sect. 3.2.3. Which sensor type is considered to be the least sophisticated, and which type is considered to be the most sophisticated? Why?
10. What is an autonomous sensor? When does it notify neighboring system(s) or IoT gateway? What is the difference between "autonomous" and "user-controller" sensors?
11. In a table, list and compare the ten characteristics of good sensors. Which characteristic you believe is the most important and why?
12. It was mentioned in Sect. 3.3 that actuators use sensor-collected and analyzed data as well as other types of data intelligence to control IoT systems. What is data intelligence? Provide two examples of data intelligence.
13. What is the definition of sensitivity and dynamic range? What are the typical units of sensitivity and dynamic range?
14. What is hysteresis? What is a typical unit of hysteresis?
15. How do touch screens operate with the presence of touch sensors?
16. In a table, list five examples of industries where pressure sensors are used. In each case, list at least one main application.

17. Some people have raised concerns about the potential invasion of privacy in RFID-enabled solutions (e.g., track the whereabouts of a person who checked out an RFID-enabled library book). Is this a major concern? How would you address it?
18. Athletic Timing: Athletic Timing is one of the most popular use cases of RFID, but often race participants never realize they are being timed using RFID technology. How does it work?
19. Describe how RFID works for laundry management. List three benefits.
20. Provide an example of how RFID works for interactive marketing.
21. How does RFID track the real-time location of assets or employees? What other technology can be used to track an employee location in real time?
22. How do retailers use Wi-Fi access point data in conjunction with video tracking to improve sales and customer experience?
23. This chapter discussed three different ways to obtain information from IoT "Things": sensors, RFID, and video tracking. In a table, compare the three technologies addressing:

    (a) Advantages
    (b) Disadvantages
    (c) Key requirements for the things
    (d) Two applications

24. What are transducers? How are they related to sensors and actuators?
25. Wind speed sensors typically involve a rotating element that is set in motion by wind. These sensor report the frequency of rotation of that moving element. An application receiving the frequency readings needs to apply a "transfer function" to translate the frequency to actual wind speed. In the weather monitoring station at Vancouver International Airport, two wind speed sensors are installed: an RM Young 05103 Wind Sensor and a Vaisala WM30 Wind Sensor. The first has the following transfer function: Wind Speed (m/s) = 0.0980 × Frequency. The second has this transfer function: Wind speed (m/s) = 0.699 × Frequency − 0.24.

    (a) If the RM Young sensor is reporting frequency of 20 Hz, and assuming both sensors are measuring the same wind speed value, then what would be the frequency reported by the Vasiala sensor?
    (b) What would be the actual wind speed measured?

References

1. A Framework for IoT Sensor Data Analytics and Visualisation in Cloud Computing Environments, University of Melbourne, Online: http://www.cloudbus.org/students/Krishnakumar-IoT-Project2011.pdf
2. Wikipedia, Online.: https://en.wikipedia.org/wiki/Sensor
3. Sensors: Online Electr. Eng. Online: http://www.electrical4u.com/sensor-types-of-sensor/

4. I. Gubbia, R. Buyyab, S. Marusica, M. Palaniswamia, Internet of things (IoT): A vision, architectural elements, and future directions. Futur. Gener. Comput. Syst. **29**(7), 1645–1660 (2013)
5. L. Patrono, A. Vilei, Evolution of wireless sensor networks towards the Internet of Things: A survey, 19th IEEE SoftCom, p. 106, Sept. 2011
6. Sukanay, A Walk Through IoT, Online: https://opentechdiary.wordpress.com/2015/07/16/a-walk-through-internet-of-things-iot-basics-part-2/
7. T. A. Kinney, Proximity sensors compared: Inductive, capacitive, photoelectric and ultrasonic, Online: http://machinedesign.com/sensors/proximity-sensors-compared-inductive-capacitive-photoelectric-and-ultrasonic
8. Sensor Basics - Different Types of Sensors with Working Principles, https://www.youtube.com/watch?v=Xs1uicalZYA
9. Thermocouple, Wikipedia, Online: https://en.wikipedia.org/wiki/Thermocouple
10. T. Agarwal, Temperature sensors – Types, Working & Operation, white paper, Online: http://www.elprocus.com/temperature-sensors-types-working-operation/
11. Future Electronics, Online: http://www.futureelectronics.com/en/sensors/humidity-dew.aspx
12. Paul Garden, Electronic design, Online: http://electronicdesign.com/communications/iot-requires-new-type-low-power-processor
13. Common Actuator Types, Online: http://www.vma.org/?ActuatorTypes
14. Actuators, The Green Book, Online: http://www.thegreenbook.com/four-types-of-actuators.htm
15. Type of Robot Actuators, Robot Platform, Online: http://www.robotplatform.com/knowledge/actuators/types_of_actuators.html
16. S. Duquet, Smart sensors, enabling detection and ranging for IoT and beyond, ladder technology magazine Elektronik Praxis, April 2015, Online: http://leddartech.com/smart-sensors
17. 50 Sensors Applications for Smarter World, Libelium, Online: http://www.libelium.com/top_50_iot_sensor_applications_ranking/
18. P. Seneviratne, Internet Connected Smart Water Sensors, September 2015, Online: https://www.packtpub.com/books/content/internet-connected-smart-water-meter
19. P. Jain, Pressure sensors, prototype PCB from $10, Online: http://www.engineersgarage.com/articles/t
20. D. Merrill, J. Kalanithi, P. Maes, Siftables: Towards sensor network user interfaces, Online: http://alumni.media.mit.edu/~dmerrill/publications/dmerrill_siftables.pdf
21. A. Alcom, RFID Can Be Hacked: Here's How, & What You Can Do To Stay Safe, October 2012, Online: http://www.makeuseof.com/tag/rfid-hacked-stay-safe/
22. B. Hoffmann, S. Moebus, A. Stang, E. Beck, N. Dragano, S. Möhlenkamp, A. Schmermund, M. Memmesheimer, K. Mann, 2006, Residence close to high traffic and prevalence of coronary heart disease. Eur. Heart J. **27** Online: http://www.ncbi.nlm.nih.gov/pubmed/17003049
23. J. Thrasher, RFID vs. Barcodes: What are the advantages?, RFID Insider, April 2013, Online: http://blog.atlasrfidstore.com/rfid-vs-barcodes
24. S. Egloff, Advantages and disadvantages of using RFID Technology in Libraries, information Technology at the University of Maryland, Online: http://terpconnect.umd.edu/~segloff/RFIDTest3/AdvantagesandDisadvantages.html
25. P. Harrison, EU considers overhauling rules for lost air luggage. Reuters, September 2009, Online: http://www.reuters.com/article/eu-aviation-baggage-idUSLS63631320090728
26. P. Sweeney, Social Media Winner's Circle at Geneva Motor Show, Social Media Today, September 2013, Online: http://www.socialmediatoday.com/content/social-media-winners-circle-geneva-motor-show-video
27. J. Thrasher, "How is RFID Used in Real World Applications?", August 2013, Online: http://blog.atlasrfidstore.com/what-is-rfid-used-for-in-applications
28. M. Nystrom, K. Holmqvist, An adaptive algorithm for fixation, saccade and glissade detection in eye tracking data. Behav. Res. Methods **42**(1), 188–204 (2010)
29. Tank Monitoring on a New Level, Online: https://www.tankutility.com/

Chapter 4
IoT Requirements for Networking Protocols

The success of the Internet is attributed, in part, to the Internet Protocol stack that offers two key characteristics:

- A normalization layer (the IP layer), which guarantees system interoperability while accommodating a multitude of link layer technologies, in addition to a plethora of application protocols. IP constitutes the thin waist of the proverbial hourglass that is the Internet's protocol stack.
- Layered abstractions that hide the specifics of a given layer from the one above or below it. Such abstractions define contracts or "slip surfaces" allowing innovations in one layer to proceed independent of the adjacent layers.

As researchers and technologists started delving into the world of IoT, it was relatively straightforward to justify the benefits of employing a similar layered architectural approach for the IoT protocol stack. However, a topic of lively debate emerged in whether the Internet Protocol stack was suited for the IoT or whether a new stack was needed. In the late 1990s and early 2000s, many researchers in the field of wireless sensor networks did not shy away from denouncing IP networking as unsuitable for that application domain.

It was deemed that the requirements of IoT were sufficiently different to warrant a white canvas approach, rather than reusing the Internet technology, which fell short of addressing the requirements in a number of areas. The decade and a half that followed witnessed an evolution of the IP stack to address many of the cited requirements for sensor networks and the shortcomings of IP technologies at the time.

In this chapter, we will discuss the key IoT requirements and their impact on each of the layers of the protocol stack. In the next chapter, we take a layer-by-layer view and discuss the industry's efforts, to date, to address these requirements. We will also discuss the gaps that remain for further study and require future solutions.

© The Author(s), under exclusive license to Springer Nature Switzerland AG 2022
A. Rayes, S. Salam, *Internet of Things from Hype to Reality*,
https://doi.org/10.1007/978-3-030-90158-5_4

4.1   Support for Constrained Devices

The devices that are to be connected to the network in the IoT span a wide gamut of capabilities and characteristics along the facets of computational power, mobility, size, complexity, dispersion, power resource, placement, and connectivity patterns. These and other device characteristics impose a set of requirements and restrictions on the network infrastructure used for interconnecting them. In particular, the devices' computational capabilities, as well as their power resources, introduce challenging requirements for IP networking technologies.

Stepping back and examining the devices that have traditionally connected to the Internet, one can easily categorize them as homogeneous in terms of being fully capable computers or peripherals (e.g., servers, desktops, laptops, printers, etc.) that have an endless source of power (e.g., mains powered or equipped with rechargeable batteries). In the IoT, this homogeneity no longer holds: on one end of the spectrum are devices with very limited processing power which scavenge energy from their environment (e.g., pressure sensors), and on the other end are devices with powerful processors, a generous amount of memory, and replenishable power sources (e.g., smartphones).

Small devices with limited processing, memory, and power resources are referred to as constrained devices. Generally speaking, a constrained device is limited in one or more of the following dimensions:

- Maximum code complexity (ROM/Flash).
- Size of run-time state and buffers (RAM).
- Amount of computation feasible in a specific period of time ("processing power").
- Available power resources.
- Management of user interface and accessibility in deployment (ability to set security keys, update software, etc.).

IETF RFC 7228 defines a taxonomy of constrained devices based on the first two dimensions above, which recognizes three classes of devices as depicted in Table 4.1.

Class 0 devices are the most severely constrained in memory and processing power. In general, such devices do not have the resources to connect to an IP network directly and will leverage the services of helper devices such as proxies or gateways for connectivity. For example, sensor motes fall under this class.

Table 4.1 Classes of constrained devices in RFC 7228

Name	Data size	Code size
Class 0	≪10 KB	≪100 KB
Class 1	~10 KB	~100 KB
Class 2	~50 KB	~250 KB

Class 1 devices are highly constrained in terms of code space and processing capacity; however they are capable of connecting to an IP network directly, without the help of gateways, as long as they are "parsimonious with state memory, code space, and often power expenditure for protocol and application usage." As such, these devices face challenges in running certain demanding IPs such as BGP, OSPF, HTTP, or Transport Layer Security (TLS) and in exchanging data using verbose data serialization formats such as XML.

Class 2 devices are less constrained when compared to the first two classes and are capable of running the same IP stack that runs on general compute nodes today. Nevertheless, these devices can still benefit from lightweight and efficient communication stacks since the resources may then be directed toward applications in lieu of networking.

Another dimension that characterizes constrained devices is power and/or energy resource constraints. These could be attributed to a number of factors such as the device size, primary mode of use, cost, operational environment, etc. Again, with this dimension, there is a spectrum of possibilities ranging from devices that harvest energy from the environment to battery-powered devices where the batteries are replaceable or rechargeable, to non-field replaceable battery-powered devices (which are discarded past the battery's lifetime), and to mains-powered devices. Energy consumption is a major issue for IoT devices. Research studies suggest that communication is over three orders of magnitude more expensive in terms of energy consumption than performing local processing functions. This is especially the case when wireless communication is used, where the radio takes the lion's share of the energy consumed by the device. To this reason, a common strategy employed by power-constrained devices is to remain in sleep mode with no network connectivity for extended periods of time and to connect only long enough to send the local data either based on periodic timers or asynchronous triggers (e.g., when new data is present or an event is detected).

To address the requirements of constrained devices, lightweight, energy-efficient, and bandwidth-conscious communication protocols are required across all the layers of the protocol stack.

4.2 Massive Scalability

Based on an estimate conducted by Cisco, about 99.4% of the physical objects in the world, which could potentially be connected to the Internet, are still unconnected. Conversely, this means that only about 10 billion out of approximately 1.5 trillion global objects are connected. The number of devices connected to the Internet surpassed 26 billion devices in 2020 (Fig. 4.1). The majority of this growth continues to be due to smart objects and "things" connecting to the Internet. This massive scalability imposes requirements on various aspects of the IoT protocol stack, in the areas of device identification and addressing, namely resolution, security, control plane (e.g., routing protocols), data plane forwarding, as well as manageability.

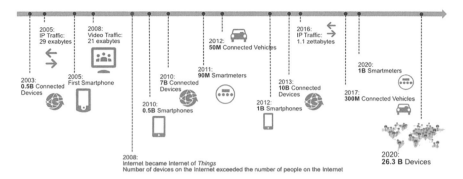

Fig. 4.1 Growth of connected devices. (Source: Cisco)

4.2.1 Device Addressing

The goal of the IoT is to build a uniform network that integrates and unifies all the communication systems between smart objects in the world. To realize the full potential of this vision, the interconnected things need to be individually address-able for ubiquitous communication between systems. In many current deployments of smart objects, the interconnection of things to the Internet, when available, is through gateways or proxies. In this sense, the connected things are proverbial second-class citizens of the Internet. Realizing the IoT vision requires that a global IP address be assigned to each one of the billions of devices that will be connected. Taking into account the fact that the IPv4 address space was completely depleted by February 1, 2011, it becomes clear that the massive scalability of the IoT will accelerate the transition of the Internet to IPv6.

4.2.2 Credentials Management

Security credentials management (e.g., shared key distribution, certificate management, etc.) poses a significant challenge in today's Internet. The addition of billions of devices to the network with IoT will only compound the problem further. Manual mechanisms currently employed for credentials management (e.g., through precon-figuration) are not going to be viable in IoT due to two reasons: the sheer number of devices and the limitations in (or complete lack of) user interfaces on constrained devices. The number of devices renders the use of pre-shared keys impractical for production deployments, especially when the devices have rudimentary user interfaces or no user interface at all.

The massive scalability of the IoT calls for lightweight, low-touch, and highly automated credentials management mechanisms.

4.2.3   Control Plane

The Internet encompasses diverse networks running different control plane protocols for the purpose of discovering topology information, communicating connectivity status or link health, signaling session or connection state, guaranteeing quality of service, and, among other things, quickly reacting to faults. These protocols maintain distributed state that is synchronized using message exchanges between peering nodes. In some cases, these peering relationships are hierarchical in nature (e.g., a client-server model) or flat (e.g., overlay peers). The behavior of the control plane functions together with the syntax and semantics of the messages exchanged defines the specifics of the control plane protocol. As the number of nodes participating in a given protocol increases, both the amount of state to be maintained by each node increases and the volume of messages required for keeping the distributed state tables in synchronization grows. Beyond a specific limit, attempts to scale a specific control plane protocol typically lead to adverse side effects on the protocol's convergence time, the node resources, and the overall network response. The scalability of the IoT calls for elastic control plane mechanisms that can accommodate the massive number of connected devices.

4.2.4   Wireless Spectrum

As the Internet of Things continues to evolve, one fact remains constant: these things require connectivity. This global network of objects, sensors, actuators, etc. must be connected to the Internet in some way, and in many cases wirelessly. The wireless spectrum is a finite resource, and the licensed portion of this spectrum is both expensive and scarce. With billions of devices coming online over the coming decade or so, many of these devices will be contending for the airwaves.

As of now, many IoT systems operate in unlicensed radio frequencies, namely, the industrial, scientific, and medical (ISM) bands, for example, the 900 MHz band for Electronic Product Code (EPC), one of the standards for radio-frequency identification (RFID); the 13.56 MHz band for near-field communications (NFC) supporting mobile payments; and the sub-125 kHz band for physical security systems (video surveillance and access control). These technologies achieve connectivity using a range of different, and in some ways competing, wireless protocol standards, such as Zigbee, Z-Wave, Bluetooth LE, and Wi-Fi, all of which were designed to work in the unlicensed spectrum. There are no spectrum bottlenecks for these bands yet, even though Wi-Fi services are approaching the point where they are maximizing the number of channels that can be fit into the allotted spectrum. However, when it comes to the licensed bands used for cellular communication (e.g., the GSM bands defined in 3GPP TS 45.005), the bottlenecks become more pronounced, especially with the accelerating growth in data traffic over cellular networks. The term "spectrum crunch" has been used in recent years to refer to this issue. There are two variables at play here: growth in the number of endpoints as well as growth in the volume of traffic per endpoint, both of which contribute to the spectrum crunch phenomenon. Research by Cisco shows that globally, mobile M2M connections grew from 495 million in 2014 to more than 3 billion in 2019, a sevenfold growth. Global mobile data traffic grew 69% in 2014 reaching 2.5 exabytes per month at the end of 2014, up from 1.5 exabytes per month at the end of 2013. Further, global mobile data traffic increased nearly tenfold between 2014 and 2019 (Fig. 4.2).

4.3   Determinism

One of the value propositions of IoT is that the technology will allow for better observation and monitoring of the physical world and will also enable the automated change of that world through closed-loop actuation. IoT opens up the door for supporting use cases that demand mission-critical networking with high requirements for real-time response as well as overall network, protocol, and device robustness. Some of these use cases emerge from industrial automation, such as monitoring systems, movement detection systems for use in process control (i.e., process manufacturing), and factory automation (i.e., discrete manufacturing). Other use cases have a much broader scope that spans mission-critical automation (e.g., rail control systems), motion control (e.g., wind turbines), vehicular networks (e.g., infotainment, power train, driver assistance), etc. With the increasing demand for connectivity and multimedia in transportation in general, use cases and applications are emerging in all elements of the vehicle from head units to rear seat entertainment modules, and to amplifiers and camera modules. While these use cases are aimed at less critical applications than industrial automation, they do share common requirements.

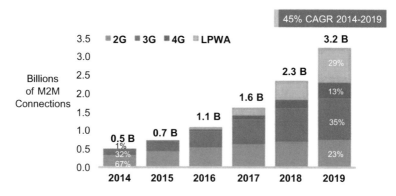

Fig. 4.2 Global machine-to-machine growth and migration from 2G to 3G and 4G. (Source: Cisco VNI Mobile, 2015)

These use cases all share the common requirement to support real-time information transfer: the time it takes for each packet to traverse a path from its source to its destination should be determined; that is, the process must be deterministic. Systems with control loops involving endpoints communicating over a network can function properly only if the networks connecting those endpoints guarantee determinism (imagine what would happen if a network delays a packet carrying a control variable for a high-speed CNC mill).

In this context, a network is said to support determinism and is thereby deemed to be a "deterministic network," if the worst-case communication latency and jitter of messages of interest are decidable based on a reasonable model of the network. A model is considered reasonable when it sufficiently represents reality for the target use cases of the networking system. Determinism does not imply speed. In control functions, both speed and determinism are required. Speed is required to attain the highest possible throughput. Determinism, on the other hand, is required to specify a level of quality for the throughput, i.e., the highest-speed throughput that is in fact usable by the application.

Deterministic Networking enables the migration of applications that have so far relied on special-purpose non-packet-based (fieldbus) technologies (e.g., HDMI, CAN bus, Profibus, etc.) to Internet Protocol technologies to support both these new applications, in addition to existing IP network applications, over the same physical network (Fig. 4.3). When applied in the context of industrial applications, this leads to what is dubbed as the "OT/IT" convergence. Operational technology (OT) refers to industrial networks, which, due to their different goals, have evolved in silo but in a manner that is substantially different from information technology (IT) networks. With OT, the focus has been on transporting fully characterized traffic flows, over a small area (e.g., plant floor), in a well-controlled environment with a bounded latency, extraordinarily low frame loss, and very narrow jitter.

Experience with custom control and automation networks, as well as proprietary audio/video networks, has shown that these applications require one or more of the following characteristics: time synchronization of all hosts and network elements (routers, bridges, etc.) and accurate in the range of 10 ns to 10 µs, depending on the application. The applications also require support for critical packet flows that need guarantees of the minimum and maximum latency end-to-end across the network. Such flows can be either unicast or multicast and can in total consume more than half of the available bandwidth of the network, thereby eliminating the possibility of relying on over-provisioning. The applications mandate packet loss ratios that are at least in the range of $1.0e-9$ to $1.0e-12$. Furthermore, the traffic for these applications cannot be subjected to throttling, congestion feedback, or stochastic network-imposed transmission delay.

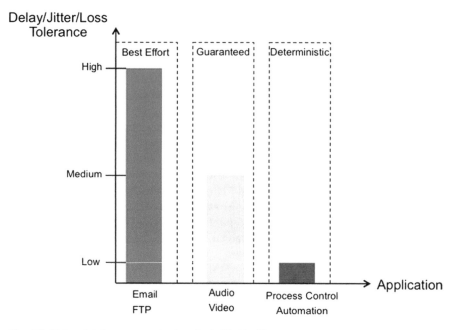

Fig. 4.3 Deterministic vs. guaranteed vs. best effort traffic

4.4   Security and Privacy

The ubiquity of IoT and its potential to extend into all aspects of human life, whether in transportation, healthcare, home automation, industrial control, etc., makes guaranteeing security and privacy paramount. With traditionally offline systems and applications being connected to the Internet, they quickly become targets for attacks that will only continue to grow in magnitude and sophistication. Such targets cover a multitude of industry segments, and the potential impact of security attacks could lead to significant damage and even loss of life.

While the threats in IoT may, at the outset, seem largely similar to those in more traditional IT environments, the potential impact of those threats is more profound. This is why threat analysis and risk assessment efforts are key in IoT to measure the impact of a security incident or breach.

A fundamental pillar in securing the IoT is around mechanisms to authenticate device identity. As was discussed in Sect. 4.1, many IoT devices are constrained devices, which lack the required processing, memory, storage, and power requirements to support state-of-the-art authentication protocols. The state-of-the-art encryption and authentication protocols are based on cryptographic suites such as Advanced Encryption Standard (AES) for confidential data transport, Rivest–Shamir–Adleman (RSA) for digital signatures and key transport, and Diffie–Hellman (DH) for key negotiations and management. While these protocols are battle-proven in deployments, they suffer from two shortcomings when it comes to applying them to IoT. The first shortcoming is that these protocols are resource hungry and generally demand high-capability compute platforms. Appropriate reengineering is required to accommodate constrained devices. The second shortcoming is that the authentication and authorization protocols are high-touch, requiring user input for provisioning and configuration. In many IoT deployments, access to the devices will be limited or impractical, thereby requiring that the initial configuration be tamper-proof throughout the usable lifespan of the devices, and such lifespan could extend to many years.

In order to address these shortcomings, new lightweight authentication and authorization protocols are required which leverage the experience of today's strong encryption/authentication algorithms but are capable of running on constrained devices.

Encryption is the cornerstone of network security protocols. The effectiveness of encryption algorithms generally decreases with time due to a number of factors including Moore's Law (availability of stronger compute to crack the encryption), public disclosure of inherent vulnerabilities with prolonged exposure to attacks, wide adoption (which increases the attack surface), etc. This creates an interesting predicament for the use of encryption in IoT: deployed devices may outlive the effectiveness of the encryption mechanisms embedded within them. For instance, a smart meter in a home can operate for 50 years, whereas the encryption protocol may lose its effectiveness in about half of that time.

Other aspects of security that need to be considered for IoT include:

- Data privacy levels and geo-fencing of data (i.e., limiting access to data to specific locales).
- Strong identities.
- Strengthening of base network infrastructure such as the Domain Name System (DNS) with DNSSEC and DHCP to prevent attacks.
- Adoption of protocols that are more tolerant to delay or transient connectivity (such as delay-tolerant networks).

Privacy is a major issue even in today's Internet. User data is collected for a multitude of purposes such as targeted advertisements, purchase recommendations, and even national security. IoT will exacerbate the importance of preserving privacy because many applications generate traceable signatures of the behavior of individuals and their physical location. Some IoT applications even involve highly sensitive personal information, such as medical records. For these types of applications, it is imperative to decouple the device from the owner's identity while still providing robust mechanisms for device ownership verification and device identity authentication. Shadowing is one mechanism proposed to achieve this. Effectively, digital shadows enable the user's objects to act on his or her behalf, storing just a virtual identity that contains information about his or her attributes. As a matter of fact, identity management in the IoT paves the way to increase security by applying a combination of diverse authentication methods for humans and machines. For instance, biometric data combined with a physical object could be used as grant access by unlocking a door.

The importance of security in IoT cannot be overstated. More details on this topic are covered in Chap. 8.

4.5   Application Interoperability

M2M deployments, in one form or another, have existed for over two decades now. However, the vision of the Internet of Things is far from being a reality, and the technology is yet to realize its full market potential. The complexity of developing, deploying, and managing IoT applications remains a key challenge for the industry. It constitutes a challenge for network operators who are trying to offer profitable services tailored to the IoT market, for application developers building vertical-specific applications, as well as for service providers who are trying to speed time to market, reduce costs, and simplify robust application deployment. This complexity drives up the cost of building IoT solutions.

The problem of complexity, and associated high cost, can be attributed in part to the closed nature of the solutions, which are developed in vertical-specific silos, thereby leading to each solution provider having to implement all the building blocks required for a minimum viable product, as opposed to reusing standard and open components. The resulting solutions are almost ubiquitously characterized by having strong coupling between application entities. Here, we use the term application entity to refer to an instance of application logic that may be implemented in hardware (analogue or digital), software, firmware, etc. Thus, an application entity denotes any IoT endpoint responsible for producing or consuming data and spans the entire gamut from a sensor/actuator to a cloud application.

The closed nature of existing IoT solutions renders them not only expensive to implement initially but also expensive and difficult to maintain and evolve over time. This is primarily because application code often needs to be updated or changed in the scenario where a device is swapped with another that is functionally equivalent albeit manufactured by a different vendor, let alone the scenario where a new device type needs to be integrated into the solution.

The above challenges lead to the requirement for application-level interoperability for the IoT. This requirement can be further broken down into requirements for abstractions and standard application programmatic interfaces (APIs) as well as requirement for semantic interoperability.

4.5.1   Abstractions and Standard APIs

Realizing the full vision of the IoT will be difficult unless the application programmatic interfaces (APIs) that control the functionalities of the devices and smart objects adhere to common standards that guarantee interoperability. To reach full API interoperability, the industry must converge on mechanisms for identifying the

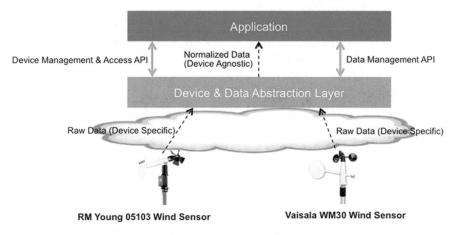

Fig. 4.4 Abstractions and APIs

data that application entities will share and methods for sharing it. APIs expose the data that enables disparate devices to be composed in innovative ways to create new and interesting workflows. With the availability of standard APIs, it is possible to introduce abstractions for common IoT functions, including:

- Device management (activation, triggering, authentication, authorization, software/firmware update, etc.)
- Data management (read, write, subscribe, notify, delete, etc.)
- Application management (start, stop, debug, upgrade, etc.)

The abstractions provide logical representations of the functions while hiding all implementation nuances and variations. They define service contracts that are governed by the syntax and semantics of the APIs and which formally specify the methods for interaction with modules supplying those functions. In other words, the use of standard APIs introduces "slip surfaces" that eliminate coupling between functionally discrete modules of a given IoT solution. This allows modules supplied by different IoT vendors to seamlessly interwork and integrate into a cohesive system. A given module can be replaced by another supplied by a different vendor as long as it subscribes to the standard API governing the associated slip surfaces between the system's building blocks (Fig. 4.4).

4.5.2 Semantic Interoperability

Semantic interoperability guarantees that application entities in the IoT can access and interpret data unambiguously. Providing unambiguous data descriptions that can be machine processed and interpreted by application entities is one of the key enablers of automated information communications and interactions in IoT.

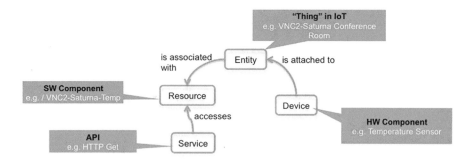

Fig. 4.5 Simple IoT ontology

Without semantic interoperability among communicating systems, sharing IoT data in a useful way is impossible. Semantic interoperability guarantees a common vocabulary that paves the way for accurate and reliable communication between applications and systems. This fluent machine-to-machine communication depends on the ability of different systems to map data to shared semantics, or meaning. If we were to use the analogy of a pyramid to visualize the different tiers of application interoperability, the base of that pyramid would be syntactic or structural interoperability: it defines the structure or format of data exchange between applications. Structural interoperability is a prerequisite; it is necessary but not sufficient for two applications to successfully work together. The top part of the pyramid is reserved to semantic interoperability. It deals with the content of the messages exchanged and their associated meaning, not just the message formats.

Semantic interoperability can be achieved in a number of ways. One is through the development of pervasive and common information models, or ontologies (Fig. 4.5), that capture the knowledge associated with a specific vertical domain. Another is through providing semantic mediators, or translators, that perform conversion of the information to a format that the application entity understands.

4.6   Summary

The Internet Protocol (IP) stack was among the factors that contributed to the success of the Internet. While this IP stack provides a strong foundation for building the IoT, a number of shortcomings need to be addressed to meet the peculiar requirements of IoT. These requirements include support for resource-constrained devices that have very limited compute capabilities and limited power; support for the massive scalability of IoT, with billions of connected devices; the need for deterministic networks to support real-time mission-critical applications; the requirement for lightweight security protocols and ensuring data privacy; and finally the requirement for application interoperability through the use of APIs and unified data semantics.

Problems and Exercises

1. What are "constrained" devices? Name their classes and characteristics.
2. What makes a network "deterministic"?
3. In what three areas does the massive scalability of IoT impact networking protocols?
4. What is the importance of standard APIs in the success of IoT?
5. Why is scalability a major requirement for IoT protocols?
6. What is an ontology? Why are ontologies applicable in the IoT?
7. Name three key IoT requirements that have impact on networking protocols.
8. What characteristics of the IP stack contributed to the success of the Internet?
9. Was the choice of the Internet as the underlying network for IoT always a given or agreed upon fact?
10. Name the various options by which IoT devices can be supplied with power.
11. Describe the characteristics of Class 0-constrained devices.
12. What is "semantic interoperability"? Why is it important in IoT?
13. How does scalability impact the network control plane? Explain the various dimensions impacted.
14. How much of the IPv4 address space is still available for allocation?
15. What common IoT functions can be abstracted through APIs in order to simplify application development and improve the time to market new IoT applications and services?
16. What types of applications can be migrated to IP technologies with the advent of Deterministic Networking?
17. Which is more expensive in terms of power consumption: Communication or local processing? What does this imply to IoT devices?
18. How does the addition of billions of devices to the Internet affect the wireless spectrum?
19. How does the complexity of developing, deploying, and managing IoT applications today affect the state of the industry?
20. What makes existing credentials management techniques inadequate for IoT?
21. What are two shortcomings of the state-of-the-art security protocols (for authentication/authorization/encryption) when applied to the IoT?

References

1. D. Estrin, R. Govindan, J. Heidemann, S. Kumar, Next century challenges: Scalable coordination in sensor networks, in *MobiCom '99: Proceedings of the 5th Annual ACM/IEEE International Conference on Mobile Computing and Networking*, (ACM, New York, 1999), pp. 263–270
2. Bormann, et al., "Terminology for Constrained-Node Networks". Internet Engineering Task Force RFC 7228. May 2014

3. V. Cantoni, L. Lombardi, P. Lombardi, Challenges for data Mining in Distributed Sensor Networks, in *18th International Conference on Pattern Recognition (ICPR'06)*, (2006), pp. 1000–1007
4. J. Bradley, J. Barbier, D. Handler, Embracing the Internet of Everything To Capture Your Share of $14.4 Trillion, Cisco Whitepaper, (2013)
5. The Zettabyte Era: Trends and Analysis, Cisco Whitepaper, (June 2016)
6. D. Evans, The Internet of Things – How the Next Evolution of the Internet is Changing Everything, Cisco Whitepaper, (April 2011)
7. "Cisco Visual Networking Index: Global Mobile Data Traffic Forecast Update, 2014–2019", Cisco Whitepaper, February 2015
8. http://www.ieee802.org/802_tutorials/2012-11/8021-tutorial-final-v4.pdf,   IEEE   802.1 Tutorial on Deterministic Ethernet, November 2012
9. N. Finn, P. Thubert, "Deterministic Networking Problem Statement", draft-finn-detnet-problem-statement-01, work in progress, (October 2014)
10. W. Steiner, N. Finn, Deterministic Ethernet: Standardization in Progress and Beyond, RATE Workshop, (December 2013)
11. P. Barnaghi et al., Semantics for the internet of things: Early progress and back to the future. Int. J. Semant. Web. Inf. Syst **8**(1) (2012)
12. Securing the Internet of Things: A Proposed Framework, Cisco Whitepaper

Chapter 5
IoT Protocol Stack: A Layered View

The IoT protocol stack can be visualized as an extension of the TCP/IP layered protocol model and is comprised of the following layers (refer to Fig. 5.1):

- Physical layer
- Link layer
- Network layer
- Transport layer
- Application Protocols layer
- Application Services layer

Note that the Application layer of the TCP/IP protocol stack is expanded into two layers in the IoT protocol stack: Application Protocols and Application Services. It is as if the proverbial "narrow waist" of the hourglass is being extended further up the stack to provide interoperability between heterogeneous "things."

5.1 Link Layer

In this section we will examine the impact of the IoT requirements on the Link layer through a combined view of the challenges that those requirements impose on networking technologies, industry efforts to address those challenges, and remaining gaps.

© The Author(s), under exclusive license to Springer Nature Switzerland AG 2022   97
A. Rayes, S. Salam, *Internet of Things from Hype to Reality*,
https://doi.org/10.1007/978-3-030-90158-5_5

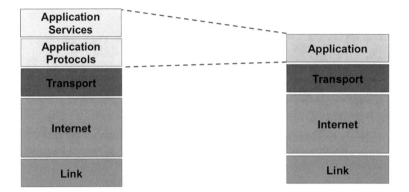

Fig. 5.1 IoT protocol stack

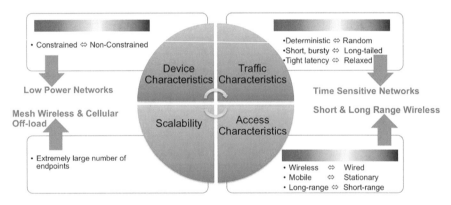

Fig. 5.2 Link layer challenges. (Source Cisco BRKIOT-2020, 2015)

5.1.1   Challenges

The challenges that the IoT presents to the Link layer of the protocol stack can be broadly categorized into the following four areas: device characteristics, traffic characteristics, access characteristics, and scalability (Fig. 5.2).

On the device characteristics front, the IoT will encompass a wide spectrum of "things" that span from fully capable (non-constrained) compute nodes to highly constrained devices. The latter typically have limited energy resources to spend on processing and communication. As discussed earlier, network communication is typically more power consuming when compared to local processing. Hence, communication technologies need to be optimized to accommodate low-power devices. Implementation of protocols at all layers of the protocol stack can affect energy consumption. However, the Link layer, in particular, has a significant impact due to the fact that this layer is responsible for the nuances of the physical transmission technology, framing, media access control, and retransmissions. For instance, it is reported that, depending on the link load, between 50% and 80% of the communication energy is used for repairing lost transmissions at the MAC layer.

The traffic characteristics of IoT endpoints vary widely depending on the application's demands and nature of devices. Some applications have relaxed requirements on packet loss, latency, and jitter (e.g., a meteorological monitoring application), whereas others have very tight availability, latency, and jitter tolerance (e.g., a jet engine control application). It is worth noting here the contrast between the meteorological monitoring and jet engine control applications: both applications may be using the same types of devices (temperature sensors, pressure sensors) and observing the same physical entities (temperature, pressure). However, it is the applications' requirements that dictate the traffic characteristics that the network must deliver. By the same token, some IoT devices generate short bursty traffic (e.g., point of sale terminal), whereas other devices generate long-tailed traffic (e.g., video camera). The dichotomy in traffic characteristics, between solutions that expect determinism and those that can withstand best-effort (random) communications, creates drivers for Link layer technologies that support deterministic and Time-Sensitive Networking.

The access characteristics of IoT endpoints become increasingly diverse as the footprint of the network grows beyond traditional IT environments, dominated by familiar local area network (LAN) and wide area network (WAN) technologies, and into new deployment environments such as industrial plant floor, oil fields, marine platforms, mines, wells, power grids, vehicles, locomotives, and even the human body. IoT devices in these environments may connect to the network using a mix of wireless and wired technologies. The devices when connected wirelessly may be either mobile or stationary and depending on the logistics of the deployment may require either long-range or short-range connectivity solutions. To accommodate this diversity, new Link layer protocols that form the foundation of field area network (FAN), neighborhood area network (NAN), and personal area network (PAN) technologies are required.

IoT scalability demands present interesting challenges for the Link layer of the protocol stack, especially for wireless technologies. On the one hand, these technologies offer a number of appealing characteristics that make them a good fit for the IoT, low upfront investments, wide geographic coverage, fast deployment, and pleasing aesthetics (no unsightly wires).

On the other hand, these technologies are susceptible to scalability issues. For instance, cellular technologies are subject to the spectrum crunch problem, which drives demand for technology optimizations and cellular off-load solutions such as Wi-Fi and femtocell. Also, wireless mesh technologies suffer from challenges such as forwarding latency and slow convergence as the diameter of the mesh scales.

5.1.2   Industry Progress

Now that we have covered the main challenges that IoT presents to the Link layer of the protocol stack, we will shift our focus to describe the industry's progress in addressing those challenges through open standard solutions.

5.1.2.1   IEEE 802.15.4

IEEE 802.15 Task Group 4 (TG4) was chartered to investigate a low data rate wireless connectivity solution with focus on very low complexity and extended battery life span that is in the range of multiple months to multiple years. The solution was meant to operate in an unlicensed, international frequency band. While initial activities of the task group focused on wearable devices, i.e., personal area networks, the eventual applications proved to be more diverse and varied. Potential applications of the solution include sensors, interactive toys, smart badges, remote controls, and home automation. As can be seen from the applications, the focus of the solution has primarily revolved around enabling "specialty," typically short-range, communication.

The resulting IEEE 802.15.4 technology is a simple packet-based radio protocol aimed at very low-cost, battery-operated devices (whose batteries last years) that can intercommunicate and send low-bandwidth data to a centralized device. The protocol supports data rates ranging from 1 Mbps to 10 kbps. The data rate is dependent on the operating frequency as well as on the coding and modulation scheme. The standard operates over several frequency bands, which vary by region:

- 169 MHz band
- 450 MHz band
- 470 MHz band
- 780 MHz band
- 863 MHz band
- 896 MHz band
- 901 MHz band
- 915 MHz band
- 917 MHz band
- 920 MHz band
- 928 MHz band
- 1427 MHz band
- 2450 MHz band

In addition, the standard supports multiple modulation schemes, including BPSK, ASK, O-QPSK, MR-FSK, MR-OFDM, and MR-O-QPSK. The transmission range varies from tens of meters up to 1 km, the latter introduced with IEEE 802.15.4g. The protocol is fully acknowledged for transfer reliability. The basic frame size is limited to 127 bytes in the original specification, and the philosophy behind that is twofold: to minimize power consumption and to reduce the probability of frame errors. However, with IEEE 802.15.4g, the maximum frame size is increased to 2047 bytes, accompanied by an increase of the frame check sequence (FCS) from 16 to 32 bits for better error protection.

The standard offers optional fully acknowledged frame delivery for transfer reliability in lossy environments (e.g., high interference). If the originator of a frame does not receive an acknowledgment after a certain time period, it assumes that the transmission failed and retransmits the frame. If an acknowledgment is still not received after multiple attempts, the originator may either terminate the transaction or continue retrying.

The IEEE 802.15.4 standard only defines the functions of the Physical and Media Access Control (MAC) layers. It serves as the foundation for several protocol stacks, some of which are non-IP, including Zigbee, Zigbee RF4CE, Zigbee Pro, WirelessHART, ISA 100.11a, and RPL.

There are two types of devices in an 802.15.4 network. The first one is the *full-function device* (FFD). It implements all of the functions of the communication stack, which allows it to communicate with any other device in the network. It may also relay messages, in which case it is dubbed as a personal area network (PAN) coordinator. The PAN coordinator is in charge of its network domain: it allocates local addresses and acts as a gateway to other domains or networks. The second type of device is the *reduced-function device* (RFD). RFDs are meant to be extremely simple devices with very modest resource and communication capabilities. Hence, they can only communicate with FFDs and can never act as PAN coordinators. The rationale is that RFDs are to be embedded into the "things." Networks can be built using either a star, mesh, or cluster tree topology (Fig. 5.3). In all three cases, every network needs at least a single FFD to act as the PAN coordinator. Networks are thus formed from clusters of devices separated by suitable distances.

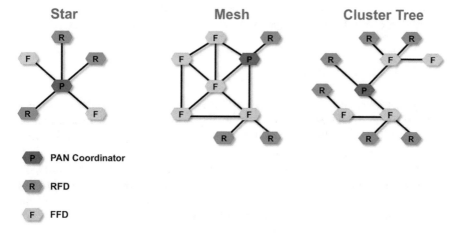

Fig. 5.3 IEEE 802.15.4 topologies

In the star topology, all devices communicate through a single central controller, namely, the PAN coordinator. This is a hub-and-spoke model: the PAN coordinator is the hub, and all other devices form spokes that connect only to the hub. The PAN coordinator is typically main powered, while the devices are most likely battery operated. Use cases that make use of this topology include smart homes (home automation), computer peripherals, personal health monitors, toys, and games. Each star network chooses a PAN identifier, which is not currently in use by any other network within the radio range. This allows each star network to operate independently of other networks.

The mesh topology (also called peer to peer) differs from the star topology in that any device can communicate with any other device as long as the two are within radio range. A mesh network can be ad hoc in formation, self-organizing, and self-healing on node or link failures. It also provides reliability through multipath routing. Use cases such as industrial control and process monitoring, wireless sensor networks (WSN), precision agriculture, security, asset tracking, and inventory management all can leverage this topology.

The cluster tree topology is a special case of a mesh network that comprised of chained clusters. In a cluster tree, the majority of the devices are FFDs. RFDs may connect to the network as leaf nodes at the end of a tree branch. As with any 802.15.4 topology, the network has a single PAN coordinator. The PAN coordinator forms the first cluster by declaring itself as the cluster head (CLH) with a cluster identifier (CID) of zero, selecting an unused PAN identifier, and broadcasting beacon frames to other neighbor devices. A device, which receives beacon frames, may request from the CLH to join the cluster. If the CLH allows the device to join, it will add the new device as a child device in its neighbor list. The newly joined device will add the CLH as its parent in its neighbor list and commence broadcasting periodic beacon frames. This allows other candidate devices to join the same cluster at that device. Once the requirements of the application or network are met, the PAN coordinator may instruct a device to become the CLH of a new cluster that is adjacent to the first. The advantage of this daisy-chained cluster structure is the ability to achieve larger coverage area at the expense of increased message latency.

5.1.2.2 IEEE 802.15.4e TSCH

IEEE 802.15.4e is the next-generation 802.15.4 wireless mesh standard. It aims to improve on its predecessor in two focus areas: lower energy consumption and increased reliability. The standard introduces a new media access control (MAC) layer to 802.15.4 while maintaining the same physical (PHY) layer. Hence, it can be supported on existing 802.15.4 hardware. Two key capabilities are added, time synchronization and channel hopping, hence the acronym TSCH. Time synchronization addresses the requirement for better energy utilization, whereas channel hopping aims at increasing the reliability of communication.

With time synchronization, time is sliced into fixed-length time slots and all nodes are synchronized. A time slot is long enough to allow a station to send a maximum transmission unit (MTU)-sized frame and receive an acknowledgment back. Time slots are grouped into slotframes of flexible width. The flexibility allows different deployments to optimize for bandwidth or for energy saving: the shorter the slotframe, the more frequently that a given time slot will be repeated, thereby giving a station more chances to transmit (i.e., higher bandwidth) but at the expense of increased energy consumption. The current time slot is globally known to all nodes in the network via an absolute slot number (ASN). The ASN is initialized to 0 and is expected to wrap around only after hundreds of years.

With channel hopping, each message transmission between nodes occurs on a specified channel offset. The channel offset is then mapped to a radio frequency using a function that guarantees that two consecutive transmissions between two nodes hop from one frequency to another within the allotted band:

$$Frequency = F\{(ASN + Channel\ Offset) \bmod nFreq\}$$

where nFreq is the number of available frequencies in the allotted band.

This enhances the reliability of communication as it is proven to be effective against multipath fading and interference. Basically, if a specific frequency is subject to fading or interference, then by changing the frequency used for communication between nodes with every new message, only a subset of the messages will be lost due to those conditions, whereas if all communication were to occur on the same frequency, then all messages between the nodes communicating over the affected frequency would be lost during the fading or interference event.

The nodes in the network all obey a TSCH schedule. The schedule is a logical two-dimensional matrix with one dimension determining the slot offset in the slotframe and the second dimension designating the channel offset in the available frequency band (Fig. 5.4). The schedule instructs each node on what it is supposed to do in a given time slot: transmit, receive, or sleep. The schedule also indicates for every communicating node its neighbor's address and the channel offset to be used for said communication. The width of the schedule is equal to the slotframe width, whereas the depth of the schedule is equal to the number of available frequencies in the allotted band. Each cell in the schedule corresponds to a unique slot offset and channel offset combination. The organization of communication in the schedule allows the network to operate using collision-free communication, by ensuring that

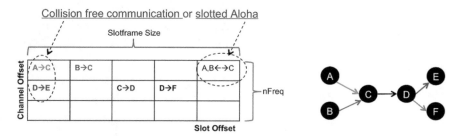

Fig. 5.4 TSCH schedule

only a single station transmits in a given cell. Alternatively, it can allow the network to operate in a slotted Aloha paradigm (i.e., carrier-sense multiple access with collision detection—CSMA/CD) by allowing multiple stations to transmit in the same cell. IEEE 802.15.4e does not define the mechanisms by which the TSCH schedule is built and leaves that responsibility to upper-layer protocols.

5.1.2.3 LPWAN

Low-power wide area networks (LPWANs) are meant to fill the gap between short-range wireless and cellular communication technologies. They are designed for low-power, long-range, and light-weight data collection IoT use cases (Fig. 5.5). Devices connecting to LPWANs will typically have a battery life of over 10 years and will require outdoor coverage of up to 20 km (12 miles) and sufficient indoor penetration. From an operational standpoint, the solutions require low service cost and endpoint complexity. In general, the LPWAN landscape spans both licensed and unlicensed spectrums.

It is not unusual to see Low-power wide area (LPWA) technology combined with LTE in solutions where high data rates are required for device (e.g., navigation, entertainment systems) and low-data rates are used in the same device for telemetry (e.g., position, direction, temperature).

There are two main LPWA technologies in the market today that dominate the landscape. They are as follows:

- *LoRaWAN (Long-Range Wide Area Network)*—An unlicensed radio technology (free) that is available for anyone to deploy much like Wi-Fi is today. Note, LoRa only provides the radio layer (link layer protocol) therefore it is combined with a network layer protocol called LoRaWAN that provided the methods and

Fig. 5.5 LPWAN positioning

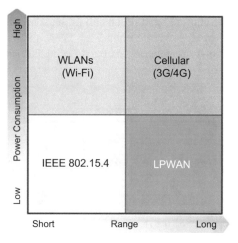

procedures for a sensor to transmit and receive packets. LoRaWAN is defined by the LoRa Alliance, an industry consortium.

- *NB-IoT (Narrow-Band IoT)*—A licensed-spectrum (paid) transport provided by service providers and defined by the 3GPP (the same organization that defines the 3/4/5G specifications).

LoRaWAN

LoRaWAN defines a communication protocol and network architecture for low-power wide area networks (LPWANs). LoRaWAN is designed to address the requirements for low-power consumption (i.e., long battery life), long range, and high capacity in LPWANs while maintaining low cost for the solution. The communication protocol used in LoRaWAN is known as LoRa. The LoRa physical layer uses chirp spread spectrum modulation. It is characterized by low-power usage while at the same time significantly increasing the communication range when compared to frequency-shifting keying (FSK), which is the modulation technique often used in legacy wireless systems. Chirp spread spectrum is not a new technique: it has been employed in military and space applications for decades because of its extended range and its robustness against interference. A key advantage of the LoRa protocol is its extended range: a single base station can cover hundreds of square miles. That is enough to provide coverage over cities. Hence, with minimal infra-structure, entire countries can be covered using LoRaWAN. In wireless communi-cation systems, the range within a given environment is determined through the link budget metric. LoRa has a link budget that is greater than any other standardized wireless communication technology today. The link budget is defined as an account-ing of all the gains and losses between a transmitter and a receiver:

$$\text{Link Budget} = \text{Transmitted Power} + \text{Gains} - \text{Losses}$$

Network Architecture

LoRaWAN employs a long-range star (or hub and spoke) architecture in order to minimize power consumption. Star architecture, in contrast to mesh architecture, eliminates the scenario where nodes receive and forward information from other nodes that is mostly irrelevant to them. In LoRaWAN, gateways act as hub nodes, whereas end devices form the spokes. End nodes are not associated with a particular gateway. Rather, when a node sends data, it is typically received by multiple gate-ways. Each of these gateways, in turn, forwards the received data toward the cloud-based network server using some backhaul[1] technology. The network server is responsible for all complex and intelligent functions: it manages the network, filters

---

[1] The backhaul can be Ethernet, Wi-Fi, satellite, or cellular.

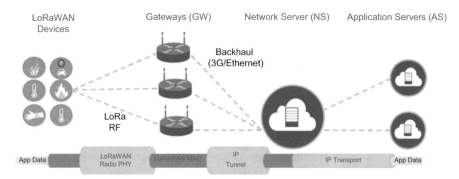

Fig. 5.6 LoRaWAN end-to-end network architecture

redundant received data, performs security verification, schedules acknowledgments through the most optimal gateway, and performs adaptive rate control, etc.

A key feature of this architecture is that no handover mechanism is required from one gateway to another to support the mobility of end nodes. Therefore, it is straightforward to enable IoT asset tracking applications. Another key feature is the built-in access redundancy, where the failure of a gateway or path toward the network server is handled by sending redundant copies of data packets (Fig. 5.6).

Device Class Capabilities

In order to address the constrained devices requirement of IoT, LoRaWAN defines three device class capabilities targeting different applications with varying needs. The classes are labeled A, B, and C. They offer a trade-off between energy consumption and downlink communication latency.

Class A devices support bidirectional communication. They include battery-powered sensors. This is the most energy-efficient device class capability and must be supported by all devices implementing LoRaWAN. The communication model is such that each uplink transmission by the end device is followed by two short downlink receive windows. The transmission schedule of the end device is dictated by its own communication requirements, albeit with a small variation in the allocated window based on a random time variance (ALOHA protocol flavor). This class of operation is suitable for applications where downlink communication from the server to the end device mostly occurs in the short window after the latter had sent an uplink transmission. Otherwise, such downlink communication must be deferred till the next scheduled uplink transmission.

Class B devices support bidirectional communication with scheduled receive slots. They include battery-powered actuators. This class offers energy efficiency with latency controlled downlink communication. The communication model for this class supports all the capabilities of Class A and in addition requires end devices to open extra receive windows at scheduled times. This is accomplished by having the end devices receive a time-synchronized beacon from the gateways, so that the applications on the servers know when the end devices are listening on these extra slots.

Class C devices support bidirectional communication with maximal receive slots. They include main powered actuators. This class is for devices that have the energy resources to afford to listen continuously. It is well suited for applications that require no latency in downlink communication. End devices in this class must continuously open receive windows when not in transmitting mode.

Scalability

LoRaWAN ensures the scalability of its long-range star network architecture through high-capacity gateways. Gateways achieve high capacity through a twofold approach, by using adaptive data rate and by employing a multichannel multi-modem transceiver. This allows the gateway to receive simultaneous messages on multiple channels from a very high volume of end devices. Several factors affect network capacity, among which the following are deemed most critical:

- Number of concurrent channels supported by the transceiver
- Data rate (i.e., time on air)
- Payload size
- Frequency of transmission of communicating nodes

Recall that LoRa uses spread spectrum modulation; hence, when different spreading factors are used, the signals end up being orthogonal to one another. The effective data rate changes with change in the spreading factor. LoRaWAN gateways capitalize on this property in order to concurrently receive multiple different data rates on the same channel. In the scenario where an end device is in the vicinity of a gateway and has a good link, there is no technical reason for it to use the lowest data rate thereby filling up the available spectrum for a longer time period than required. If this device was to shift to a higher data rate, its time on air will be shortened, thereby freeing up more time for other devices to transmit. It is worth noting that in order for adaptive data rate to work, the uplink and downlink need to be symmetrical, with sufficient downlink capacity. These features all contribute to making a LoRaWAN network scalable.

However, the duty-cycle limitation in the ISM bands may arise as a limitation to the scale of LoRaWAN networks. As an example, the maximum duty cycle of the EU 868 ISM band is 1%. This results in a maximum transmission time of 36 s in each hour for each end device in a sub-band.

Energy Efficiency

Energy efficiency is achieved in LoRaWAN through the use of the ALOHA method of communication: nodes are asynchronous and only communicate when they have data ready to be sent, whether scheduled or event driven. This alleviates the need for end devices to frequently wake up and synchronize with the network or check for messages. Such synchronization is one of the primary contributors to energy consumption in wireless networks.

Energy efficiency is also achieved through the use of adaptive data rate, where transmission power is varied according to link quality. When adaptive data rate is enabled, the network collects metrics on a number of the most recent transmissions from a node. These metrics include the frame counter, signal-to-noise ratio (SNR), and the number of gateways that have received each transmission. Based on these metrics, the network then calculates if it is possible to increase the data rate or lower the transmission power. If possible, the network will lower the transmission power to save energy and cause less interference.

Security

LoRaWAN defines two layers of security: one at the Network layer and one at the Application layer. Network security is responsible for ensuring the authenticity of the node in the network, whereas the Application layer security guarantees that the user's application data is inaccessible to the network operator. LoRaWAN uses AES encryption with key exchanges based on the IEEE EUI64 identifier.

Three different security keys are defined: network session key, application session key, and application key. The network session key is used for securing the interactions between the end node and the network. It helps in checking the validity of the messages. The application session key is used for payload encryption/decryption. These two session keys are unique per device, per session. When a device is dynamically activated, these keys are regenerated upon every activation, whereas, if the device is statically activated, these keys remain the same until changed by the operator. Devices which are dynamically activated use the application key in order to derive the two session keys in the course of the activation procedure. In general, it is possible to have either a default application key that is used to activate all devices or a customized key per device.

Regional Variations

Due to differences in spectrum allocations and regulatory requirements between regions, the LoRaWAN specification varies slightly from region to region. These variations affect the following: frequency band, number of channels, channel bandwidth, transmission power, data rate, link budget, and spreading factor.

Challenges

LoRaWAN relies on the acknowledgment of frames in the downlink for reliability. This, in turn, causes capacity drain. Therefore, in general, application should try to minimize the volume of acknowledgments in order to avoid this drain. This raises an open question regarding the feasibility of very large-scale and ultrareliable applications using LoRaWAN.

Also, the uncoordinated deployment of LoRaWAN gateways and alternate LPWAN technologies in large urban centers may lead to a decrease in network capacity due to collisions in the ISM bands. This, in addition to the duty-cycle regulation for these bands, poses potential challenges for large-scale LoRaWAN deployments.

NB-IoT

In June 2016, 3GPP completed the standardization of Narrow Band IoT (NB-IoT), a radio access technology with a spectrum bandwidth that can go as small as 180 kHz and with higher modulation rates compared to LoRaWAN. 3GPP had started NB-IoT under the name "Cellular System Support for Ultra-low Complexity and Low Throughput Internet of Things (CIoT) " with the goal of finding a solution that would be competitive in the Low-Power Wide Area segment, which at that time was largely defined by unlicensed spectrum technologies.

NB-IoT has its roots in LTE, albeit its operation is kept as simple as possible in order to reduce device costs and minimize battery consumption. In order to do so, it removes many features of LTE, including handover mechanisms, channel quality monitoring measurements, carrier aggregation, and dual connectivity. It uses the same licensed frequency bands used in LTE, and employs QPSK modulation. There are different frequency band deployments, which are stand-alone, guard-band, and in-band deployment. There are 12 subcarriers of 15 kHz in downlink using OFDM and 3.75/15 kHz in uplink using SC-FDMA.

Network Architecture

The core network architecture of NB-IoT is based on the 3GPP's Evolved Packet Core (EPC), with simplifications and optimizations that were designed specifically for IoT use cases focusing on communication between an IoT device and an application in the external network (cloud/Internet). This is achieved using a combined node called C-SGN (CIoT Serving Gateway Node) which serves the combined functionality of the Mobility Management Entity (MME)/Serving Gateway (SGW) and of Packet Data Network Gateway (PGW) in the original EPC architecture. Figure 5.7 depicts the architecture.

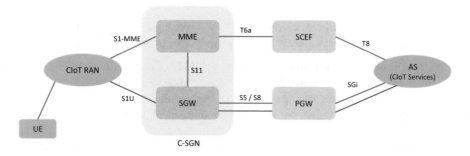

Fig. 5.7 NB-IoT network architecture

For NB-IoT, 3GPP has introduced in addition to the IP Packet Data Network (PDN), a non-IP PDN. This is to handle IoT devices where the packets used for communication are unstructured from the Evolved Packet System standpoint. While an IP based PDN is established through the regular attach procedure, a non-IP type PDN can be accomplished by one of two mechanisms:

- Delivery using SCEF.
- Delivery using point to point SGi tunnel (via PGW) based on UDP/IP where PGW acts as transparent forwarding node via transferring Non-IP data between UE and the AS (Fig. 5.7).

Each application shall have its own PDN and APN to differentiate the bearer. The APN configuration in the HSS helps the network to decide whether Non-IP data is sent via SCEF or PGW.

SCEF stands for Service Capability Exposure Function. It provides a means to securely expose the services and capabilities provided by the 3GPP network and hence enables enterprises to develop applications that may benefit from the transport network information. SCEF is primarily used for Non-IP data delivery provided:

- The Application server (AS) registers itself with the SCEF for a particular device followed by SCEF informing the Home Subscriber Server (HSS) about the registration request.
- The device has a PDN connection/bearer available between itself and SCEF (via MME) for non-IP data delivery.

In general, a device does not need to know whether a PDN connection is obtained via a SCEF or a PGW. In case of the former, an association between the AS and SCEF needs to be established to enable transfer of non-IP data. SCEF also helps in monitoring device events or state and performing application specific actions based on the device trigger or SCS/AS request.

Device Categories

Table 5.1 provides a summary of device categories as they relate to CIoT.

Scalability

NB-IoT allows mobile network operators to support high scale deployments, with up to 60K devices per cell, by employing a number of optimizations:

Control Plane CIoT optimization: In the original EPC architecture, the S1U path (refer to Fig. 5.7) is required to transfer data. This path is established every time the device (UE) needs to send data. In IoT applications that are expected to transfer small amounts of data per day or per month, establishment of frequent data radio bearers and consequently S1U path is a signaling overhead. To overcome this, data transfer to MME can take place over control plane/Signaling Radio

Table 5.1 CIoT device categories

	Release-8	Release-12	Release-13	Release-13
	Cat. 1	Cat. 0	Cat. M1	Cat. NB1
Downlink peak	10 Mbps	1 Mbps	1 Mbps	200 kbps
Uplink peak rate	5 Mbps	1 Mbps	1 Mbps	144 kbps
Number of antennas	2	1	1	1
Duplex mode	Full duplex	Half duplex	Half duplex	Half duplex
UE receive bandwidth	20 MHz	20 MHz	1.4 MHz	200 kHz
UE transmit power	23 dBm	23 dBm	20 dBm	23 dBm
Use case	Voice services for emergency in elevators, smart Grid Management	Cat0 is the interim solution prior to Cat-M. Cat0 is used for replacing Cat1 but cannot replace voice use cases	Environment monitoring, vehicle tracking	Smart metering, smart buildings, home automation

Bearer (SRBs) as Network Access Stratum (NAS) PDUs. This data is further sent by MME to SGW or SCEF depending on the PDN type.

User Plane CIoT optimization: In scenarios where large data transfer is required, such as remote installation or device software update, Data Radio Bearers (DRBs) are used. The existing procedure of S1U establishment consumes significant signaling resources due to frequent recurring UE inactivity timer expiry. This is why 3GPP TS 36.003 introduces the suspension of the Radio Resource Control (RRC) context at eNB until the next data request. A "resume id" is shared by radio base-station (eNB) to the device during RRC connection release and indicates to the latter to store its context information while suspending SRBs and DRBs. This RRC context can later be resumed by the device by simply sending its "resume ID" to the eNB.

Attach without PDN connectivity: This is a new capability to allow devices to remain attached without PDN connection. It is useful for devices which seldom transfer data and stay inactive most of the time. The device can stay attached without PDN but SMS service is available for any data transmission. The SMS could also be used to trigger the device to initiate a PDN connection.

APN rate control: Since many IoT devices use minimal data and hence cannot be charged based on data usage, Access Point Name (APN) rate control is used to decide the maximum number of packets to/from device per time unit (day, month, etc.). This upper cap or the limit is decided by the network operator and is based on the general data consumption by the IoT application. APN rate control comes into the picture only for devices attached with PDN.

eSIM: eSIM is a global specification by GSMA that enables remote SIM provisioning of any mobile device. This is not based on a regular SIM card rather using
embedded SIM (also called eUICC) which can accommodate multiple SIM profiles, having their respective operator and subscriber data. This allows remote
provisioning and migrating the SIMs to a different operator/network over the air,
thereby providing significant operational efficiency for large scale IoT
deployments.

Energy Efficiency

NB-IoT provides mechanisms for efficient energy consumption, namely:

Power saving mode: This is a device mechanism to conserve energy and support
extended battery life. When enabled, the device and the network can negotiate
the sleep and active state duration for transfer and reception of data. However,
the final values are determined by the network and no re-attach procedures are
required when the device becomes active again. Mobile network operators are
expected to use store-and-forward approach during power saving mode so that
stored messages can be forwarded to the device when it becomes active. The
amount of storage capacity to be reserved for storing the messages is decided by
the operator.

Extended Discontinuous Reception (eDRX): eDRX is an extension of an existing
feature to save more energy and allows the device to sleep for an extended period
of time. During sleep time, the device does not listen for any paging or control
channels. While power saving mode can effectively reduce power consumption
for devices that originate messages, e-DRX could do the same for devices that
terminate messages.

Security

NB-IoT inherits LTE's secure authentication, signalling protection, user identity
confidentiality, data integrity, and encryption capabilities. To protect the mobile
operator's network from misbehaving devices, NB-IoT supports PLMN rate control. It allows the network to measure and protect itself by enabling a rate control on
the data traffic being carried in NAS PDUs in UL/DL and hence is not applicable to
user plane optimization.

Comparison of LoRaWAN and NB-IoT

Table 5.2 illustrates the technical differences between LoRaWAN and NB-IoT in
both implementation and attributes. In short, only Service Provider networks can
deploy NB-IoT, whilst LoRa/LoRaWAN can be deployed by both Service Providers
and private enterprises.

Table 5.2 LoRaWAN and NB-IoT comparison

Attribute	LoRaWAN	NB-IoT
Frequency/spectrum	Unlicensed	Licensed
Bandwidth	500 kHz–125 kHz	180 kHz
Max data rate	50 kbps	200 kbps
Range	5 km (urban) 20 km (rural)	1 km (urban) 10 km (rural)
Base station architecture	Device TX to multiple base stations	Devices TX to single base
Power efficiency	Very high	High
Max messages per day	Unlimited	Unlimited
Protocol	Asynchronous	Synchronous
Interference immunity	High	Low
Allows private network	Yes	No
Standardization	LoRa Alliance	3GPP
Modulation	CSS	QPSK

5.1.2.4 IEEE 802.11ah

The popularity of IEEE 802.11 wireless technologies (Wi-Fi) has grown steadily over the years in home, business, as well as metropolitan area networks. The technology, however, cannot sufficiently address the requirements of IoT, due to the following two reasons:

- High power consumption for client stations: Wi-Fi has the reputation of not being very power efficient, due to the need for client devices to wake up at regular intervals to listen to AP announcements, waste cycle in contention processes, etc.
- Unsuitable frequency bands: Wi-Fi currently uses the 2.4–5 GHz frequency bands, which are characterized by short transmission range and high degree of loss due to obstructions. A common solution to this is the use of repeaters, but those add to the power consumption of the solution and add to the network's complexity.

To address these issues, IEEE 802.11 formed Task Group "ah." The 802.11ah group was chartered to develop a wireless connectivity solution that operates in the license-exempt sub-1 GHz bands to address the following IoT requirements: large number of constrained devices, long transmission range, small (approximately 100 bytes) and infrequent data messages (inter-arrival time larger than 30 s), low data rates, and one-hop network topologies. The solution is intended to provide a transmission range of up to 1 km in outdoor areas with data rates above 100 kbps while maintaining the current Wi-Fi experience for fixed, outdoor, point-to-multipoint applications. From a design philosophy perspective, the solution optimizes for lower power consumption and extended range at the expense of throughput, where applicable. In addition, the solution aims for scalability by supporting a large number of devices (up to 8191) per Wi-Fi access point.

IEEE 802.11ah introduces new PHY and MAC layers. The new layers are designed for scalability, extended range, and power efficiency. Compared to existing Wi-Fi technologies which operate in the 2.4–5 GHz range, the use of the sub-1 GHz band provides longer range through improved propagation and allows better penetration of the radio waves through obstructions (e.g., walls).

However, one of the challenges in the use of the sub-1 GHz spectrum is that its availability differs from one country to the next, with large channels available in the USA, whereas many other regions only have a few channels. This led the 802.11ah group to create several channel sizes: 1, 2, 4, 8, and 16 MHz channels based on the needs and regulatory domains of different countries. It also led the group to define operation over several frequency bands, which vary by region:

- Europe: 868–868.6 MHz
- Japan: 950–958 MHz
- China: 314–316, 390–434, 470–510, and 779–787 MHz
- Korea: 917–923.5 MHz
- USA: 902–928 MHz

IEEE 802.11ah will support data rates ranging from 150 kbps up to 340 Mbps. The supported modulation schemes include BPSK, QPSK, and 16 to 256 QAM.

In order to address the IoT requirements of low-power consumption and massive scalability, the emerging 802.11ah introduces several enhancements to Wi-Fi technology that can be categorized into three functional areas:

- Providing mechanisms for client stations to save power through longer sleep times and reducing the need to wake up.
- Improving the mechanisms by which a client station accesses the medium by providing procedures to allow the station to know when it will be able to, or will have to, access the channel.
- Enhancing the throughput of a client station that accesses the channel, by reducing the overhead associated with current IEEE 802.11 exchanges through reducing frame headers, as well as simplifying and speeding management frames exchanges.

In what follows, we will describe a number of those enhancements in more detail.

Short MAC Header

To enhance throughput, 802.11ah adds support for a shorter MAC header compared to the current 802.11 standard. Information contained in the QoS and HT control fields (the latter introduced to the MAC header with 802.11n) are moved to a signal (SIG) field in the PHY header. The other non-applicable parts of the header are suppressed, e.g., no duration/ID fields, since there is no virtual clear channel assessment (CCA). The new header is 12 bytes shorter than the standard 802.11n header. Following the same logic, the acknowledgment (ACK) frame is replaced with a null data packet, which only contains the PHY header (no MAC header, no FCS). That frame is sent at a special reserved modulation and coding scheme (MCS) to make it recognizable. MCS is a simple integer assigned to every permutation of modulation, coding rate, guard interval, channel width, and number of spatial streams.

Large Number of Stations

To enable support for a large number of client stations, 802.11ah extends the Association Identifier (AID), which is limited to 2007 in the current 802.11 standard, by creating a hierarchical identifier with a virtual map, bringing the number up to 8191.

Speeding Frame Exchanges

In current 802.11 frame exchanges, a client station first has to contend for the medium, then transmit its frames, and then wait for an acknowledgment from the access point (AP). If the client station expects a response, it has to stay awake, while the AP contends for the medium and then sends. The client station finally sends an acknowledgment. With the 802.11ah speed frame exchange mechanism, the dialog can occur within a single transmission opportunity (TXOP): the client station wakes up, contends for the medium, and sends the frame to the AP, and the AP immediately replies after just a short inter-frame gap, allowing the client station (e.g., sensor) to immediately go back to sleep mode after receiving the answer, saving on uptime wasted in inter-frame and two-way acknowledgments.

Relay

Client stations often need to exchange information with one another, going through one or more intermediary APs when a direct connection is not available. In such exchanges, the client stations are forced to stay awake for the entire duration of the dialog. This process is greatly optimized with 802.11ah relay coupled with speed frame exchange. The client station wakes up and sends a frame to the AP, asking the latter to relay. The client station can then immediately go back to sleep/power-saving mode. The AP may relay the frame through another AP or deliver it directly to the destination. This model is appealing due to a number of reasons: the AP is usually main powered and has enough resources to buffer the frame until the destination client station wakes up. The same process can be repeated for the response message, allowing both client stations to optimize power consumption when they are not actively sending or receiving. This also eliminates the need for the client stations to synchronize wake/sleep cycles.

Target Wake Time

With target wake time (TWT), the AP can inform client stations when they will gain the right to access the medium. A client station and an AP can exchange initial frames expressing how much access the former needs. Then, the AP can assign a target wake time for the station, which can be either aperiodic or periodic (thus eliminating the need for the client station to have to wake up to listen to TWT values). Outside of the TWT, the client station can sleep and does not have to wake up to listen to any messages, not even beacon frames. At those target wake times (TWTs), the AP can send a null data packet paging (NDP) that tells the client station about the AP buffer status. This allows the AP to smoothly deliver buffer content to all client stations one after the other, instead of having all stations wake up at beacon time.

Grouping

Client stations can be grouped based on their location, using a group identifier assignment that relies on their type or other criteria. The AP then announces which groups are allowed to be awake for the next time period and which groups can go back to sleep mode because they will not be allowed to access the channel. This saves battery power on the sleeping groups, as these do not have to listen to the traffic. This logic brings a form of time division multiplexing (TDM) to Wi-Fi, by allowing transmission to each group based on time periods.

Traffic Indication Map (TIM) and Paging Mechanism

802.11ah introduces a traffic indication map (TIM) and page segmentation mechanism, by which an AP splits the TIM virtual bitmap into segments and each beacon only carries one segment. This allows IoT devices to wake up only to listen to the TIM matching their segment number. 802.11ah also introduces the concept of TIM stations (that need to get TIM info and therefore wake up at regular intervals) and non-TIM stations (that do not expect to receive anything and therefore can sleep beyond TIMs and do not need to wake up unless they need to send).

Restricted Access Windows

The AP can define a restricted access window (RAW), which is a time duration composed of several time slots. The AP can inform client stations that they have the right to send or receive only during certain time slots within the window, in order to distribute traffic evenly. The AP would use the RAW parameter set (RPW) to determine and communicate these slots and transmission or reception privileges. A client station that has traffic to send upstream but for which the AP does not have traffic to send downstream can send a request message to indicate to the AP that it needs a slot upstream.

5.1.2.5   Comparison of Wireless Link Layer Protocols

The table below summarizes key characteristics of the wireless IoT link layer protocols discussed in this chapter:

Protocol	Range	Data rate	Topology	Application	Power consumption
IEEE 802.15.4	Up to 1 km	1 Mbps to 10 Kbps	Mesh	Personal area network/ home network	Very low
LPWAN	Up to 20 km	Up to 50 Kbps	Star	Wide area network	Low
IEEE 802.11ah	Up to 1 km	>100 Kbps	Star	Metropolitan block	Medium

5.1.2.6   Time-Sensitive Networking

The requirements for Time-Sensitive Networking originate from real-time control applications such as industrial automation and automotive networks. These requirements contribute to some of the most prominent gaps that current Internet technologies need to address at the Link layer to realize the vision of IoT. In the case of industrial automation, the networks are relatively large (in the order of one to several kilometers) and may include up to 64 hops for a factory and up to 5 hops within a work cell (e.g., robot). The network needs to accommodate, in addition to real-time control traffic, other long-tailed traffic such as video or large file transfers. One of the key requirements for such networks is precise time synchronization, in the order of ±500 ns within a work cell and ±100 μs factory wide. Another key requirement is deterministic delay, which is not to exceed 5 μs within a work cell and 125 μs factory wide. Last but not least, a fundamental requirement for such networks is high availability as it is critical for the safety of the operators. This translates to a requirement for redundant paths with seamless or instantaneous switchover time, not to exceed 1 μs. In the case of automotive networks, the physical size of the deployments is relatively small, but the number of ports required is large: as an example, the network may span 30 m over 5 hops with over 100 devices connected (sensors, radar, control, driver-assist video, information, and entertainment audio/ video). A key requirement for these networks is support for deterministic and very small latency, less than 100 μs over 5 hops using 100 Mbps links. Another important requirement is high availability to ensure driver and passenger safety.

The above networks have typically been based on non-IP technologies. Connectivity has traditionally been achieved using some fieldbus technology such as DeviceNet, Profibus, and Modbus. Each of these technologies conforms to specific power, cable, and communication specifications, depending on the supported application. This has led to the situation where multiple desperate networks are deployed in the same space and has driven the need to have multiple sets of replacement parts, skills, and support programs within the same organization. With IoT, it will be possible to unite these separate networks into a converged network infra-structure based on industry standards. A candidate set of technologies to provide the Link layer functions of this converged network infrastructure is the IEEE 802 family of local area network (LAN)/metropolitan area network (MAN) protocols. One of the more popular technologies in the IEEE 802 family of protocols is Ethernet. Ethernet is by far the most widely deployed LAN technology today, connecting more than 85% of the world's local area networks (LANs). More than 300 million switched Ethernet ports have been installed worldwide. Ethernet's ubiquity can be attributed to the technology's simplicity, plug-and-play characteristics, and ease of manageability. Furthermore, it is low cost and flexible and can be deployed in any topology. Ethernet and the IEEE 802 family of protocols have steadily evolved over the years, with the IEEE Audio-Video Bridging (AVB) task group focusing on standards for transporting latency-sensitive traffic over bridged networks, primarily for multimedia (audio and video) streaming applications. These standards provide a foundation on which to build Time-Sensitive Networking technologies for IoT. They provide architecture for managing different classes of time-sensitive traffic through a set of in-band protocols. In particular, IEEE 802.1AS defines a profile for the Precision Timing Protocol (PTP), which provides time synchronization of distributed end systems over the network with accuracy better than ±1 µs. IEEE 802.1Qav defines forwarding and queuing rules for time-sensitive traffic in Ethernet. It specifies two traffic classes, class A and class B, with maximum latency guarantees of 2 ms and 50 ms, respectively. Traffic that does not belong to one of these two classes is considered to be "best effort," which includes all legacy Ethernet traffic. Traffic shaping and transmission selection are performed using a credit-based shaping algorithm: traffic is organized by priority, according to its class, and transmission of a frame in one of the above two classes is only allowed when credits are available for the associated class. Upper and lower bounds on the credit-based shaper limit the bandwidth and burstiness of the streams. Furthermore, IEEE standard 802.1Qat (part of IEEE 802.1Q-2011) defines a signaling protocol for dynamic registration and resource reservation of new streams, which provides per-hop delays in the order of 130 µs on 1 Gbps Ethernet links.

These standards, however, fall short in a number of areas: First, IEEE 802.1AS can take up to 1 s to switch to a new grandmaster clock (GMC) in the case of failure of the primary GMC. For real-time control applications, it is required to have the switchover time be in the order of 250 ms or less. Also, it is highly desirable to support multiple concurrently active GMCs for high availability. Second, per-hop switch delays need to be reduced by almost two orders of magnitude. Third, path selection and reservation for critical streams need to be made faster and simpler in order to accommodate high-scale deployments with thousands of streams.

As discussed previously, network high availability is of paramount importance in real-time IoT applications. Ethernet has historically, and for a long period of time, relied on the Spanning Tree Protocol (STP) in order to support redundancy and failure protection. However, in the past decade or so, requirements for massively scalable Ethernet networks in data center and metropolitan area network (MAN) deployments have resulted in the evolution of the Ethernet plane toward the use of the Intermediate System-to-Intermediate System (IS-IS) protocol, as defined in IEEE 802.1aq-2012 (Shortest Path Bridging) and IEEE 802.1Qbp-2014 (Equal Cost Multiple Path). IS-IS provides mechanisms for topology discovery and setup of redundant paths. It also includes mechanisms for network reconfiguration in the case of failures with reasonable delays (better than STP). These standards, however, are still lacking in the following areas: There are no standardized mechanisms to engineer paths with nonoverlapping or minimally overlapping links and nodes. Also, there are no mechanisms that provide extremely fast (i.e., instantaneous) switchover in the case of failures. Finally, there are no mechanisms for redundant (simultaneous) transmission of streams along nonoverlapping paths.

The IEEE Time-Sensitive Networking TSN task group was formed in November 2012, by renaming the Audio/Video Bridging (AVB) task group, with the goal of addressing the gaps highlighted above. Under that umbrella, work on three emerging standards commenced: 802.1Qca Path Control and Reservation, 802.1Qbv Enhancements for Scheduled Traffic, and 802.1CB.

IEEE 802.1Qca

This emerging standard extends the use of IS-IS to control Ethernet networks beyond what is defined in IEEE 802.1aq Shortest Path Bridging. It provides explicit path control, bandwidth, and stream reservation and redundancy (through protection or restoration) for data streams. It proposes the use of IS-IS for topology discovery and to carry control information for scheduling and time synchronization. The new protocol will enable the use of non-shortest paths and will provide explicit forwarding path (explicit tree—ET) control. Path calculation and determination will be done through a Path Computation Element (PCE), the latter being defined by the IETF PCE workgroup. The PCE is an application that computes paths between nodes in the network based on a representation of its topology. In 802.1Qca, IS-IS is currently being proposed as the protocol to convey the topology information from the Ethernet network to the PCE. The PCE may be centralized and reside in a dedicated server or in a network management system (NMS), or it may be distributed and embedded in the network elements (e.g., routers or bridges) themselves.

Figure 5.8 shows an example Ethernet network controlled by a single PCE residing in end station X. This end station is connected to SPT Bridge 11. The PCE peers with the bridge using IS-IS to learn the topology. The PCE can compute explicit trees based on, for example, bandwidth or delay requirements, and communicates them using IS-IS extensions to the bridges (Fig. 5.8).

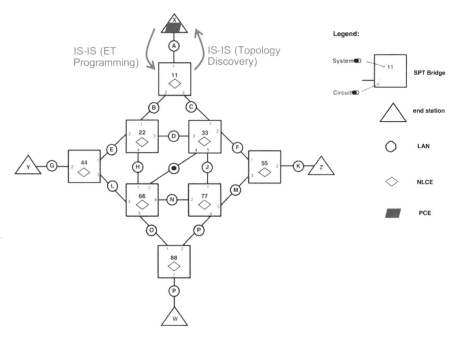

Fig. 5.8 Example IEEE 802.1Qca network

IEEE 802.1Qbv

The IEEE 802.1Qbv standard will provide real-time control applications with performance assurances for network delay and jitter over "engineered" LANs while maintaining coexistence with IEEE 802.1Qav/Qat reserved streams and best-effort traffic on the same physical network. Engineered LANs are so-called because traffic transmission schedules for the network can be designed offline. These pre-configured schedules assign dedicated transmission slots to each node in the network, for the purpose of preventing congestion and enabling isochronous communication with deterministic latency and jitter. The emerging standard will define time-aware shaping algorithm that enables communicating nodes to schedule the transmission of messages based on a synchronized time. It is proposed that priority markings carried in the frames will be used to distinguish between time-scheduled, reserved stream (credit based), and best-effort traffic.

Figure 5.9 depicts the traffic queue architecture for a bridge port that implements this emerging standard. A transmission gate is associated with each traffic queue; the state of the transmission gate determines whether or not queued packets can be selected for transmission on the port. Global Gate Control logic determines what set of gates are open or closed at any given point of time. A packet on a queue cannot be transmitted if the transmission gate, for that queue, is in the closed state or if the packet size is known and there is insufficient time available to transmit the entirety of that packet before the next gate-close event associated with that queue (Fig. 5.9).

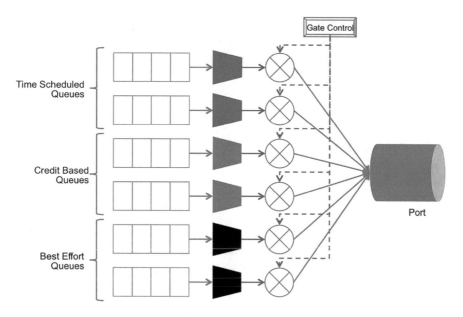

Fig. 5.9 IEEE 802.1Qbv time-based queuing

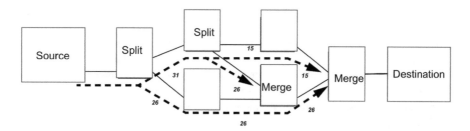

Fig. 5.10 IEEE 802.1CB seamless redundancy

IEEE 802.1CB

In order to maximize the availability and reliability of the network, IEEE 802.1CB proposes mechanisms that will enable "seamless redundancy" over 802.1Qca networks. With seamless redundancy, the probability of packet loss is reduced by sending multiple copies of every packet of a stream. Each copy is transmitted along one of a multitude of redundant paths. Duplicate copies are then eliminated to reconstitute the original stream before it reaches its intended destination.

This is effectively done by tagging packets with sequence numbers to identify and eliminate the duplicates and by defining new functions for bridges, a split function, responsible for replicating packets in a stream, and a merge function responsible for eliminating duplicate packets of a stream (Fig. 5.10).

IEEE 802.1CB proposes introducing a new tag to the 802.1Q frame, the redundancy tag, which includes a 16-bit sequence number. The emerging standard recognizes that alternate tagging mechanisms are possible, for example, through the use of multiple protocol label switching (MPLS) pseudowires [RFC4448] or using IEEE 802.1AE MacSec.

5.2   Internet Layer

5.2.1   Challenges

Many IoT deployments constitute what is referred to as low-power and lossy networks (LLNs). These networks comprised of a large number (several thousand) of constrained embedded devices with limited power, memory, and processing resources. They are interconnected using a variety of Link layer technologies, such as IEEE 802.15.4, Bluetooth, Wi-Fi, or power-line communication (PLC) links. There is a wide scope of use cases for LLNs, including industrial monitoring, building automation (HVAC, lighting, access control, fire), connected homes, healthcare, environmental monitoring, urban sensor networks (e.g., smart grid), and asset tracking. LLNs present the following five challenges to the Internet layer of the protocol stack:

Nodes in LLNs operate with a hard, very small bound on state. As such, Internet layer protocols need to minimize the amount of state that needs to be kept per node for routing or topology maintenance functions. The design of LLN routing protocols needs to pay close attention to trading off efficiency for generality, as most LLN nodes do not have resources to spare.

Typically, LLNs are optimized for saving energy. Various techniques are used to that effect, including employing extended sleep cycles, where the embedded devices only wake up and connect to the network when they have data to send. Thus routing protocols need to adapt to operate under constant topological changes due to sleep/wake cycles.

Traffic patterns within LLNs include point-to-point, point-to-multipoint, and multipoint-to-point flows. As such, unicast and multicast considerations should be taken into account when designing protocols for this layer.

LLNs will typically be employed over Link layer technologies characterized with restricted frame sizes; thus routing protocols for LLNs should be adapted specifically for those Link layers.

Links within LLNs may be inherently unreliable with time-varying loss characteristics. The protocols need to offer high reliability under those characteristics.

Internet layer protocols in LLN have to take the above issues and challenges as design requirements. The protocol design should take into account the link speeds and the device capabilities. For example, if the devices are battery powered, then protocols that require frequent communication will deplete the nodes' energy faster. As described above, LLNs are inherently lossy: a characteristic that is typically unpredictable and predominantly transient in nature. The design of the Internet layer protocols must account for these characteristics. In conventional networks, these protocols react to loss of connectivity by quickly reconverging over alternate routing paths. This is to minimize the extent of data loss by routing around link, node, or other failures as quickly as possible (e.g., MPLS fast reroute mechanism strives for reconvergence within 50 ms). In LLNs, such quick reaction to failures is undesirable due to the transient nature of loss in these networks. As a matter of fact, it would lead to instability and unacceptable control plane churn. Instead, the protocols should follow a paradigm of underreacting to failures in order to dampen the effect of transient connectivity loss, combined with confidence-monitoring model to determine when to trigger full reconvergence. The varying link quality levels in LLNs have direct bearing on protocol design, especially with regard to convergence characteristics and time. In traditional networks, global reconvergence is triggered to minimize the convergence time, whereas in LLNs local reconvergence is preferred, where the traffic is locally redirected to an alternate next hop during transient instabilities. This is to minimize the effect of routing instabilities that may lead to overall network oscillations or forwarding loops. Another consideration for LLNs is the dynamic nature of link and node metrics used in route computation. There are so many dynamic factors in LLNs, such as link quality deteriorating due to interference, node switching from mains power to battery power, momentary CPU overload on a node, etc. These factors cause node and link metrics to be time varying in nature, and the routing protocols must be able to handle that.

Existing routing protocols such as OSPF, IS-IS, etc. in their current form do not satisfy the routing requirements imposed by the above challenges (Fig. 5.11).

5.2.2   Industry Progress

5.2.2.1   6LowPAN

As discussed previously, one of the challenges imposed by IoT on the Internet layer is the adaptation of this layer's functions to Link layer technologies with restricted frame size. A case in point is adapting IP, and specifically the scalable IPv6, to the IEEE 802.15.4 Link layer. The base maximum frame size for 802.15.4 is 127 bytes, out of which 25 bytes need to be reserved for the frame header and another 21 bytes for link layer security. This leaves, in the worst case, 81 bytes per frame to cram the IPv6 packet into. What add to the problem are two issues: first, the IPv6 packet header, on its own, is 40 bytes in length, and second, IPv6 does not perform segmentation and reassembly of packets; this function is left to the end stations or to lower layer protocols. Even though 802.15.4 g increases the maximum frame size to 2047 bytes, it is still highly desirable to be able to compress IPv6 packet headers over this Link layer. For the aforementioned reasons, the IETF defined IPv6 over low-power

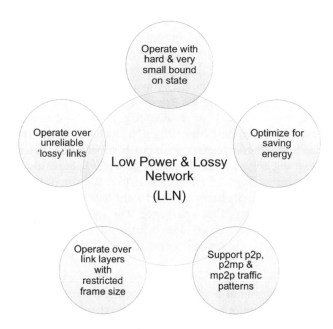

Fig. 5.11 IoT challenges for the Internet layer. (Source Cisco BRKIOT-2020, 2015)

Fig. 5.12 6LowPAN
Adaptation layer

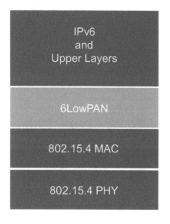

wireless personal area networks (6LowPAN). 6LowPAN is defined in RFC6282. It is an adaptation layer for running IPv6 over 802.15.4 networks (Fig. 5.12). 6LowPAN provides three main functions: IPv6 header compression, IPv6 packet segmentation and reassembly, and layer 2 forwarding (also referred to as mesh under). With 6LowPAN, it is possible to compress the IPv6 header into 2 bytes, as most of the information is already encoded into the Link layer header.

6LowPAN introduces three headers for each of the three functions that it provides. Those headers are compression header, fragment header, and mesh header.

One or more of these headers may be available in any given packet depending on which functions are applied (Fig. 5.13).

6LowPAN defines new mechanisms to perform IPv6 neighbor discovery (ND) operations including link layer address resolution and duplicate address detection.

A recurring issue when adapting IPv6 to any Link Layer technology is support for a single broadcast domain, where a host can reach any number of hosts within the subnet by sending a single IP datagram. Accommodating a single broadcast domain within a 6LoWPAN network requires Link layer routing and forwarding functions, often referred to as mesh under, since the multi-hop mesh topology is abstracted away from the IP layer to appear as a single network segment. However, the IETF has not specified a mesh-under routing protocol for 6LoWPAN. Hence, this constitutes a technology gap, especially for IoT applications that can benefit from or that rely on intra-subnet broadcast capabilities.

Even though the scope of 6LoWPAN was originally focused on the IEEE 802.15.4 Link layer, the technology has very limited dependency on 802.15.4 specifics, thereby allowing other link technologies (e.g., power-line communication— PLC) to utilize the same adaptation mechanisms. Consequently, the term "6LoWPAN networks" is often generalized to refer to any Link layer mesh network built on low-power and lossy links leveraging 6LoWPAN mechanisms.

5.2.2.2   RPL: IPv6 Routing Protocol for Low-Power and Lossy Networks

The routing over low-power and lossy networks (ROLL) workgroup in IETF has defined in RFC 6550 an IPv6 routing protocol for LLNs, known as RPL.[2] RPL is a distance-vector routing protocol. The reason for choosing a distance-vector protocol, as opposed to a link-state paradigm, is primarily to address the requirement of minimizing the amount of control-plane state (memory) that needs to be maintained on the constrained nodes of LLNs. Link-state routing protocols build and maintain a link-state database of the entire network on every node and hence tend to be heavier on memory utilization compared to distance-vector algorithms. RPL

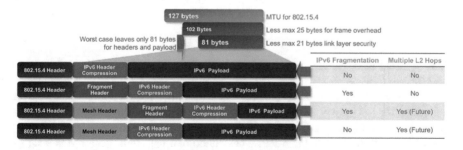

Fig. 5.13   6LowPAN header stack. (Source: Cisco BRKIOT-2020, 2015)

---

[2] Pronounced as "ripple".

computes a destination-oriented directed acyclic graph (DODAG) based on an objective function and a set of metrics and constraints. In the context of routing, a directed acyclic graph (DAG) is formed by a series of nodes and links. Each link connects one node to another in a directed fashion such that it is not possible to start at a node N and follow a directed path that cycles back to node N. A destination-oriented DAG is a DAG that includes a single root node. The DODAG is a logical topology built over the physical network for the purpose of meeting specific criteria and carrying traffic subject to certain requirements. These criteria and requirements are captured in the objective function, metrics, and constraints. The objective function captures the goal behind setting up a specific topology. Example objective functions include minimizing latency of communication or maximizing the probability of message delivery. Metrics are scalar values that serve as input parameters to the best-path selection algorithm. Example metrics include link latency or link reliability or node energy level. Constraints refer to conditions that would exclude specific nodes or links from the topology if they do not meet those constraints, such as exclude battery-powered nodes or avoid unencrypted links. RPL supports dynamic metrics and constraints, where the values change overtime and the protocol reacts to those changes.

In a RPL network, a given node may be a member of different logical topologies, or DODAGs, each with a different objective. This is supported through the notion of RPL "instances." An RPL instance is a set of DODAGs rooted at different nodes, all sharing the same objective function (Fig. 5.14).

The DODAG root is typically a border router that connects the LLN to a backbone network. It is always assigned a rank of 1. RPL calculates ranks for all nodes connected to the root based on the objective function. The rank value increases moving down from the root toward leaf nodes. The rank indicates the node's position or coordinates in the graph hierarchy.

RPL has two characteristics that render it well suited for LLNs: First, it is a proactive protocol, i.e., it can calculate alternate paths as part of the topology setup, as opposed to reactive protocols which rely on exchanging control plane messages after a failure occurs to determine backup paths. Second, RPL is underreactive: it prefers local repair to global reconvergence. Failures are handled by locally choosing an alternate path, which makes the protocol well suited for operation over lossy links.

5.2.2.3 6TiSCH

As discussed previously, IEEE 802.15.4 TSCH defines the medium access control functions for low-power wireless networks with time scheduling and channel hopping. TSCH can fit as the Link layer technology in an IPv6-enabled protocol stack for LLNs, with 6LoWPAN and RPL. The functional gap in the solution is a set of entities that can take control of defining the policies to build and maintain the TSCH schedule, matching that schedule to the multi-hop paths maintained by the RPL

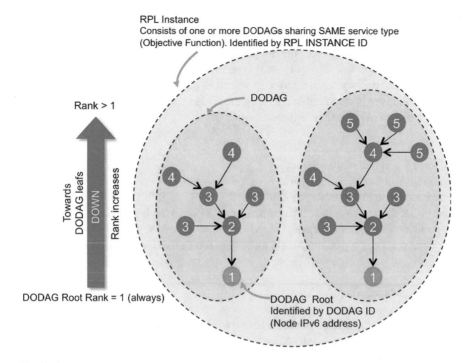

Fig. 5.14 RPL instances and DODAGs

routing protocol and adapting the resources allocated between adjacent nodes to traffic flows.

As such, an adaptation layer is required in order to run the IPv6 stack on top of IEEE 802.15.4 TSCH. The IETF has recently formed the 6TiSCH workgroup in order to address this technology gap and define what is referred to as the "6top" adaptation layer. This adaptation layer is sandwiched in between the 802.15.4 link layer and the 6LoWPAN adaptation layer. Its goals are to address the following issues:

Network Formation

The adaptation layer must control the formation of the network. This includes two functions: the mechanisms by which new nodes securely join the network and the mechanisms by which nodes that are already part of the network advertise its presence.

Network Maintenance

After the network is formed, the adaptation layer needs to maintain the network's health and ensure that the nodes stay synchronized. This is because a TSCH node must have a time-source neighbor to which it can synchronize at all times. The adaptation layer is responsible for assigning those neighbors to the nodes, to guarantee the correct operation of the network.

Topology and Schedule Mapping

The adaptation layer needs to gather basic topological information, including node and link state, and provide this information to RPL, so the latter can compute multi-hop routes. Conversely, the adaptation layer needs to ensure that the TSCH schedule contains cells corresponding to the multi-hop routes calculated by RPL.

Resource Management

The adaptation layer is responsible for providing mechanisms by which neighboring nodes can exchange information regarding their schedule and negotiate the addition or deletion of cells. Note that a cell maps to a transmission/reception opportunity, and, hence, constitutes an atomic unit of resource in TSCH. The number of cells to be assigned between two neighbor nodes should be sized proportionately to the volume of traffic between them.

Flow Control

While TSCH defines mechanisms by which a node can signal to its neighbors when it can no longer accept incoming packets, it does not, however, specify the policies that govern when to trigger those mechanisms. Hence, it is the responsibility of the adaptation layer to specify mechanisms for input and output packet queuing policies, manage the associated packet queues, and indicate to TSCH when to stop accepting incoming packets. The adaptation layer should also handle transmission failures, in the scenario where TSCH has attempted to retransmit a packet multiple times without receiving any acknowledgment.

Determinism

The adaptation layer is responsible for providing deterministic behavior for applications that demand it. This includes providing mechanisms to ensure that data is delivered with guaranteed upper bounds on latency and possibly jitter, all while maintaining coexistence between deterministic flows and best-effort traffic.

Scheduling Mechanisms

It is envisioned that multiple different scheduling mechanisms may be employed and even coexist in the same network. This includes centralized mechanisms, for example, where a Path Computation Element (PCE) takes control of the schedule, in addition to distributed mechanisms where, for instance, neighboring nodes monitor the amount of traffic and adapt the number of cells autonomously by negotiation of the allocation or deallocation of cells as needed. The adaptation layer needs to provide mechanisms to allow for all these functions.

Secure Communication

TSCH defines mechanisms for encrypting and authenticating frames, but it does not define how the security keys are to be generated. Hence, the adaptation layer is responsible for generating the keys and defining the authentication mechanisms by which a new node can join an existing TSCH network. The layer is also expected to provide mechanisms for the secure transfer of signaling (i.e., control) as well as application data between nodes.

The envisioned 6TiSCH protocol stack is depicted in Fig. 5.15. RPL will be the routing protocol of choice for the architecture. As the work in IETF progresses, there may be a need to define a new 6TiSCH-specific objective function for RPL. For the management of devices, the architecture will leverage the Constraint Application Protocol Management Interface (COMI), which will provide the data model for the 6top adaptation layer management interface. Centralized scheduling will be carried out by the Path Computation Element (PCE). The topology and device capabilities will be exposed to the PCE using an extension to a Traffic Engineering Architecture and Signaling (TEAS) protocol. The schedule computed by the PCE will be distributed to the devices in the network using either a light-weight Path Computation Element Protocol (PCEP) or an adaptation of Common Control and Measurement

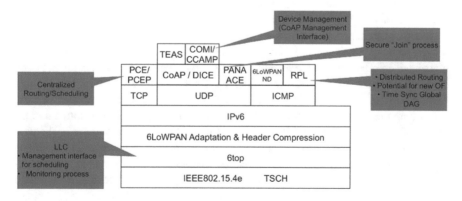

Fig. 5.15 6TiSCH protocol stack

Plane (CCAMP) formats. The Datagram Transport Layer Security in Constrained Environments (DICE) can be used in the architecture to secure CoAP messages. Also, the Protocol for Carrying Authentication for Network Access (PANA) will secure the process of a node joining an existing network.

5.3   Application Protocols Layer

Application protocols are responsible for handling the communication between Application Entities, i.e., things, gateways, and applications. They typically support the flow of data (e.g., readings or measurements) from things to applications and the flow of command or control information (e.g., to trigger or actuate end devices) in the reverse direction. These protocols define the semantics and mechanisms for message exchanges between the communicating endpoints.

The landscape of the application protocols layer in IoT is currently crowded with competing protocols and standards, each having its own set of strengths and weaknesses and with no clear path toward convergence being agreed upon by the industry yet. In this section, we will discuss the characteristics and attributes of the protocols in this layer as they pertain to IoT and will highlight, where applicable, the requirements and challenges that IoT applications impose on these protocols.

5.3.1   Data Serialization Formats

Applications protocols vary in the data serialization formats used to encode information into messages. One of the challenges in IoT data serialization formats is mapping between the formats used in constrained devices and those used by applications in the World Wide Web. These applications should be able to interpret the data from IoT devices with minimal format translations and a priori knowledge. Hence, the formats should be general and compatible with Web technologies. Popular data serialization formats on the Web include XML, JSON, and EXI.

Another challenge in IoT data serialization formats is the impact they have on device resource utilization, especially in terms of energy consumption. Data formats have an effect on device resource usage in two facets: in their local processing demands and their communication efficiency. The local processing demands include both the processing required to serialize memory objects into data encoded in messages and the processing required to parse the encoded messages into memory objects. The communication efficiency is a function of the compactness of the data serialization format and its efficiency to encode information in the least amount of message real estate. Both of these facets, namely, local processing and communication, have a direct impact on the energy consumption of the IoT device. Research in wireless sensor networks suggests "communication is over 1000 times more expensive in terms of energy than performing a trivial aggregation operation." Therefore, the data serialization formats for IoT application protocols should be chosen such that they require minimal processing and communication demands.

A third challenge in IoT data serialization formats is the impact they have on network bandwidth utilization. This ties back to the compactness of the format and its encoding efficiency, as discussed above. The more verbose that the data format is, the more message space that it will consume on the wire to carry the same amount of information, which leads to less efficient use of network bandwidth. For IoT, especially when devices are connected over low-bandwidth wireless links, the data serialization format of application protocols should be chosen carefully to maximize the use of the available bandwidth.

5.3.2   Communication Paradigms

Application protocols support different communication patterns. These patterns enable varying paradigms of interaction between IoT applications and devices.

5.3.2.1   Request/Response Versus Publish/Subscribe

The request/response paradigm enables bidirectional communication between endpoints (Fig. 5.16). The initiator of the communication sends a request message, which is received and operated upon by the target endpoint. The latter then sends a response message to the original initiator. This paradigm is well suited for IoT deployments that have one or more of the following characteristics:

- The deployment follows a client-server architecture.
- The deployment requires interactive communication: both endpoints have information to send to the other side.

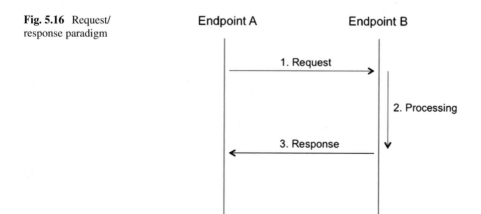

Fig. 5.16 Request/response paradigm

• The receipt of information needs to be fully acknowledged (e.g., for reliability).

However, not all IoT deployments have the above characteristics. In particular, in many scenarios, all what is required is one-way communication from a data producer (e.g., a sensor) to a consuming entity (the application). For this, the request/response paradigm is sub-optimal due to the overhead of the unneeded messages running in the reverse direction. This is where the publish/subscribe pattern comes in (Fig. 5.17).

The publish/subscribe paradigm, often referred to as pub/sub, enables unidirectional communication from a publisher to one or more subscribers. The subscribers declare their interest in a particular class or category of data to the publisher. When the publisher has new data available from that class, it pushes it in messages to interested subscribers. Besides the obvious proclamation that this paradigm optimal for IoT applications requires one-way communication, the pub/sub model is well suited for IoT deployments that can benefit from the following characteristics:

• Loose coupling between the communicating endpoints, especially when compared with the client-server model.
• Better scalability by leveraging parallelism and the multicast capabilities of the underlying transport network.

5.3.2.2 Blocking Versus Non-blocking

Application protocols can offer IoT endpoints blocking or non-blocking messaging service.

In the blocking mode, the endpoint originating a request must wait to get a response to its request, after the requested operation has finished on the other endpoint. This involves potentially long or unknown wait times (where a pending request has not been responded to) for the originator.

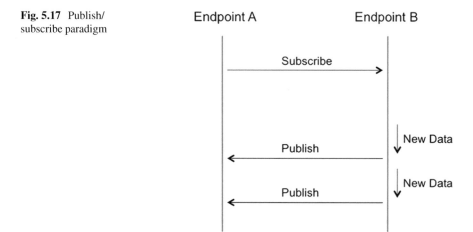

Fig. 5.17 Publish/subscribe paradigm

In the non-blocking mode, the endpoint originating a request does not wait until the other endpoint has fully serviced the request. Rather, it expects a prompt acknowledgment of the request together with a specified reference, so that the originator can retrieve the outcome of the requested operation at a later point of time.

In the synchronous case, the originator of a request is not able to receive asynchronous messages, i.e., all exchanges of information between the originator and the receiver need to be initiated by the originator. The later retrieval of the result of a requested operation is through the exchange of request/response messages between the originator and the receiver.

In the asynchronous case, the originator of a request is able to receive notification messages, i.e., the receiver can send an unsolicited message to the originator at an arbitrary time to report the requested operation. The mechanisms for the notification to the originator are the same as in the case of a notification after a subscription.

5.3.3   QoS

Application protocols should provide mechanisms for fine-grained control over the real-time behavior, dependability, and performance of IoT applications by means of a rich set of QoS policies. These policies should provide control over local resources and the end-to-end properties and characteristics of data transfer. The local properties controlled by QoS relate to resource usage, whereas the end-to-end properties relate to the temporal and spatial aspects of data communication.

5.3.3.1   Resource Utilization

Application protocols should provide QoS policies to control the amount of memory and processing resources that can be used by the application protocol for data transmission and reception. These policies include:

Resource Limits Policy

This policy allows control of the amount of message buffering performed by a protocol implementation, as this impacts the amount of memory consumed by that protocol. Such controls are particularly important for embedded applications running on constrained devices.

Time Filter Policy

This policy allows applications to specify the minimum inter-arrival time between data samples. Samples that are produced at a faster pace are not delivered. This policy allows control of both network bandwidth and memory and processing power for applications which are connected over limited bandwidth networks and which might have limited computing resources.

5.3.3.2   Data Timeliness

Application protocols should provide a set of QoS policies that allow control of the timeliness properties of distributed data. Specifically, the QoS policies that are desirable are described below:

Deadline Policy

This QoS policy allows an application to define the maximum inter-arrival time for data. Missed deadline can be notified by the protocol to the application.

Latency Budget Policy

This QoS policy provides a means for the application to communicate to the application protocol the level of urgency associated with a data communication. The latency budget specifies the maximum amount of time that should elapse from the instance when the data is transmitted to the instance when the data is placed in the queue of the associated recipients.

5.3.3.3   Data Availability

Application protocols should provide the following QoS policies to allow control of data availability:

Durability Policy

This QoS policy provides control over the degree of persistence of the data being transmitted by the application. At one end of the spectrum, it allows the data be configured to be volatile, while at the other end, it allows for data persistency. It is worth noting that data persistence enables time decoupling between the producing and the consuming endpoint by making the data available for late-joining consumers or even after the producer has disconnected.

Life Span Policy

This QoS policy allows control of the interval of time for which a data sample will be valid.

History Policy

This QoS policy provides a means to control the number of data samples that have to be kept available for the recipients. Possible values are the last sample only, the last N samples, or all the samples.

5.3.3.4   Data Delivery

Application protocols should provide QoS policies to allow control of how data is delivered.

Reliability Policy

This QoS policy allows the application to control the level of reliability associated with data diffusion. The possible choices are reliable and best-effort distribution. With reliable distribution, the application protocol must ensure message delivery and handle acknowledgments and retransmissions without direct application involvement.

Transport Priority

This QoS policy allows the application to take advantage of transports that are capable of sending messages with different priorities. Application protocols are responsible for interacting with the underlying transport layer in order to map this QoS policy to the right underlying transport network QoS markings (e.g., IP DSCP, TOS, or PCP).

5.3.4   RESTful Constraints

Some application protocols adhere to a set of constraints defined by the representational state transfer (REST) architectural paradigm. REST is a distributed client-server software architecture style that was coined by Roy Fielding after he analyzed the design principles that contributed to the success of the Hypertext Transfer Protocol (HTTP) employed in the World Wide Web. Fielding concluded on a set of constraints that collectively define the REST architectural style and yield a system that is simple, scalable, and reliable.

The formal REST constraints are as follows:

Client-Server Communication Model

This allows for separation of concerns where the server focuses on functions such as data storage, whereas clients focus on the user interface and user state. Uniform interfaces separate the clients from the servers. This allows for independent development of servers and clients as long as they honor the same interface.

Stateless Communication

The server must not store any client context that persists between requests. Session state is maintained by the client, which passes all the information necessary to service a particular request in the request itself. In other words, requests are self-contained from a server perspective.

Cacheable Communication

Responses from the server may be cacheable by clients and intermediate nodes. This improves the scalability and performance of the system by partially or completely eliminating some client–server interactions.

Layered Architecture

To allow for better scalability, the system comprised of a layered architecture that includes clients, servers, and potentially multiple intermediate nodes interspersed between them. Clients may be in communication with intermediate nodes or directly with servers without ordinarily being able to identify a difference between the two.

Uniform Interfaces

All interactions between clients and servers (or intermediate nodes) are governed by uniform interfaces. These interfaces use the notion of "resources." A resource is an abstraction for server-side information and associated native data representation. Resources have unique identifiers (e.g., URIs in Web systems). When a server communicates with a client, it transfers an external representation of the resource to the client (hence the name representational state transfer). REST interfaces are representation centric. Hence, a small set of operations (also called verbs), which are uniform across all use cases, can be used in the interface. Usually, this set of verbs is referred to as CRUD for create, read, update, and delete. In REST interfaces, there is no out-of-band contract that defines the types of actions that can be initiated by a client. Rather, this information is discovered dynamically by the client from prior server interactions through hypermedia (i.e., by hyperlinks within hypertext). This characteristic of the interface is known as hypermedia as the engine of application state (HATEOAS).

Code on Demand

Client functionality may be extended or modified by the server through the transfer of executable pieces of code that can be executed on the client side (e.g., scripts or applets). This is an optional REST constraint known as "code on demand."

5.3.5 Survey of IoT Application Protocols

5.3.5.1 CoAP

The Constrained Application Protocol (CoAP) was standardized by the IETF Constrained RESTful Environments (CORE) workgroup as a lightweight alternative to HTTP, targeted for constrained nodes in low-power and lossy networks (LLNs). The need for a lighter-weight version of HTTP can be appreciated by examining, for example, the number of messages that need to be exchanged between a client and a server to perform a simple Get operation on a resource: first there are three TCP SYN messages exchanged to bring up the TCP session, followed by the HTTP Get request from the client, then the HTTP response from the server, and finally two messages to terminate the TCP session. Hence, a total of seven messages are required just to fetch a resource. CoAP reduces this overhead by using UDP as a transport in lieu of TCP. CoAP also uses short headers to reduce message sizes.

Similar to HTTP, CoAP is a RESTful protocol. It supports the create, read, update, and delete (CRUD) verbs but in addition provides built-in support for the publish/subscribe paradigm via the new observe verb. CoAP optionally provides a mechanism where messages may be acknowledged for reliability and provides a bulk transfer mode. CoAP was standardized as RFC 7252. Furthermore, there is an ongoing work in the IETF to define mechanisms for dynamic resource discovery in CoAP via a directory service.

5.3.5.2 XMPP

The Extensible Messaging and Presence Protocol (XMPP) was originally designed for instant messaging, contact list, and presence information maintenance. It is a message-centric protocol based on the Extensible Markup Language (XML). Due to its extensibility, the protocol has been used in several applications, including network management, video, voice-over IP, file sharing, social networks, and online gaming, among others. In the context of IoT, XMPP has been positioned for smart grid solutions, for example, as depicted in RFC 6272. XMPP originally started as an open-source effort, but the core protocol was later standardized by the IETF in RFC 6120 and 6121. Moreover, the XMPP Standards Foundation (XSF) actively develops open extensions to the protocol.

The native transport protocol for XMPP is TCP. However, there is an option to run XMPP over HTTP.

5.3.5.3   MQTT

The Message Queue Telemetry Transport (MQTT) protocol is a lightweight pub-
lish/subscribe messaging protocol that was originally designed by IBM for enterprise telemetry. MQTT follows a client-server architecture where clients connect to
a central server (called the broker). The protocol is message oriented, where messages are published to an address, referred to as a topic. Clients subscribe to one or
more topics and receive updates from a client that is publishing messages for this
topic. In MQTT, topics are hierarchical (similar to URLs), and subscriptions may
use wildcards. MQTT is a binary protocol, and it uses TCP transport. The protocol
is being standardized by the Organization for the Advancement of Structured
Information Standards (OASIS).

The protocol targets endpoints where "a small code footprint" is required or
where network bandwidth is limited; hence it could prove useful for constrained
devices in IoT.

5.3.5.4   AMQP

The Advanced Message Queuing Protocol (AMQP) originates from financial sector
applications but is generic enough to accommodate other types of applications.
AMQP is a binary message-oriented protocol. Due to its roots, AMQP provides
message delivery guarantees for reliability, including at least once, at most once,
and exactly once. The importance of such guarantees can be easily seen in the context of financial transactions (e.g., when executing a credit or debit transaction).
AMQP offers flow control through a token-based mechanism, to ensure that a
receiving endpoint is not overburdened with more messages than it is capable of
handling. AMQP assumes a reliable underlying transport protocol, such as TCP.

AMQP was standardized by OASIS in 2012 and then by the International
Standards Organization (ISO) and the International Electrotechnical Commission
(IEC) in 2014. Several open-source implementations of the protocol are available.
AMQP defines a type system for encoding message data as well as annotating this
data with additional context or metadata. AMQP can operate in simple peer-to-peer
mode as well as in hierarchical architectures with intermediary nodes, e.g., messaging brokers or bridges. Finally, AMQP supports both point-to-point communication
and multipoint publish/subscribe interactions.

5.3.5.5   SIP

The Session Initiation Protocol (SIP) handles session establishment for voice, video,
and instant messaging applications on IP networks. It also manages presence (similar to XMPP).

SIP invitation messages used to create sessions carry session descriptions that
enable endpoints to agree on a set of compatible media types. SIP leverages

elements called proxy servers to route requests to the user's current location, authenticate and authorize users for services, implement call-routing policies, and provide features. SIP also defines a registration function that enables users to update their current locations for use by proxy servers. SIP is a text-based protocol and can use a variety of underlying transports, TCP, UDP, or SCTP, for example. SIP is standardized by the IETF as RFC 3261.

5.3.5.6 IEEE 1888

IEEE 1888 is an application protocol for environmental monitoring, smart energy, and facility management applications. It is a simple protocol that supports reading and writing of time-series data using the Extensible Markup Language (XML) and the simple object access protocol (SOAP). The data is identified using Universal Resource Identifiers (URIs). The latest revision of the protocol was standardized by the IEEE Standards Association in 2014.

5.3.5.7 DDS RTPS

Distributed Data Service Real Time Publish and Subscribe is a data-centric application protocol that, as its name indicates, supports the publish/subscribe paradigm. DDS organizes data into "topics" that listeners can subscribe to and receive asynchronous updates when the associated data changes. DDS RTPS provides mechanisms where listeners can automatically discover speakers associated with specific topics. IP multicast or a centralized broker/server may be used to that effect. Multiple speakers may be associated with a single topic and priorities can be defined for different speakers. This provides a redundancy mechanism for the architecture in case a speaker fails or loses communication with its listeners.

DDS RTPS supports very elaborate QoS policies for data distribution. These policies cover reliability, data persistence, delivery deadlines, and data freshness. DDS RTPS is a binary protocol, and it uses UDP as the underlying transport. The latest version of the protocol was standardized by the Object Management Group (OMG) in 2014. Table 5.3 provides a summary of the protocols discussed in this section.

5.4 Application Services Layer

5.4.1 Motivation

M2M deployments have existed for over two decades now. However, what has characterized these deployments is a state of fragmentation: vertical solutions are implemented in silos with proprietary communication stacks and very tight coupling

Table 5.3 Survey of IoT application protocols

Protocol	Functions	Primary use	Transport	Format	SDO
CoAP	REST resource manipulation via CRUD Resource tagging with attributes Resource discovery through RD	LLNs	UDP	Binary	IETF
XMPP	Manage presence Session establishment Data transfer (text or binary)	Instant messaging	TCP HTTP	XML	IETF XSF
MQTT	Lightweight pub/sub messaging Message queuing for future subscribers	Enterprise telemetry	TCP	Binary	OASIS
AMQP	Message orientation, queuing and pub/sub Data transfer with delivery guarantees (at least once, at most once, exactly once)	Financial services	TCP	Binary	OASIS
SIP	Manage presence Session establishment Data transfer (voice, video, text)	IP telephony	TCP, UDP, SCTP	XML	IETF
IEEE 1888	Read/write data into URI Handling time-series data	Energy and facility management	SOAP/ HTTP	XML	IEEE
DDS (RTPS)	Pub/sub messaging with well-defined data types Data discovery Elaborate QoS	Real-time distributed systems (military, industrial, etc.)	UDP	Binary	OMG

between applications and devices. The paradigm can be best described as "one application-one device." The application code is exposed to all the device specifics under this modus operandi. This, in turn, creates complexity and increases the cost of the solution's initial development and ongoing maintenance. For instance, if the operator of a deployment wanted to replace a defective device with another from a different manufacturer, parts of the application source code would have to be rewritten in order for the replacement device to be integrated into the solution. By the same token, adding new types of devices to the solution cannot be performed without application source code changes. Furthermore, the networks interconnecting the devices and the applications are in many case closed proprietary systems, and interconnecting those networks requires application gateways that are complex and expensive. These issues constitute a major current gap in IoT. What is required is a layer of abstraction that fits in between the applications and the devices, i.e., things, and enables the paradigm of "any application-any device" (Fig. 5.18).

1 Application – 1 Device   Any Application – Any Device

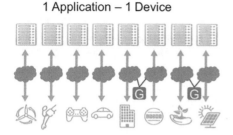

Fig. 5.18 Application to device coupling

In other words, this abstraction layer provides a common set of services that enables an application to interface with potentially any device without understanding a priori the specifics and internals of that device. This abstraction layer is referred to as the Application Services layer in our model of the IoT protocol stack. It provides seamless interoperability between applications and devices and promotes nimble development of IoT solutions.

From a business perspective, the emergence of this new layer is driven, in part, by communication service providers (CSPs) looking at using IoT to gain additional revenue from their networks. Key to this revenue will be differentiating beyond providing simple IP connectivity. CSPs know well the value of IoT is in the data, not the way it is transported. To unlock this value, the Application Services layer aims to turn the network to a common platform to enable diverse IoT applications. This common platform will be built across an ecosystem of heterogeneous devices and will enable CSPs to monetize IoT data access, storage, management, and security.

5.4.2   Industry Progress

In 2012, the European Telecommunications Standards Institute (ETSI) published the first release of its M2M service layer standard defining a standardized platform for multiservice IoT solution. Later that year, seven standards development organizations (TIA and ATSI from the USA, ARIB and TTC from Japan, CCSA from China, ETSI from Europe, and TTA from Korea) launched a global organization to jointly define and standardize the common horizontal functions of the IoT Application Services layer under the umbrella of the oneM2M Partnership Project (http://www.onem2m.org). The founders agreed to transfer and stop their own overlapping IoT application service layer work.

In what follows, we will discuss the ETSI M2M and oneM2M efforts in more details.

5.4.2.1   ETSI M2M

The network architecture adopted by the ETSI M2M effort draws heavily on existing technologies. The architecture comprised of three domains: M2M device
domain, network domain, and application domain (Fig. 5.19). The M2M device
domain provides connectivity between things and gateways, e.g., a field area network or personal area network. Devices are entities that are capable of replying to
request for data contained within those entities or capable of transmitting data contained within those entities autonomously. Gateways ensure that end devices (which
may not be IP enabled) can interwork and interconnect with the communication
network. Technologies in the M2M device domain include IEEE 802.15.4, IEEE
802.11, Zigbee, Z-WAVE, PLC, etc.

The network domain includes the communication networks, which interconnect
the gateways and applications. This typically includes access networks (xDSL,
FTTX, WiMax, 3GPP, etc.) as well as core networks (MPLS/IP). The application
domain includes the vertical-specific applications (e.g., smart energy, eHealth,
smart city, fleet management, etc.) in addition to the Service Capabilities layer
(SCL), a middleware layer that provides various data and application services. The
main focus of the ETSI M2M standards is on defining the functionality of the
SCL. The SCL provides functions that are common across different applications
and exposes those functions through an open API. The goal is to simplify application development and deployment through hiding the network specifics.

The functions of the SCL may reside on entities deployed in the field such as
devices and gateways or on entities deeper in the network (e.g., servers in a data
center). This gives rise to three flavors of SCL, depending on its placement: device
SCL (D-SCL), gateway SCL (G-SCL), and network SCL (N-SCL). While the three
flavors of SCL do share some common functions, they also differ due to the

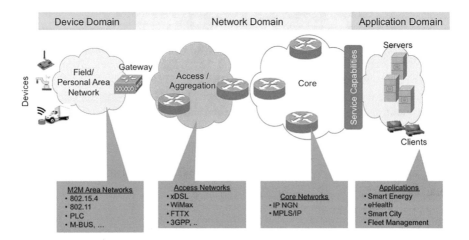

Fig. 5.19 ETSI M2M network architecture

different operations that need to be carried out by devices, gateways, and network nodes (servers). In general, the SCL provides the following functions:

- Registration of devices, applications, and remote SCLs
- Synchronous and asynchronous data transfer
- Identification of applications and devices
- Group management for bulk endpoint addressability and operations
- Security mechanisms for authentication, authorization, and access rights control
- Remote device management (through existing protocols)
- Location information

ETSI M2M adopted a RESTful architecture style where all data in the SCL is represented as resources. This includes not only the data generated by the devices but also data representing device information, application information, remote SCL information, access rights information, etc. Resources in the SCL are uniquely addressable via Universal Resource Identifiers (URIs). Manipulation of the resources is done through a RESTful API, which provides the CRUD primitives (C, create; R, read; U, update, D, delete). The API can be bound to any RESTful protocol, such as HTTP or CoAP. ETSI technical specification TS 102 921 specifies the API binding to HTTP and CoAP protocols.

Resources within the SCL are organized in a well-specified hierarchical structure known as the resource tree (Fig. 5.20). This provides a number of advantages: it provides a data mediation function, describes how resources relate to each other, allows traversal and query of data in an efficient manner, and speeds up the development of platforms. The resource tree of an SCL includes:

- Location of other SCLs in the network (in other devices or GWs)
- List of registered applications
- Announced resources on remote elements
- Access rights to various resources
- Containers to store actual application data

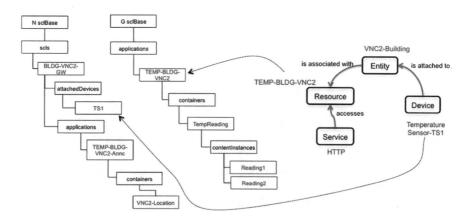

Fig. 5.20 Example ETSI M2M resource tree

In addition to the different flavors of SCL, ETSI M2M defines the following types of entities: application and devices. Applications are further categorized as network applications (NA), gateway applications (GA), or device applications (DA) depending on whether they run in the network domain, on a gateway or embedded on a device, respectively. Devices are categorized into those that support the ETSI SCL functions (known as D devices) and those that do not support these functions (known as D devices).

ETSI M2M defines a number of reference points, or interfaces, between interacting entities. These reference points define the semantics of the interactions, and associated API, between the entities. In particular, the following three reference points are defined:

- mIa: defines the interactions between a network application and the N-SCL. Allows the application to register with the SCL and access resources on it.
- mId: defines the interactions between a device application, on the one hand, and a D-SCL or G-SCL on the other. Allows the application to register with the SCL and access resources on it.
- dIa: defines the interactions between the N-SCL, on the one hand, and the D-SCL or G-SCL on the other. Allows the various SCL instances to register with one another and access their respective resources.

The ETSI M2M architecture supports backward compatibility with devices that do not support the ETSI reference point functions. This compatibility is achieved through gateways that communicate with the legacy devices via their own proprietary mechanisms and handle the translation of the data into the resource tree. ETSI does not define the specifics of how the translation should be performed (Fig. 5.21).

Irrespective of the underlying physical network topology, the ETSI model defines a strict two-level hierarchy with N-SCL at the top level and G-SCL or D-SCL at the bottom level. The daisy chaining of SCLs in deeper hierarchies is not defined or supported.

The ETSI M2M functional architecture is defined in technical specification TS 102 690.

5.4.2.2   oneM2M

The oneM2M standards consider any IoT deployment to be comprised of two domains: the field domain and the infrastructure domain (Fig. 5.22). The field domain includes things (e.g., sensors, actuator, etc.) and gateways, whereas the infrastructure domain includes the communication networks (aggregation, core) as well as the data centers. From a functional perspective, each of these domains includes three flavors of entities: an application entity, a common services entity, and a network services entity.

The application entity implements the vertical-specific application logic. It may reside on one or multiple physical nodes in the deployment. Examples of an application entity would be a home automation application or a smart parking application.

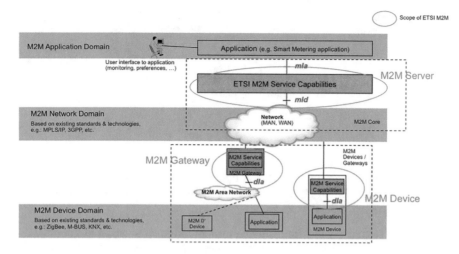

Fig. 5.21   ETSI M2M system architecture

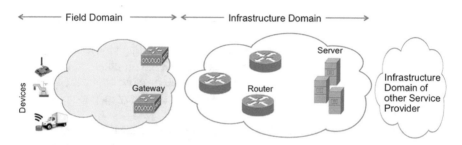

Fig. 5.22   oneM2M domains

The common services entity is a middleware layer that sits in between applications (application entity) and the underlying network services (network services entity) (Fig. 5.23). The common services entity (CSE) provides the following set of common functions to applications:

- Identity management: Identification of applications entities and CSEs.
- Registration: Includes registration of application entities and CSEs.
- Connectivity handling: This ensures efficient, reliable, and scalable use of the underlying network.
- Remote device management: This includes configuration and diagnostic functions.
- Data exchange: Supports storing and sharing of data between applications and devices, in addition to event notification.
- Security and access control: Provides control over access to data (who can access what and when, etc.).
- Discovery: Provides discovery of entities as well as data and resources.

Fig. 5.23 oneM2M
common services entity

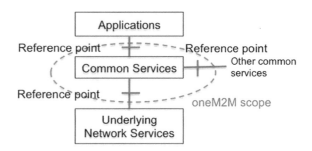

- Group management: Support of bulk operations and access.
- Location: Provides an abstraction for managing and offering location information services.

The CSE is, more or less, logically equivalent to the ETSI M2M SCL.

The network services entity provides value-added services to the CSE, such as QoS, device management, location services, and device triggering.

The oneM2M reference architecture identifies five different types of logical nodes: application-dedicated nodes, application service nodes, middle nodes, infrastructure nodes, and none-oneM2M nodes. These nodes may map to one or more physical devices in the network or may have no corresponding physical mapping.

Application-dedicated nodes (ADNs) are oneM2M compliant devices (i.e., things) with restricted functionality: they include one or more application entities but no CSE. From a physical mapping perspective, ADNs may map to constrained IoT devices.

Application service nodes (ASNs) are fully featured oneM2M compliant devices. They include a CSE in addition to one or more application entities. From physical mapping standpoint, they map to (typically non-constrained) IoT devices.

Middle nodes (MNs) host a CSE. A middle node may or may not include application entities. There could be zero, one, or many middle nodes in the network. MNs physically map to gateways in the network.

Infrastructure nodes (INs) host the CSE and may or may not host any application entities. The CSE on the IN includes functions that do not typically exist in any other CSE in the network. There is a single infrastructure node per domain per service provider in the oneM2M architecture.

Non-oneM2M Nodes are legacy devices that interwork with the oneM2M architecture. This provides backward compatibility of oneM2M with existing systems (similar to D devices in the ETSI M2M architecture).

As with ETSI M2M, oneM2M follows a RESTful architecture style where all data is modeled as resources, albeit oneM2M does not define a static resource structure like the ETSI resource tree. Instead, the standard provides means by which resources can be linked together (through resource links). Client applications can discover the resource organization dynamically. In this regard, the oneM2M approach complies with the HATEOAS (Hypermedia as the Engine of Application

State) REST constraint discussed in Sect. 5.3.4, because it does not assume that the clients have any a priori knowledge of the resource organization (Fig. 5.24).

Similar to ETSI M2M, oneM2M defines a set of reference points or interfaces between interacting entities. The oneM2M standard defines the following four reference points:

- Mca: Defines the interactions between application entities and CSE.
- Mcn: Defines the interactions between the CSE and the underlying network service entity.
- Mcc: Defines the interactions between two CSEs in the same service provider domain.
- Mcc': Defines the interactions between two CSEs across service provider domain boundary.

A number of notable differences between the reference points defined by ETSI M2M and those defined by oneM2M are worth highlighting:

First, ETSI M2M defines two different reference points for interactions between applications and the middleware as well as between devices and the middleware (mIa and mId interfaces, respectively), whereas oneM2M collapses both interfaces into the Mca reference point.

Second, the Mcn reference point is unique to oneM2M and has no equivalent in the ETSI standard. This interface enables the middleware to access network service functions. For example, it can be used to signal information from the service layer to the transport layer to request QoS and prioritization for M2M communi-

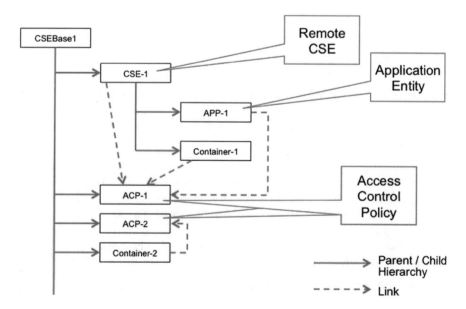

Fig. 5.24 oneM2M resource organization

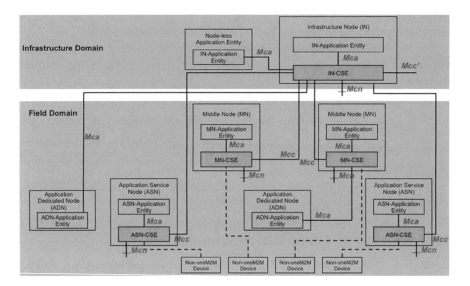

Fig. 5.25 oneM2M functional architecture

cation, for transmission scheduling, to signal indication for small data transmission, for device triggering, etc.

The interface may also be used to extract information from the underlying transport layer, for example, to fetch data related to the location of M2M devices or gateways (Fig. 5.25).

5.4.3   Technology Gaps

While ETSI and oneM2M have made strides in defining standard APIs and common application services for IoT, several gaps remain.

First, in terms of search and discovery capabilities, the IoT Application Services layer should provide support for:

- Mechanisms by which devices as well as applications can automatically discover each other as well as discover middleware/common services nodes.
- Mechanisms by which applications can search for devices with specific attributes (e.g., sensors of particular type) or context (e.g., within a specific distance from a location).
- Mechanisms by which applications can search for data based on attributes (e.g., semantic annotations) or context (e.g., spatial or temporal).

Both ETSI and oneM2M define basic mechanisms for resource search based on metadata or text strings. However, these are rudimentary capabilities and do not provide the contextual search functions that will be needed for IoT. Furthermore, no mechanisms for device or gateway auto-discovery are provided by either standard. It is assumed that the various instances of the middleware (SCL in case of ETSI and CSE in case of oneM2M), which need to communicate with each other, have a priori knowledge of their respective IP addresses. The same assumption holds between application endpoints and other entities (devices or middleware instances) that they need to communicate with.

Second, with regard to data encoding, interpretation, and modeling, the Application Services layer should encompass:

- Mechanisms that render IoT data understandable to applications without a priori knowledge of the data or the devices that produced it.
- Mechanisms that enable application interaction at a high level of abstraction by means of physical/virtual entity modeling.
- Mechanisms that enable data management services to host the semantic description of IoT data that is being handled.
- Framework for defining formal domain-specific semantic models or ontologies, including but not limited to defining an upper-level ontology for IoT.

ETSI's effort stopped at defining opaque containers for holding data. The interpretation of that data was outside the scope of what was standardized. OneM2M went one step further by providing an attribute to link the data container to an ontology reference (URI). However, no formal effort has been undertaken to define any ontologies or define any associated framework for tying semantic systems with the rest of the architecture, beyond this simple linkage.

5.5 Summary

In this chapter we started with an overview of the IoT protocol stack, and then we examined each of the Link layer, Internet layer, Application Protocols layer, and Application Services layer in details. For each of these layers, we examined the IoT challenges and requirements impacting the protocols, which operate at that respective layer, and discussed the industry progress and gaps.

In the course of the discussion on the Link layer, we covered IEEE 802.15.4, TCSH, IEEE 802.11ah, and Time-Sensitive Networking (TSN). In the Internet layer, we discussed 6LowPAN, RPL, and 6TiSCH. In the Application Protocols layer, we surveyed a subset of the multitude of available protocols. Finally, in the Application Services layer, we covered the work in ETSI M2M and oneM2M on defining standard application middleware services.

Problems and Exercises

1. What is the difference between IEEE 802.15.4 full-function device (FFD) and reduced-function device (RFD)?
2. IEEE 802.11ah and IEEE 802.15.4 both provide a low-power wireless protocol. What are the main differences between the two?

3. Why does IEEE 802.1Qca use IS-IS as the underlying protocol and not some other routing protocols such as OSPF or BGP?
4. What are three functions provided by the 6LowPAN adaptation layer?
5. Is RPL a link-state or distance-vector routing protocol? Why did the IETF ROLL workgroup decide to go with that specific flavor of routing protocols?
6. What are the constraints that characterize the RESTful communication paradigm?
7. What is the Application Services layer in the IoT protocol stack? What services does it provide?
8. What are the functions of the Service Capabilities layer (SCL) in the ETSI M2M architecture?
9. What are functions of the common services entity (CSE) in the oneM2M architecture? How do they compare to those of ETSI's SCL?
10. Why do the IoT application services architectures under standardization all follow the RESTful paradigm?
11. A temperature sensor that supports CoAP has an operating range of 0–1000 °F reports a reading every 5 s. The sensor has a precision of 1/100 °F. The sensor reports along with every temperature reading a time stamp using the ISO 8601 format (CCYY-MM-DDThh:mm:ss).

    (a) If the current temperature measured by the sensor is 342.5 °F, construct the payload of a CoAP message with the reading encoded in XML and then in JSON.
    (b) Assuming that the sensor consumes 3 nJ per byte (character) transmitted over a wireless network, calculate the total energy required to transmit each message. Which of the two encoding schemes (XML or JSON) is more energy efficient? By what percentage?

12. Compare the bandwidth utilization for the XML vs. JSON messages of Question 11 in bits per second assuming UTF-8 text encoding is being used.
13. An IoT water level monitoring application requires updates from a sensor periodically, using the command/response paradigm. The application triggers a request every 1 s. The roundtrip propagation delay between the application and the sensor is 12 ms. The sensor consumes 3 ms on average to process each request. The application consumes 2 ms to send or receive any message. If the application blocks on every request to the sensor, how much of its time budget can be saved by redesigning the application to use the publish/subscribe communication model in lieu of the command/response approach?
14. A utility company uses IPv6-enabled smart meters running in an IEEE 802.15.4 mesh. If the mesh is operating at 1 Mbps without 6LoWPAN IPv6 header compression, what is the throughput of the smart metering application in the worst-case scenario?
15. An automotive parts manufacturer is looking to upgrade the network that controls their computer numerical control (CNC) mill. At full speed, the mill can cut into solid steel at a rate of 1 inch per second. The manufacturer's quality assurance (QA) guideline mandates that the dimensions of any part produced

must be accurate within ±1/100 inch. In order to meet the QA guideline, what is the maximum jitter that needs to be guaranteed by the new deterministic network that connects the mill to the controlling computer?

16. Given the following IEEE 802.15.4 mesh running the RPL protocol. The numbers indicated next to each link is the associated latency. If the objective function is to minimize the communication latency to the Internet, what will be the topology computed by RPL?

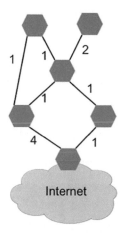

17. An automation engineer is looking to deploy a deterministic network in a sheet metal factory. The control system in charge of safety expects a message from the embedded application of a heating element controller every 50 ms, otherwise it immediately shuts down the production line. The network in question has on average a delay of 1 ms per link and 2 ms per node. What is the maximum number of hops that can separate the control system from the heating element controller?

18. Why does channel hopping improve the reliability of wireless sensor networks?

19. An application protocol supporting a time filter policy support for client applications must not deliver messages at a rate higher than what the client application is willing to consume. What are common strategies to achieve this?

20. Which Application layer protocol would you choose for deploying an IoT solution for a financial institution? Why?

References

1. J. Yick et al., Wireless sensor network survey. Comput. Netw **52**(12), 2292–2330 (2008)
2. M. Sichitiu, Wireless Mesh Networks: Opportunities and Challenges, Wireless World Congress, 1–6, 2005
3. IEEE 802.15.4–2011, September 2011

4. R. Krasteva et al., Application of wireless protocols Bluetooth and ZigBee in telemetry system development. Prob. Eng. Cybern. Robot **55**, 30–38 (2005)
5. N. Garg, M. Yadav, A review on comparative study of Bluetooth and ZigBee, Proceedings of the Second International Conference on Advances in Electronics, Electrical and Computer Engineering, EEC 2013
6. IEEE 802.15.4g-2012, April 2012
7. T. Adame et al., IEEE 802.11ah: The Wi-Fi Approach for M2M Communications, IEEE Wireless Communications, December 2014
8. IEEE draft standard 802.11ah Draft 4
9. M. Teener. IEEE 802 Time Sensitive Networking: Extending Beyond AVB
10. "Industrial Ethernet: A Control Engineer's Guide," Cisco Whitepaper
11. D. Pannell, Audio Vidor Bridging Gen 2 Assumptions, July 2011
12. IEEE 802.1Qca Draft 2.0, April 2015
13. P. Meyer et al., Extending IEEE 802.1 AVB with time-triggered scheduling: A simulation study of the coexistence of synchronous and asynchronous traffic, IEEE Vehicular Networking Conference (VNC), At Boston, Massachusetts, 2013
14. IEEE 802.1Qbv Draft 2.3, April 2015
15. IEEE Standard 802.15.4e-2012
16. T. Watteyne et al., Using IEEE 802.15.4e Time-Slotted Channel Hopping (TSCH) in the Internet of Things (IoT): Problem Statement, IETF RFC 7554, May 2015
17. X. Su et al., Enabling Semantics For The Internet of Things – Data Representation and Energy Consumption, Internet of Things Finland, January 2013
18. Z. Shelby et al., The Constrained Application Protocol (CoAP), IETF RFC 7252, June 2014
19. Z. Shelby et al., CoRE Resource Directory, draft-ietf-core-resource-directory, work in progress, March 2016
20. Baker & Meyer, Internet Protocols for the Smart Grid, IETF RFC 6272, June 2011
21. J. Rosenberg et al., SIP: Session Initiation Protocol, IETF RFC 3261, June 2002
22. T. Watteyne et al., Reliability through frequency diversity: why channel hopping makes sense, PE-WASUN '09 Proceedings of the 6th ACM symposium on Performance evaluation of wireless ad hoc, sensor, and ubiquitous networks, Pages 116–123, October 2009
23. LoRa Alliance, Technical Marketing Workgroup 1.0, "LoRaWAN What is it? A technical overview of LoRa and LoRaWAN," November 2015
24. LoRaWAN Adaptive Data Rate: https://www.thethingsnetwork.org/wiki/LoRaWAN/ADR
25. F. Adelantado et al. Understanding the Limits of LoRaWAN, IEEE Communications Magazine, January 2017
26. Rashmi Sharan Sinha, Yiqiao Wei, Seung-Hoon Hwang, A survey on LPWA technology: LoRa and NB-IoT, ICT Express, Volume 3, Issue 1, 2017, Pages 14-21
27. Kumar, V., Jha, R.K. & Jain, S. NB-IoT Security: A Survey. Wireless Pers Commun 113, 2661–2708 (2020).

Chapter 6
Fog Computing

6.1 Defining Fog Computing

In order to define Fog computing, a recap of the concept of Cloud computing is in order. Cloud computing refers to a model that provides users with on-demand access of a shared pool of computing resources over a network. These resources can be quickly provisioned and released through a self-service model. One of the key characteristics of the Cloud computing model is the notion of resource pooling, where workloads associated with multiple users (or tenants) are typically collocated on the same set of physical resources. This guarantees the economy of scale of the Cloud computing model. Hence, essential to Cloud computing is the use of network and compute virtualization technologies. Cloud computing provides elastic scalability characteristics, where the amount of resources can be grown or diminished based on user demand.

Fog computing, or in short Fog, refers to a platform for integrated compute, storage and network services that is highly distributed and virtualized. This platform can extend in locality from IoT end devices and gateways all the way to Cloud data centers, but is typically located at the network edge (Fig. 6.1). Fog augments Cloud computing and brings its functions closer to where data is produced (e.g., sensors) or needs to be consumed (e.g., actuators). Fog is not an alternative to Cloud computing, rather the two synergistically interplay in order to enable new types and classes of IoT applications that otherwise would not have been possible when relying on Cloud computing stand-alone.

© The Author(s), under exclusive license to Springer Nature Switzerland AG 2022   153
A. Rayes, S. Salam, *Internet of Things from Hype to Reality*,
https://doi.org/10.1007/978-3-030-90158-5_6

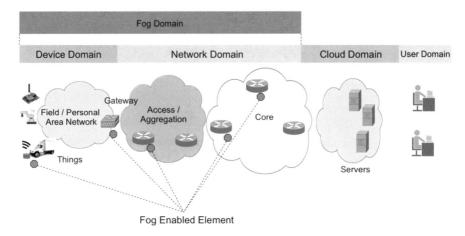

Fig. 6.1 Fog and Cloud

6.2 Drivers for Fog

There are several IoT requirements that act as the drivers for the Fog architecture. These will be discussed next.

6.2.1 Data Deluge

It has been claimed that 5 exabytes of data have been generated from the dawn of humanity to 2003.[1] Now this much data is generated every 2 days[1], and the rate is only increasing. The billions of devices that are projected to be connected to the Internet will only exacerbate the data deluge problem. At heart of the issue is the question of whether the state of the art will evolve fast enough to handle the imminent explosion of data? There are two technology evolution curves at play here: one represents the evolution of compute and storage technologies, which is governed by Moore's Law, and the second represents the growth of bandwidth at the network edge, which is covered by Nielsen's Law. Moore's Law stipulates that compute and storage technologies will double in capability/capacity every 18 months. Nielsen's Law, on the other hand, projects that the bandwidth at the network edge doubles every 24 months. Acknowledging that there is a positive correlation between the growth of compute and storage technologies and the growth in data volume, it is conceivable to foresee an IoT future where data will be produced at rates that far outpace the network's ability to backhaul the information, from the network edge where it is produced by the billions of Things, to the Cloud where it will ultimately

---

[1] As quoted by Eric Schmidt, Executive Chairman of Google

need to be processed and potentially stored. This disparity between the data volume and the available bandwidth is best exemplified with the analogy of attempting to push a golf ball through a straw. Luckily, Moore's Law is not only a culprit by contributing, in part, to the problem but is also a key enabler to the solution: it can be leveraged to augment the functions of the network itself with compute and storage capabilities at the edge. This allows the network to perform processing, analysis and storage of data in lieu of blindly pushing all data up to the Cloud. With that, Cloud computing is brought closer to the data sources, the Things, which gives rise to the notion of Fog computing. *Cloud* becomes *Fog* when it is closer to *Things*, pun intended.

6.2.2   Rapid Mobility

Certain IoT use cases require support for rapid mobility of Things, for example, sensors on a speeding vehicle communicating with road-side infrastructure or a passenger commuting on a train. Due to rapid mobility, network conditions may vary frequently, due to signal fading, interference, or other conditions. This may even lead to severe service degradation or intermittent loss of connectivity to the Cloud. Another consideration is the characteristics of the communication path to the Cloud: bandwidth and/or latency limitations may have adverse side effects on the operation of the IoT application. Multiple variables will typically be at play to contribute to these characteristics, including radio coverage, interference, and the amount of resources shared with other mobile nodes.

To guarantee the quality of service and reliability required by the application, especially when dealing with mobility over extended geographic distances, the Cloud infrastructure needs to be augmented with compute and storage functions that move with the mobile Things. The mobility of these functions may be either physical or virtual. In the former case, the compute and storage is physically situated with the moving Thing, whereas in the latter, these functions maintain close proximity by shadowing and following the Thing albeit in the network edge. In this capacity, Fog augments the Cloud to achieve the required pervasiveness and reliability required by rapid mobility in IoT.

6.2.3   Reliable Control

IoT applications that focus on closed-loop control and actuation often share the following characteristics: the data input space and the processing logic required to produce the control decision have intensive computational and considerable storage demands. The sensing and actuating devices are typically constrained devices, and therefore need to off-load the storage and compute functions to external systems or infrastructure. In many cases, these control applications require very low latency for correct operation. In a subset of the scenarios, connectivity to the Cloud may be either too expensive (e.g., satellite links connecting sensors deployed in oilfields) or unreliable due to rapid mobility patterns.

The combination of the above characteristics makes it unpalatable to rely on Cloud computing to support reliable real-time control with fixed latency. This is where Fog computing can complement the Cloud to address that IoT application niche.

6.2.4   Data Management and Analytics

A class of IoT applications characterized with the confluence of very large scale, in terms of the number of devices generating data, widespread geographic footprint where these devices are deployed, vast amounts of data that need to be collected, aggregated, processed, and exposed to consuming entities, as well as real-time analytics or closed-loop control. For such class of applications, a data management and analytics platform that can handle the scale and performance requirements is needed. Experience with large-scale information and communication systems has proven that distributed systems built on hierarchical division of functions provide the elasticity required while maintaining key performance metrics. Such systems typically exploit locality of data for their most basic functions. In other words, they tend to minimize the amount of data required from remote sources for critical functions. Interactions between widespread entities are typically confined to system wide functions. For data management and analytics, this operating paradigm is even more relevant because the IoT data often needs to be operated on within a context, which is well known at the edge of the network, close to the data sources, and is often lost or is irrelevant as the data travels deeper in the network and into the Cloud. Take as an example an ambient noise sensor in a Smart City application, which is constantly measuring noise levels and streaming the recorded data. Backhauling all the data to the Cloud is both unnecessary and inefficient, especially when compared with an alternate design where a local analytics function situated close to the sensor filters readings below a specified threshold (depending on the context associated with where the sensor is deployed) and only propagates to the Cloud interesting readings above that threshold, e.g., to alert city personnel.

Fog computing, in concert with Cloud computing, provides the necessary compute and storage infrastructure required to support such distributed and hierarchical data management and analytics.

6.3   Characteristics of Fog

The Fog and the Cloud both comprised of the same three building blocks: compute, storage, and networking. However, there are multiple characteristics that uniquely shape the Fog and distinguish it from the Cloud:

Table 6.1 Summary comparison of Cloud and Fog computing

Requirement	Cloud Computing	Fog Computing
Latency & Jitter	High/medium	Low
Location of service	Within Internet	Network Edge
Distance between data sources/consumers	Multiple Hops	Single Hop
Location awareness	No	Yes
Geo-distribution	Centralized (Data Center)	Distributed
Number of nodes	Large	Larger
Support for mobility	No	Yes
Data analytics	Data at Rest	Data in Motion
Connectivity	Wire-line	Wireless

First are the network edge location, location awareness, and low latency. Fog locates the services close to the data sources and consumers where it is possible to enrich the data with location context and operate on it with minimal latency.

Second is geographical and architectural distribution. This is in stark contrast to the Cloud model where are all services are centralized in the data center.

Third is the extremely large number of nodes. While the Cloud drives demand for massively scalable data centers (MSDC), the Fog pushes the envelope further on scalability.

Fourth is mobility of nodes and endpoints. The data sources, consumers, compute or storage resources can all be mobile.

Fifth is real-time interaction. In the Fog, the focus is on real-time analysis of streaming data as opposed to batch processing. Fog requires analysis of data in motion as opposed to data at rest.

Sixth is predominance of wireless access. In the Cloud, connectivity relies on wire-line technologies, predominantly Gigabit Ethernet (10 Gbps, 40 Gbps, and soon 100 Gbps). Whereas the Fog will be mostly connected over wireless links, both because of the impracticality of running wires everywhere, as well as to support the mobility requirements.

Seventh is the heterogeneity of resources. In the Cloud, a given data center is managed by a single business entity, which goes about deploying homogeneous resources in order to minimize complexity and operational costs. With the Fog, the architecture is federated over resources managed by different business entities. Hence, these resources will vary widely in capabilities, form factors, and operating environment.

Table 6.1 summarizes the main facets of difference between Cloud and Fog computing.

6.4   Enabling Technologies and Prerequisites

The realization of the vision of Fog computing relies on a number of technologies that provide enabling building blocks and are key prerequisites for the architecture. These include lightweight compute virtualization, network mobility, orchestration, and application enablement technologies. In what follows, we will discuss each of those technologies in more detail.

6.4.1   Virtualization Technologies

Inherent to Fog computing is the ability to locate compute functions close to data producers and/or consumers. This assumes the availability of lightweight compute virtualization technologies that allow workloads to be instantiated, as needed, on Fog nodes. The latter act as shared compute resources among potentially a multitude of IoT applications.

Virtualization technologies combine or partition computing resources to present one or more operating environments using techniques such as hardware and software partitioning or aggregation, hardware emulation, resource sharing or time multiplexing, etc. Virtualization provides a number of advantages: It enables consolidation of both hardware and applications, thereby eliminating the expense associated with procuring and managing under-utilized infrastructure. It also enables sandboxing, i.e., providing application with secure isolated execution environments. Virtualization also provides the flexibility of multiple simultaneous operating systems over the same hardware infrastructure. It eases the migration of software stacks and allows the packaging of applications as stand-alone appliances. Furthermore, virtualization enables the portability and mobility of applications from one hardware or physical location to another with ease.

Virtualization technologies generally differ in the abstraction level at which operate: CPU instruction set level, hardware abstraction layer (HAL) level, operating system level.

Virtualization at the CPU instruction set level allows an "emulator" to provide to an application the illusion of running on one processor architecture, whereas the real hardware actually belongs to a different architecture. It is the job of the emulator to translate the guest instruction set (offered to the application) to the host instruction set (used by the actual hardware).

Virtualization at the hardware abstraction layer level involves a virtual machine manager, or hypervisor, which is a software layer that sits above the physical hardware (sometimes referred to as "bare metal") and provides a virtualized view of all its services. The hypervisor can create multiple virtual machines (VMs) on top of the bare metal. The VMs can be running different operating systems. Applications can run within their respective operating systems and are completely oblivious to the underlying virtualization.

Virtualization at the operating system level relies on virtualization software that runs on top of or as a module within the operating system. It provides an abstraction of the kernel-space system calls to user-space applications, in addition to security and sandboxing capabilities to prevent one application from causing collateral damage to another.

Other higher levels of virtualization are possible, such as library and application level virtualization, but these are not relevant for the purpose of this discussion.

6.4.1.1   Containers and Virtual Machines

Both Containers and Virtual Machines are popular virtualization constructs employed in Cloud Computing today. Each of the two technologies has its own set of advantages and trade offs. Virtual Machines (VMs) are a virtualization technology at the Hardware Abstraction Layer level. VMs provide an abstraction of a compute platform's hardware and software resources, complete with all the drivers, full operating system and needed libraries. Containers, on the other hand, are a virtualization technology at the operating system level. They include portions of the operating system and select libraries: the minimal pieces that are absolutely required to run the application. Containers share the same operating system and, where applicable, common libraries. Due to this, Containers are lighter weight when compared to VMs, both in terms of their memory and processing requirements. As a result, given a specific hardware (e.g., a server) with a fixed resource profile, it is possible to support more Containers than VMs running concurrently. This gives Containers

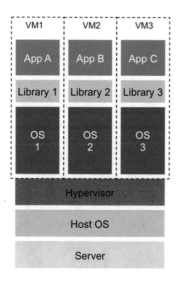

Virtual Machine Paradigm   Container Paradigm

Fig. 6.2 VMs and Containers

a clear scalability advantage over VMs, not only for Cloud computing but also for the Fog. In fact, the compact memory footprint for Containers gives them another advantage in the Fog context: they are faster to migrate from one hosting node to another, a matter which characterizes them with the nimbleness required to support rapid mobility (Fig. 6.2).

However, the lightweight nature of Containers comes with a set of trade offs: since Containers share the same underlying operating system, it is not possible to use them to deploy applications that require disparate operating system environments, or different OS versions, on the same physical hardware. Such restriction does not apply to Virtual Machines, since they include their own copy of the operating system. Another trade off associated with the shared operating system in Containers is the security implications: there is potential for an application in a Container to be subjected to security threats due to malicious or misbehaving code running in another Container on the same operating system. With Virtual Machines, the security threat is smaller in comparison, because the attack surface is minimized due to the fact that each VM has an independent operating system instance. Therefore, an application in one VM is better sandboxed and isolated from applications or code running in another VM.

Linux, the leading open operating system platform, supports both Virtual Machines and Containers. Both Kernel-based Virtual Machines (KVM) and Linux Containers (LXC) are available in the standard distribution.

Containers and VMs both provide the capability to sandbox Fog applications from one another and to control their resource usage. In addition to these relatively low-level functions, Fog requires a framework for the packaging, portability, sharing, and deployment of applications. One such framework that has been gaining popularity in the industry is Docker, which will be discussed next.

6.4.1.2 Docker

Docker is an open source project that provides a packaging framework to simplify the portability and automate the deployment of applications in Containers. Docker introduces scripts composed of a series of instructions that automate the deployment process from start to finish. These scripts are referred to as "Dockerfiles." Docker defines a format for packaging an application and all its dependencies into a single portable object. The portability is guaranteed by providing the application a runtime environment that behaves exactly the same on all Docker-enabled machines. Docker also provides tooling for container version tracking and management. In addition, it provides a community for sharing useful source code among developers.

6.4.1.3   Application Mobility

Virtualization technologies decouple the application software from the underlying compute, storage, and networking resources. As such, it enables unrestricted workload placement and mobility across geographically dispersed physical resources. For instance, multiple hypervisors support different flavors of Virtual Machine migration, including "cold" migration and "live" migration. In the former case, a VM that is either powered down or suspended is moved from one host to another. In the latter, a VM that is powered on and operational is moved across hosts, without any interruption to its operation. The VM mobility solution takes care of moving the VM's memory footprint, and if applicable, any virtual disk/storage from the old to the new hardware. In order to ensure seamless mobility in the case of "live" migration, the VM retains its original Internet Protocol (IP) and Medium Access Control (MAC) addresses. This ensures that any clients or services that are in communication with the migrating VM can continue to reach it using the same communication addresses. The successful orchestration of such seamless live migration requires the underlying network infrastructure to support mobility. This will be the topic of the next section.

6.4.2   Network Support for Mobility

As previously discussed, rapid mobility is one of the drivers for Fog computing. To ensure uninterrupted operation of the IoT application, the network infrastructure that is providing the underlying communication fabric for the Fog deployment must support seamless mobility of the communicating endpoints.

Networking systems rely on the address of the endpoints in order to deliver messages to their intended recipients. Depending on the technology at hand, the address either connotes the identity or the location of the endpoint. For example, Media Access Control (MAC) addresses are identity addresses, because they are burnt into the machine and uniquely identify it on a network. Internet Protocol (IP) addresses, on the other hand, are typically used as location addresses because they indicate the geographic locality of the endpoint. In some contexts, IP addresses are used as identity addresses as well, for example, in wireless mobile IP applications.

Applications that are deployed in a virtualization construct, such as a Virtual Machine, can perform seamless mobility. With seamless mobility, the application's MAC and IP addresses remain unchanged as the associated VM moves from one physical server node to another. The network infrastructure needs to handle the application mobility event and update the forwarding information on the routers and/or switches to deliver the messages correctly to the right physical server that is now hosting the VM. In order to do this, the network infrastructure needs to treat the VM's IP and MAC addresses as identity addresses, and correlate them with dynamic

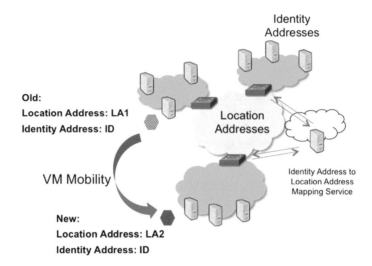

Fig. 6.3 Identity vs. location addresses with application mobility

location addresses that get updated automatically as the VM moves from one locality to another. In order to properly scale the solution, the knowledge of identity addresses should be confined to the edge of the network, whereas the core of the network performs forwarding solely based on the location addresses. This is achieved by relying on tunnels established between the edge nodes of the network to forward the end-host traffic over the core. The tunnel encapsulation uses location addresses and hides identity addresses from the core network nodes. The correlation between identity addresses and location addresses is established through a mapping service provided by the network infrastructure. In a way, this is similar to how the post office mail forwarding service works: If a person moves her home, then she informs the post office in order to update the association of her name (identity address) from an old home address (old location address) to a new home address (new location address), in order to guarantee uninterrupted delivery of mail (packets) (Fig. 6.3).

The industry has recently been working on defining networking solutions to support seamless VM mobility, primarily driven by Enterprise mobility, Data Center, and Cloud use cases. The solutions generally differ in how the mapping service (for identity to location address) is implemented: some proposals use a centralized server for the mapping service, whereas others rely on a distributed control protocol. These solutions can be leveraged by Fog computing. We will discuss two of the most promising solutions: Ethernet Virtual Private Network (EVPN) and Locator/identifier Separation Protocol (LISP).

6.4.2.1 EVPN

Ethernet Virtual Private Network (EVPN) is an overlay technology that allows Layer 2, and even Layer 3, virtual private networks to be created over a shared Internet Protocol (IP) or Multiprotocol Label Switched (MPLS) transport network. EVPN was standardized by the IETF in RFC 7432. EVPN uses the Border Gateway Protocol (BGP) in order to build the forwarding tables on the participating network elements. Given that EVPN is an overlay technology, only network elements that are at the edge of the network need to support it, and core network elements are oblivious to the fact that EVPN is running in the network. The edge nodes, which run EVPN, are known as EVPN Provider Edge (PE) nodes. PE nodes learn the MAC and IP addresses of connected hosts, from the access side, either by snooping on the host traffic in the data-plane (similar to how Ethernet bridges learn addresses) or by running some control protocol (e.g., the Address Resolution Protocol—ARP). The PE nodes then build a database of the local addresses and advertise these addresses to remote PEs using BGP route messages. Remote PEs, which receive the BGP route messages, build their own forwarding databases where they associate the MAC and IP addresses (identity addresses) of the hosts with the next hop address (location address) of the PE that advertised the route. Host traffic packets received by ingress PE nodes are tunneled (using IP or MPLS encapsulation) over the core network to egress PE nodes, where the tunnel encapsulation is removed, and the original host packets are forwarded to their intended destination(s) (Fig. 6.4).

To handle application mobility, EVPN introduces new BGP messages and dedicated protocol machinery. These mechanisms provide a solution for two issues: first, updating the network infrastructure with the new identity address to location

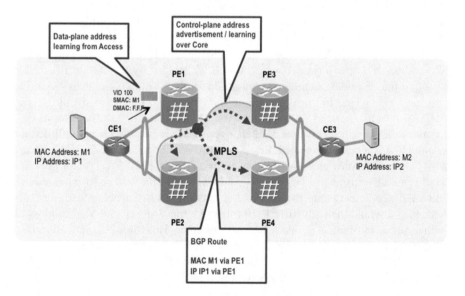

Fig. 6.4 Ethernet virtual private network (EVPN) architecture

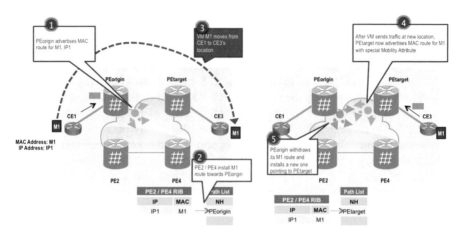

Fig. 6.5 Mobility in EVPN

address mappings, and second, guaranteeing optimal forwarding to the default IP gateway after mobility. These two issues and how they are addressed with EVPN will be discussed next.

Updating the Identity to Location Address Mappings

When an application running in a VM starts sending traffic, the EVPN PE that is servicing the physical server on which the VM is hosted will receive this traffic and learn the application/VM IP and MAC addresses. This PE, call it PE$_{origin}$, will then advertise the VMs addresses in BGP to all the remote PEs in the virtual private network instance. The remote PEs will then update their forwarding tables to indicate that the VM IP and MAC addresses are reachable via PE$_{origin}$. Now, assume that the VM moves to a new physical server, which is serviced by a different PE, call it PE$_{target}$. If the PE nodes continue to send traffic for the VM to PE$_{origin}$, then this traffic will not be delivered to the VM because the latter is no longer on the old server. EVPN solves this issue as follows: when the VM starts sending traffic from its new location, PE$_{target}$ will receive the packets over its access interfaces and will deduce that the VM is locally connected. PE$_{target}$ would also recognize that the VM's IP and MAC addresses were previously learnt from a remote PE, PE$_{origin}$, via a previous BGP route advertisement. Hence, PE$_{target}$ deduces that the VM must have moved, and so it needs to update the rest of the network with the new location of the VM. PE$_{target}$ would then advertise BGP routes for the VM's IP and MAC addresses with a special attribute to indicate the mobility event. This route is sent to all remote PEs, including PE$_{origin}$. When PE$_{origin}$ processes the BGP route message, the special attribute indicates to it that the VM has moved, so PE$_{origin}$ withdraws its previously advertised BGP route for that VM's addresses. This handshake mechanism results in all the PEs converging on using PE$_{target}$ as the new next hop (location address) for the VM traffic (Fig. 6.5).

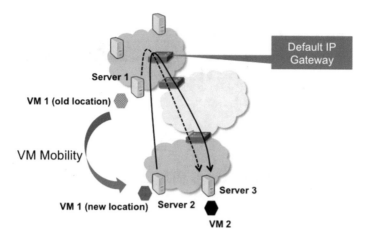

Fig. 6.6 Default IP Gateway problem with VM mobility

Default IP Gateway Problem

As a VM moves from one physical server to another, both its memory (RAM) and disk image are maintained unchanged. This means that the VM's configuration remains unmodified. The configuration includes, among other things, the address of the Default IP Gateway that the VM should use in order to forward network traffic to remote nodes. Typically, the Default IP Gateway should be in close topological proximity to the server that is hosting the VM, in order to guarantee optimal forwarding of network traffic originating from the VM. However, with VM mobility, the VM may land on a new host server that is topological distant from the original Default IP Gateway. In such a case, network traffic sourced by the VM will most likely follow a sub-optimal forwarding path to its destination.

For example, consider the network of Fig. 6.6, where VM1 is in communication with VM2 (hosted on Server 3). VM1 is originally hosted on Server 1, and its network traffic that is destined to VM2 initially follows an optimal forwarding path through the Default IP Gateway (the dotted black line). When this VM moves from its initial location to a new location on Server 2, the network traffic will start following a sub-optimal path from Server 2, via the same default gateway, to Server 3 (the solid black line).

To address this problem, EVPN delegates the Default IP Gateway function to the edge of the network (the PE nodes), and enables all the PEs to act as a distributed logical default gateway for hosts that are attached over the PE access interfaces. When a host sends an ARP request for the Default IP Gateway IP address, the EVPN PE intercepts the ARP message and responds to it with its own MAC address. The default gateway IP address is the same across all the participating EVPN PEs. This is specifically to cater for the fact that the VM retains its configured default gateway address after a mobility event (Fig. 6.7).

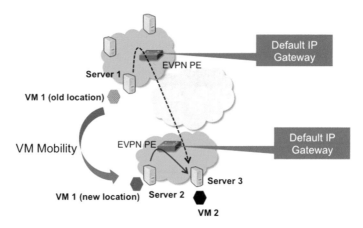

Fig. 6.7 EVPN Default Gateway solution

This approach solves the problem by ensuring that the default gateway is always in topological proximity to the VM after it moves from one physical host to another.

6.4.2.2 LISP

Locator/Identifier Separation Protocol (LISP) is an overlay networking solution that allows complete decoupling of the addressing structure of end hosts from that of the network infrastructure. LISP formally defines two namespaces for IP addresses: Endpoint Identifiers (EIDs) and Routing Locators (RLOCs). EIDs are identity addresses associated with end hosts, whereas RLOCs are location addresses primarily assigned to routers. LISP dedicates an entire system for the directory service that performs the mapping between EIDs and RLOCs, and provides two approaches by which that system can be implemented: a distributed approach that relies on BGP over an Alternative Logical Topology (ALT), and a centralized approach that uses a dedicated database for the mapping known as Dedicated Database Tree (DDT). LISP is standardized in IETF RFC 6830.

Network elements that sit at the edge of a LISP network are known as Ingress Tunnel Routers (ITRs) and Egress Tunnel Routers (ETRs). The ITR receives traffic from end hosts and is responsible for encapsulating the traffic within a tunnel to be transported over the LISP network. The ETR decapsulates the tunneled traffic and forwards the original end-host packets to their destinations. ITRs and ETRs are identified based on their RLOCs. In order to determine which ETR to forward the traffic to, the ITR consults with a Map Resolver to resolve the RLOC of the ETR associated with the destination EID of the traffic. The Map Resolver is responsible for identifying which Map Server to direct the query to in order to determine the RLOC associated with a given EID. The Map Server is a database that holds all EID/ETR associations. It may be deployed on a pair of devices or a full-blown hierarchy of devices for large-scale implementation (LISP-DDT). Each ETR registers

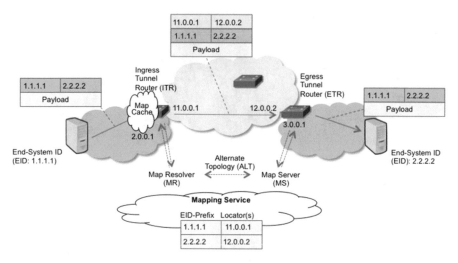

Fig. 6.8 LISP architecture

with the Map Server the EID address space that it is authoritative for. When triggered in the data-plane by a packet destined to a remote EID, the ITR issues a "Map-Request" towards the Map Resolver. The latter forwards it to the right Map Server, which in turn forwards the request to the authoritative ETR. This ETR replies to the requesting ITR with a "Map-Reply" message that contains the list of the RLOCs having the capability to reach the requested EID, with their characteristics in terms of priority of usage and weighted load partitioning (Fig. 6.8).

To handle application mobility, LISP introduces specific protocol mechanisms. These mechanisms provide a solution for the two issues discussed in the previous section: first, updating the network infrastructure with the new identity address to location address mappings, and second, guaranteeing optimal forwarding to the default IP gateway after mobility.

Updating the Identity to Location Address Mappings

Mobility is enabled on an ETR by configuring the node with the list of the mobile IP subnets (EIDs) that the ETR is to support. This ETR then becomes the local Default IP Gateway for these mobile EIDs. When an application, with its unique EID, moves into the LISP site, the first packet that it will send to its local Default IP Gateway will trigger the mobility detection on the ETR. The ETR then registers this specific EID with the Map Server. The latter, in turn, deregisters the EID from the previous authoritative ETR. What remains is to update the map caches of all the ITRs that have communicated with the application prior to its move, as those ITRs will have stale entries to the RLOC of the old authoritative ETR. This function is performed by the old authoritative ETR itself, which upon receiving any data traffic

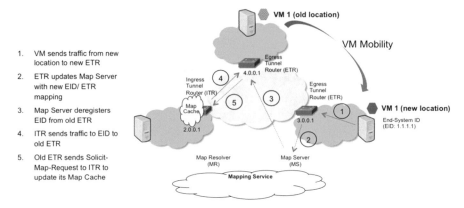

1. VM sends traffic from new location to new ETR

2. ETR updates Map Server with new EID/ ETR mapping

3. Map Server deregisters EID from old ETR

4. ITR sends traffic to EID to old ETR

5. Old ETR sends Solicit-Map-Request to ITR to update its Map Cache

Fig. 6.9 LISP mobility

for the EID that has moved, sends back a "Solicit-Map-Request" message to the originating ITR. This message instructs the ITR to refresh its cache (Fig. 6.9).

Default IP Gateway Problem

LISP solves the Default IP Gateway Problem by ensuring that every site has a default gateway configured for the same prefix. This gateway must use the same (virtual) IP and MAC Addresses in order to guarantee that the traffic originating from the moved VM follows an optimal path out of the local LISP Tunneling Router rather than being forwarded to another site. First Hop Redundancy Protocols (e.g., VRRP) must be configured with identical gateway and MAC addresses in all sites, and their packets must not be allowed to leak beyond a given site. This way, when a VM moves it will always find the same default gateway regardless of its location.

6.4.3   Fog Orchestration

Orchestration, in the context of Fog computing, refers to the process of automating the various workflows that perform the full lifecycle management of the Fog infrastructure. This includes the provisioning and management of its three components (compute, network, storage) and associated resources. For illustration, tasks such as deploying, debugging, patching, and updating applications or operating systems, setting up network connectivity between application entities and reserving bandwidth, as well as allocating and expanding disk space are all examples of workflows that fall under orchestration.

Orchestration is a complex task in Fog environment as it involves components spread across heterogeneous systems and distributed across multiple locations. Due to the Fog's multi-tiered hierarchical organization, it requires a hierarchically

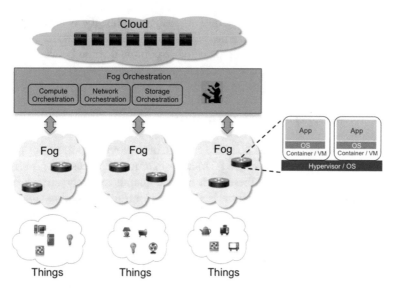

Fig. 6.10 Fog orchestration

organized Orchestration plane that supports dynamic policies and interplay with Cloud orchestration (Fig. 6.10).

Fog orchestration differs from Cloud orchestration in three different facets: Topology, Things Connectivity, and Network Performance Guarantees.

6.4.3.1   Topology

Cloud orchestration systems that are available today make assumptions about the network: the physical layout of the topology (3-tiered, 4-tiered, Fat Tree, etc.), the abundance of available bandwidth, and the fact that the network elements are capable devices and therefore have no restrictions on the size of the routing tables. While these assumptions are valid in the Cloud, they do not hold true in the Fog. Fog topologies are ad-hoc best-fit affairs. They have heterogeneous interconnects as well as dynamically varying bandwidth, latency and reliability characteristics. Fog orchestration software has to deal with isomorphic topologies that are directly connected to Things.

6.4.3.2   Things Connectivity

With Fog, the orchestration software needs to be able to deploy applications, which need direct access to Things (e.g., legacy applications), on Fog nodes that are physically connected to these specific Things. To enable the communication between the applications and their Things, specialized device drivers need to be initialized on the

Fog nodes by the orchestration system. Furthermore, applications may require data from remote Things, in which case the orchestration software needs to dynamically establish network overlays to facilitate network communication between the applications and those remote Things.

6.4.3.3 Network Performance Guarantees

Orchestration systems for the Cloud are capable of deploying applications on nodes that can offer the right performance guarantees in terms of processing power, memory, and disk space. For Fog, these performance guarantees alone are not enough. Another dimension of complexity arises due to Control Applications that require network performance guarantees, in terms of upper bounds on latency and jitter, in their communication with Things. In order to support these control applications in the Fog, the orchestration system needs to be able to incorporate network latency and jitter into the application placement and scheduling algorithms. Mobility complicates this further, as the placement decisions need to be recalculated with changing conditions.

6.4.4 Data Management

6.4.4.1 Data in Motion

There are vast amounts of data crossing the network every day. However, those bits and bytes provide a wealth of information about actions, time, location, and devices. By gathering and combining pieces of information together it is possible to start seeing patterns, and gain greater insights. In other words, it is possible to gain knowledge. And it is through knowledge that we, as humans, can learn and apply wisdom, leading to better outcomes (Fig. 6.11).

Fig. 6.11 DIKW pyramid

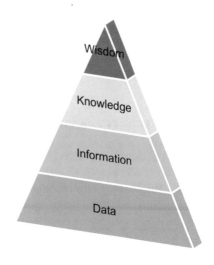

New data sources are being created and added to the network every day. From a video camera in a transit bus, a tire pressure sensor in a truck, a jet engine, to a smart meter attached to a house. These devices are creating a constant stream of data. Very soon, the data generated by the IoT will make up the majority of all information available on the Internet and will change the face of Big Data. It will not be possible to store all this data and analyze it later. The real-time nature of these new sources of data requires that their output be evaluated in motion and in meaningful way. The value of data is often dictated by time—being at its highest value when it is first created. Actionable insights can be extracted and acted upon, as data is generated, to create advantage here and now or even predict the future. Mastery of data—moving from data to wisdom—has the potential to improve various aspects of our personal and business life. Organizations can make better decisions, provide enhanced experiences, and achieve competitive advantage.

Most of the new data that will be generated in the IoT is real-time data that fits into a broad category called Data in Motion. This refers to the constant stream of sensor-generated data that defies traditional processes for capture, storage, and analysis.

Historically, in order to find actionable insights, enterprises have focused their analytics or business intelligence applications on data captured and stored using traditional relational data warehouses or "enterprise historian" technologies.

However, the limits of this approach have been tested by the increase in volume of this so-called Data at Rest. The challenges inherent in collecting, searching, sharing, analyzing, and visualizing insights from these ever-expanding data sets have led to the development of massively parallel computing software running on tens, hundreds, or even thousands of servers. As innovative and adaptive as these Big Data technologies are, they still rely on historical data to find the proverbial needle in the haystack.

As the IoT gathers momentum, the vast number of connections will trigger a flood of data, at an even more accelerated pace. While this new Data in Motion has huge potential, it also has a very limited shelf life. As such, its primary value lies in it being analyzed soon after it is created—in many cases, immediately after it is created. Hence, the traditional data management paradigm where raw data is stored first and analyzed later does not fit the temporal nature of IoT data. A new paradigm for handling Data in Motion is required, where data is analyzed as soon as it is generated and then optionally stored if required. The analysis can involve one or more of the following: aggregation, reduction/filtering, categorization/classification, contextualization, dimensioning, compression, pattern matching, normalization, and anonymization. All of these functions can be applied in micro-services that are hosted in the Fog (Fig. 6.12).

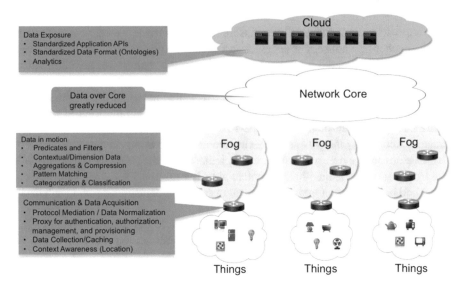

Fig. 6.12 Data management in the Fog

6.4.4.2   Search Technologies and Engines

With the availability of massive amounts of data, the need arises for reliable and effective mechanisms of searching for information that is useful and relevant. Search technologies have made great strides since the inception of the World Wide Web. However, these technologies, and the engines that utilize them, target static or slowly changing web data, and are generally lacking when dealing with the constantly streaming data in IoT.

IoT requires a solution for distributed data search, where queries can be propagated throughout the Fog domains. The solution can be logically organized into two planes: Things Plane and Search Plane (Fig. 6.13). The Things Plane encompasses the physical Things, Network and Compute nodes in the Fog. The Search Plane is a logical view of the various Fog nodes that support the distributed search functionality together with the network overlay that enables communication between them. Such overlay could be implemented, for instance, using a Federation Message Bus. Search queries are injected into the Search Plane at some Fog node, and propagate throughout the Search Plane. Special considerations are required to ensure that such propagation does not lead to traffic storms that overwhelm the network or the Fog nodes. Furthermore, mechanisms are required to limit the search scope, or radius, order to guarantee scalability and relevance of returned results. One approach would be to rely on Wave algorithms, such as the Echo algorithm, for query distribution and perform tree-based aggregation of partial results. These algorithms typically result in very low latency, have a low overhead and generally scale to hundreds of thousands of nodes.

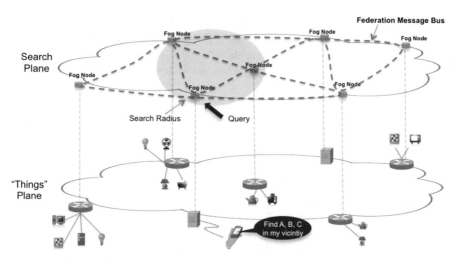

Fig. 6.13 Data search in Fog

As discussed in Chap. 5, both the ETSI and oneM2M standards define basic mechanisms for data search based on metadata. However, these mechanisms only allow elementary search procedures based on string matching between requests and the resource metadata. This provides a syntactic search capability with binary (yes/no) outcomes based on exact matches. Exact matches are highly unlikely in real-world IoT deployments with heterogeneous devices and Things from different vendors and providers. As such, effective search mechanisms should allow for "fuzzy matches," with partial correspondence between the request and the available data. Such mechanisms, ideally, would provide a measure of the semantic similarity between the original request and the retrieved results. To achieve this, Semantic Web technologies could be applied to the IoT: the IoT data can be enriched with semantic-based annotations that reference shared domain conceptualizations, and the search mechanisms can utilize Semantic Matching techniques to perform the ranking of potential results. Ruta et al. [15] propose such a framework that utilizes and enhanced version of CoAP as the underlying protocol for the Federation Message Bus.

6.4.5 More Gaps Ahead

Clouds are deployed in data centers, where network topologies are well defined and the infrastructure is physically secured with solid walls and cages. Network input and output between applications deployed in the data center and the outside world (e.g., Internet) are mediated through security appliances, such as Firewalls, which provide applications with a well-incubated environment under which they can operate. Furthermore, network bandwidth is abundant and it is relatively easy to change the network physical topology. With Fog, applications may be logically grouped together but not necessarily part of the same physical set up. The first gap to address is providing an orchestration system that enables the connection of applications deployed on Fog nodes to other applications, which are part of the same group, but are on desperate Fog nodes, as well as to applications that are in the Cloud. These connections could be over bandwidth-constrained links that cannot be changed due to the physical realities of the deployment. In light of this, open questions remain as to whether the Fog nodes need to replicate the entire functionality of the data center, including server, switch, and gateway functions (Data-Center-in-a-Box) or whether these functions should be distributed across multiple nodes and assembled together logically through the notion of "service chains." Another open gap is security: Fog nodes may be mounted in the field or on top of a light pole, so anyone could potentially gain physical access to them, attach wires, and compromise the security of the application or the network connectivity. New mechanisms of anomaly and tampering detection are needed. Yet another gap is in how would Fog nodes talk to Things: should that be through direct electric connectivity (e.g., PCI bus) or via the networking stack. Furthermore, in order for applications to leverage Fog, a high-level programming model is required which simplifies the development of large-scale distributed software. Such model provides simplified programming abstractions and supports dynamic application scaling at runtime.

6.5   Summary

In this chapter we introduced the concept of Fog computing and discussed its relationship to Cloud computing. The various IoT requirements driving the need for Fog were covered. We also discussed the prerequisites and enabling technologies for Fog, in terms of virtualization technologies, network mobility technologies, orchestration, and data management technologies.

Problems and Exercises

1. Will Fog Computing replace Cloud Computing? Why or why not?
2. What is the definition of Fog Computing?
3. What are the characteristics that uniquely distinguish Fog from Cloud Computing?
4. What makes Containers lighter-weight virtualization constructs compared to Virtual Machines? Why is this attribute of Containers important for Fog?
5. What are the two problems that all network mobility solutions aim to address?
6. Why can't traditional data management and analytics techniques be applied to IoT?
7. What three functions should a Fog Orchestration solution address and solve?
8. What is "data in motion"?
9. Why are semantic search mechanisms important for IoT?

10. Consider the following Fog domain shown in the Figure below. For each Fog
    node, the diagram shows the number of virtual CPUs (vCPU) and RAM avail-
    able. Also, the communication latency from each node to a remote sensor
    (labeled R_1 through R_4) is captured.

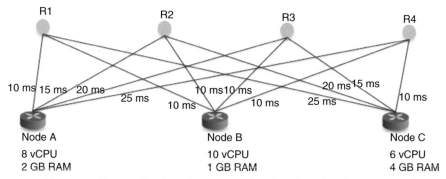

R1   R2   R3   R4

10 ms 15 ms 20 ms   10 ms10 ms   20 ms 15 ms
            25 ms   25 ms   10 ms
                    10 ms   10 ms

Node A   Node B   Node C
8 vCPU   10 vCPU   6 vCPU
2 GB RAM   1 GB RAM   4 GB RAM

There are five applications that need to be placed on the Fog nodes, and each
application has specific demands for CPU, RAM, and communication as
depicted in the table below:

Application	CPU demand (vCPU)	RAM demand (GB)	Communications demand
1	5	0.5	R_1 (<12 ms)
2	2	1	R_2
3	1	0.25	R_3
4	1	1	R_4

Find the optimal placement of the five applications on the three Fog nodes
such as to minimize the communication latency between each application and
the sensor that it needs to connect to.

11. A Fog domain is using EVPN to support workload mobility. The topology of
    the domain is as shown in the figure below. Every BGP speaker requires approx-
    imately 10 ms to process a BGP message, including any transmission/reception
    delay. A VM moves from the Melville server farm to the Granville server farm.

    (a) If N1, N2, and N3 form a BGP route-reflector (RR) cluster (i.e., fully
        meshed BGP sessions) and each of PEb, Peg, and PEm have a BGP session
        with their directly attached RR, how long would it be before all other appli-
        cations are capable of communicating with the VM in its new location
        assuming it takes 20 ms for GARP messages to be received and processed
        by the PE connected to the new server?
    (b) If N1, N2, and N3 are MPLS core routers, rather than route-reflectors, how
        does the above convergence time change?

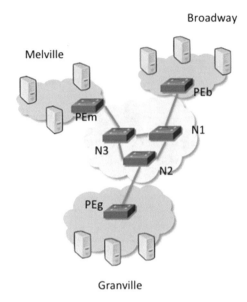

12. An IT administrator is trying to decide on whether to use Linux container or Virtual Machine for an interactive location-based interactive marketing application. Each instance of the application requires 200 MB of RAM to run, including all dependencies/libraries. The Linux distribution she is considering has a runtime memory footprint of 800 MB. A given application instance needs to move frequently in the Fog domain, to maintain close proximity to a target customer and deliver an immersive HD video/audio experience. Assume that the wireless links interconnecting the Fog nodes operate at 100 Mbps.

   (a) In the best-case scenario, how long would it take for the memory image of the application to move from one Fog Node to another in the case where the application runs in a Virtual Machine?
   (b) Repeat (a) for the case where the application runs in a Linux container?
   (c) Which virtualization construct should the IT administrator pick for her application and why?

13. A smart parking application is implemented in the future city of Metrotown using Fog computing. Fred is looking for parking in Metrotown's downtown shopping district. His car is capable of communicating automatically with the city infrastructure to locate available parking. The Fog domain in Metrotown is such that Fog nodes are placed roughly 50 m apart, on street lighting poles. The car's embedded application is searching for parking availability within a 1 km radius from the current vehicle's location. Assume that the Fog domain is using the Echo algorithm to search for data. If node processing latency and link propagation latency are 2 ms and 1 ms respectively, how long would it be before the search request has reached all nodes in the Fog domain?

14. A Fog orchestration system is responsible for the mobility of workloads among three Fog nodes dispersed in three locations: Coal Harbor, Yaletown and West End. The choice of a server for a given workload is a function of the CPU load of that server and the network communication latency from the server to the client. The orchestrator assigns a score between 0 and 1 to each server based on its CPU load, with a score of 1 for servers having less than 25% utilization, a score of 0.5 for servers with utilization between 25% and 75%, and a score of 0.25 for utilization above 75%. The orchestrator ranks the servers based on network latency and assigns them a score between 0 and 1 linearly depending on their rank in the ordered list, with a score of 0 assigned to the server with the highest latency and a score of 1 assigned to the server with the least latency. Assume that a user on her smartphone is roaming between the three locations. The network latency from her phone to the Coal Harbor Fog node is 200 μs, to the West End Fog node is 300 μs and to the Yaletown Fog node is 250 μs. The average CPU utilization for the servers is 80% for Coal Harbor, 13% for Yaletown and 50% for West End Fog nodes.

   (a) If the Fog orchestrator is configured to give equal weight to communication latency as server CPU load, which server would the orchestrator select?
   (b) If the communication latency carries twice the weight of the server CPU load, what would be the server that the orchestrator selects?

15. Explain the difference between the three different levels of virtualization: CPU instruction set level, hardware abstraction layer (HAL) level, operating system level.
16. What distinguishes LISP from other networking solutions that support mobility?
17. Describe Nielsen's Law. How does it relate to Moore's Law? What are the implications for IoT?
18. How is network connectivity different in the Fog from the Cloud?
19. How does rapid mobility impact communicating IoT applications?
20. When you conduct a search on your favorite Web search engine, is the search conducted over the Internet in real time? Will this model work for IoT?

References

1. Nielsen's Law of Internet Bandwidth, J. Nielsen, April 5, 1998, Online: http://www.nngroup.com/articles/law-of-bandwidth/
2. Yannuzzi M. et al., "Key Ingredients in an IoT Recipe: Fog Computing, Cloud Computing and more Fog Computing", IEEE 19th International Workshop on Computer Aided Modeling and Design of Communication Links and Networks (CAMAD), December 2014.
3. Bonomi F., et al., "Fog Computing and its Role in the Internet of Things", SIGCOMM 2012.
4. Mell P., Grance T., "The NIST Definition of Cloud Computing", National Institute of Standards and Technology Special Publication 800-145, September 2011.
5. Nanda S., et al., "A Survey on Virtualization Technologies". http://www.ecsl.cs.sunysb.edu/tr/TR179.pdf

6. Docker, www.docker.com
7. Kondo T., et al., "A Mobility Management System for the Global Live Migration of Virtual Machine across Multiple Sites", Computer Software and Applications Conference Workshops (COMPSACW), 2014 IEEE 38th International, July 2014.
8. Sajassi A., et al., "BGP MPLS-Based Ethernet VPN", IETF RFC 7432, February 2015.
9. Rekhter Y., et al., "Network-related VM Mobility Issues", draft-ietf-nvo3-vm-mobility-issues, work in progress, June 2014.
10. Farinacci D., et al., "The Locator/ID Separation Protocol (LISP)", IETF RFC 6830, January 2013.
11. Hertoghs Y., Binderberg M., "End Host Mobility Use Cases for LISP", draft-hertoghs-lisp-mobility-use-cases, work in progress, February 2014.
12. Morales C., "A Vision for Fog Software and Application Architecture", Fog Computing Expo, November 2014.
13. Uddin M., et al. "Graph Search for Cloud Network Management", Network Operations and Management Symposium (NOMS), 2014 IEEE, May 2014.
14. Chang, E.J.H., "Echo Algorithms: Depth Parallel Operations on General Graphs", Software Engineering, IEEE Transactions on (Volume:SE-8 , Issue: 4), July 1982.
15. Ruta, M., et al. "Resource Annotation, Dissemination and Discovery in the Semantic Web of Things: a CoAP based Framework", IEEE International Conference on Green Computing and Communications and IEEE Internet of Things and IEEE Cyber, Physical and Social Computing, 2013.

Chapter 7
IoT Services Platform: Functions and Requirements

IoT is expected to connect billions of sensors, devices, and applications over the Internet. One of the most critical prerequisites for successful, scalable, and effective IoT solutions is a Services Platform that provides abstraction across the multitude of diverse devices and data sources in addition to allowing for the management and control of a range of systems and processes. The operation of this platform requires a comprehensive and diverse set of requisites to gather relevant data, analyze it, and create actionable insights.

The Services Platform must surpass vertical solutions by integrating all essential technologies and required components into a common, open, and multi-application environment. The functions of the IoT Services Platform include the ability to deploy, configure, troubleshoot, secure, manage, and monitor IoT devices. They also include the ability to manage applications in terms of software/firmware installation, patching, starting/stopping, debugging, and monitoring. The Services Platform also provides capabilities that simplify application development through a core set of common application services that include data management, temporary caching, permanent storage, data normalization, policy-based access control and exposure. In addition to these, the Services Platform may offer some advanced application services, which include support for business rules, complex event processing, data analytics, and closed loop control. Figure 7.1 shows examples of key IoT Services Platform Functions. A more detailed and structured list will be provided in Sects. 7.2–7.12.

As can be seen from the list above, many of the capabilities of the IoT Services Platform represent what can be loosely categorized as "management functions." These, however, are different from traditional network management. Traditional network-level management functions were originally defined, in the early 1980s, by the Open Systems Interconnection (OSI) Systems Management Overview (SMO) standard as FCAPS: Fault, Configuration, Accounting, Performance, and Security. A decade later, the Telecommunications Management Network (TMN) of ITU-T, advanced the FCAPS as part of the TMN recommendation on Management

© The Author(s), under exclusive license to Springer Nature Switzerland AG 2022
A. Rayes, S. Salam, *Internet of Things from Hype to Reality*,
https://doi.org/10.1007/978-3-030-90158-5_7

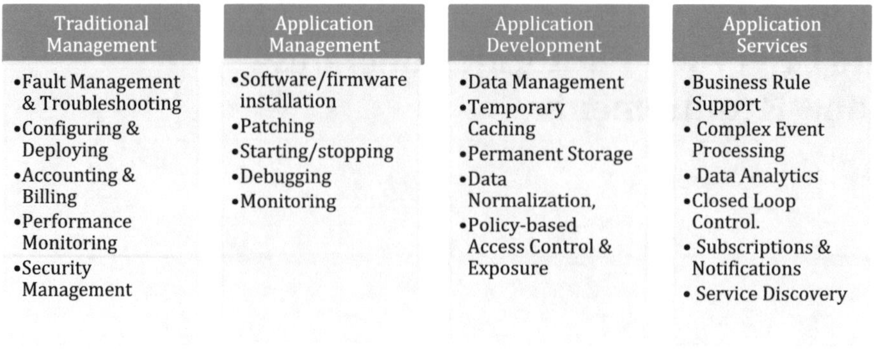

Traditional Management	Application Management	Application Development	Application Services
•Fault Management & Troubleshooting •Configuring & Deploying •Accounting & Billing •Performance Monitoring •Security Management	•Software/firmware installation •Patching •Starting/stopping •Debugging •Monitoring	•Data Management •Temporary Caching •Permanent Storage •Data Normalization, •Policy-based Access Control & Exposure	•Business Rule Support • Complex Event Processing • Data Analytics •Closed Loop Control. • Subscriptions & Notifications • Service Discovery

Fig. 7.1 Examples of key IoT Services Platform functions

Functions. The term FCAPS is often used in network management books as a useful way to break down the multipart network management functions.

While FCAPS still apply, the overall management functions of IoT solutions are more multifaceted than traditional networks. This is due to the following factors:

- IoT solutions include new devices (e.g., sensors, white-label gateways, and white-label switches). Some of these devices are inexpensive and generally lack the type or level of instrumentation required for traditional management functions.
- IoT solutions utilize relatively recent technologies (e.g., tracking exact location of IoT device using GPS triangulation) that were not considered by traditional management solutions.
- IoT solutions support more than two dozen access protocols (as was mentioned in Chaps. 4 and 5). The network management for each protocol may vary.
- IoT solutions support multiple verticals, each of which has different sets of management, quality of service, and grade of service requirements.
- IoT solutions utilize a new Fog layer with new and challenging network, compute and storage management requirements.
- Finally, many enterprises and service providers are expected to outsource and, in many cases, multisource key parts of the network and/or management functions. This requires additional, mostly new, capabilities such as secure integration that spans connecting workflows between multiple services providers.

This chapter describes the essential functions of the IoT Services Platform, as shown in Fig. 7.2. It focuses on identifying key capabilities with minimum emphasis on the relationship between the functions or their access protocol interfaces. Such relationship and protocols were addressed in the IoT Protocol Stack Chaps. 4 and 5.

Before introducing the main functions of the IoT Services Platform, we will first revisit the key components of IoT solutions that consist of IoT Device elements, IoT Network elements, IoT Services Platform, and IoT Applications as shown in Fig. 7.3.

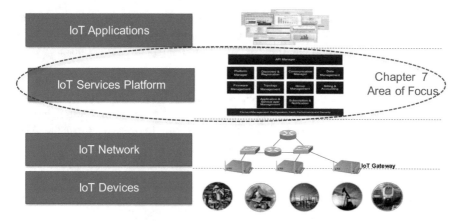

Fig. 7.2 Areas of focus for this chapter

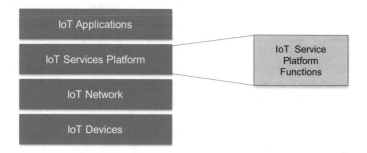

Fig. 7.3 Key components of IoT solution

- **IoT Device Entities**: IoT devices include sensing devices, actuators, and gateways. The main functions of the gateways are (1) collecting and aggregating information from the devices, (2) on-site filtering and simple correlation of collected information, (3) transferring correlated data to the network layer, and (4) taking action on the devices (e.g., shutting power off) based on commands from higher layers.
- **IoT Network Entities**: IoT network entities provide services from the underlying network to the Services platform. They include super-gateways, access routers, switches, and possibly element management servers with specific network management functions.
- **IoT Services Platform Entity**: The IoT Services Platform sometimes referred to as "IoT Platform" or "The IoT Application Services Platform," of any IoT solution. It is responsible for monitoring and controlling IoT elements in the IoT Device and Network Layers. It also allows the creation of direct integration between physical devices (e.g., sensors, actuators, gateways) and computer-based application systems to improve efficiency, accuracy, and economic benefit.

- IoT Services Platform entity receives information from IoT Device and Network Entities, and provides services to the Application Entities. More importantly, it provides network-level and often service-level management functions as will be discussed in this chapter.
- **IoT Application Entities**: Application entities receive information from the Services Platform and provide services and business level functions. These functions are typically vertical dependent. Examples of Application Entities include an IoT-based Automated Parking application, an IoT-based Hurricane Alert System application, etc.

7.1   IoT Services Platform Functions

Without a doubt, the IoT Services Platform constitutes the linchpin of successful IoT solutions. It is responsible for many of the most challenging and complex tasks of the solution. The IoT Services Platforms include numerous fundamental functions to ensure proper and secure deployment and comprehensive supervision and control. In this chapter, we will identify key IoT Services Platform functions by grouping related requirements together and by utilizing recent IoT standards such as those devised by oneM2M[1] and European Telecommunications Standards Institute (ETS) standards bodies. More information on the IoT standards was provided in Chap. 5 (Sect. 5.4.2).

The overall functions of the IoT Services Platform can be categorized into the following 11 key areas:

1. Platform Manager
2. Discovery and Registration Manager
3. Communication Manager
4. Data Management and Repository
5. Firmware Manager
6. Topology Manager
7. Group Manager
8. Billing and Accounting Manager
9. Cloud Service Integration Function/Manager
10. API Manager
11. Element Manager: Configuration Management, Fault Management, Performance Management and Security Management

Figure 7.4 shows the IoT Services Platform functions. It does not constrain the multiplicity of the entities nor the relationships among them.

---

[1] OneM2M is the global standards initiative for Machine-to-Machine Communications and the Internet of Things.

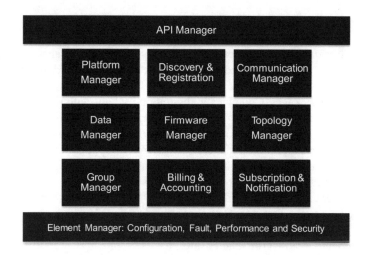

Fig. 7.4 Common IoT Services Platform functions

7.2 IoT Platform Manager

The IoT Services Platform Manager, also known as IoT Service Platform's Management Entity in some standards, is responsible for managing the IoT Service Platform internal modules and interfaces. It works with the Communication Manager (Sect. 7.4) and the Element Manager (Sect. 7.6) to monitor, configure, troubleshoot, and upgrade the Services Platform modules. It is really the "manager of managers" responsible for providing the overall management of the entire Services Platform functions.

The Platform Manager is used for the overall control and management of the common management functions. It allows the system administrator, or an application in the Application Layer, to manage IoT Services Platform components and interfaces. This includes initiating an action (e.g., discovery) and receiving results (e.g., discovered elements) within a specific amount of time.

The Platform Manager is expected to have a full user interface, allowing the system administrator to initiate requests and review reports, and providing interfaces to receive and send information. It must be noted that user and application's authorization (specifying access rights level) and authentication (verifying the user's credentials) is a top requirement.

The Platform Manager may be a physical system/server or virtual system with functions distributed among the common management components.

The IoT Platform Manager is responsible for:

- Performance Monitoring and Fault Management of the Services Platform functions. This includes continuous monitoring, troubleshooting, fault identification, fault correction, and diagnostics. This requires constant collection of logs, performance and fault parameters from the platform functions (e.g., system logs, alarms).

- Lifecycle software management allowing the IoT Platform Manager to manage any software packages related to the above Services Platform functions. This includes upgrading, updating, installing, uninstalling/removing, and downloading software packages. Complete configuration backups with roll-back capabilities must be supported (Why? See Problem 24).
- Configuring any of the platform functions when they are first installed. This includes the configuration of the services offered to Application Entities.
- Supporting multiple levels of IoT Platform Managers operating in a hierarchical environment. For instance, supporting two Platform Managers, representing two separate networks, and a third "Supper Platform Manager" with full read and write access to the first two. Consequence, Platform Managers should have the ability to establish relationships among each other including establishing parent–child and Read–Write relationships.

The concept of Super Platform Manager is needed to address high availability requirements.

7.3   Discovery: Entities, Services, and Location

Discovery is the process of identifying and transferring information regarding existing IoT entities and/or resources with their locations. Accurate discovery is essential for most IoT management tasks such as asset management, network monitoring, network diagnostics and fault analysis, network planning, capacity expansion, high availability, and others.

One of the key discovery requirements is for IoT entities (e.g., sensors, gateways, routers) to uniquely identify themselves via a common registration process. Hence, each entity needs to be uniquely identifiable through its embedded computing system. It also needs to be able to interoperate within the existing IoT infrastructure via IoT access protocols as we defined in Chap. 5.

An essential requirement for discovery is entity registration. In this section, we will first introduce the registration function and then provide the key requirements for discovery.

7.3.1   Registration

IoT device registration can be defined as the process of delivering the device information to the Management Entity (or to another server) in order for IoT devices to communicate and exchange information. Most IoT devices will be identified and tracked by their IP addresses. However, as we mentioned in Chap. 2, not all IoT devices are IP-enabled. In such case, devices (e.g., basic sensors) may be tracked by their local (typically non-unique) addresses (e.g., local identifier) in combination with their corresponding gateway IP address. Gateways are expected to have unique IP addresses and are responsible for providing a means to uniquely identify their associated sensors and actuators.

In order for the IoT registration process to work, the following key capabilities are necessary:

- IoT devices must have the capability to register to an associated Platform Manager entity. This procedure may be self-registration (preferred solution) where a new IoT device identifies itself to the management entity as soon as it joins the IoT network or identifies itself during the discovery process as will be discussed in the next section. The registration requirements must be addressed in all IoT domains, i.e.,

  - Ability for new sensors and actuators to register themselves with their associated gateways.
  - Ability for new gateways to register themselves with their associated Platform Manager entities.
  - Ability for Platform Managers to register themselves with a super (or another) Platform Manager(s) as defined by the network administrator.

- Once the registration is complete,

  - The IoT Platform Manager must be able to access the IoT gateway and retrieve information (i.e., Read Access is granted). In other words, IoT gateways must grant full access privilege to the associated IoT Platform Manager(s). Hence, all resource information must be available to the IoT Platform Manager.
  - The IoT gateways must be able to access their associated sensors and actuators and retrieve information. In this case, sensors and actuators resource information must available to the associated IoT gateway(s).
  - Super IoT Platform Manager(s), if present, must be able to access their corresponding IoT Platform Managers and retrieve information. Hence, all resource information must be available to the super management entities where applicable.

7.3.2   Discovery

Based on some filtering criteria (typically specified by a management entity such as the Platform Manager, IoT Gateway, or a northbound application) in the discovery request, the discovery function is responsible for discovering, identifying, and retuning matching information regarding entities and/or resources. The discovery function sends matching information to the requester's system. The discovery request may include the IP or MAC address (obtained from device registration), set of addresses, or range of IP addresses of the resource where the discovery is to be performed. Full discovery, without any specified addresses, may also be supported. In such case, all entities (based on some filtering criteria in the discovery request) are discovered. Example: Discover all entities in a given enterprise network.

In IoT, the location of the physical entities (e.g., sensors, gateways) is also essential. The discovery function also supports obtaining geographical location information.

It is assumed, therefore, that IoT entities have the capability of identifying, storing, and updating their geographical location information. This may be accomplished with a GPS module in the entity, a location server responsible for tracking and storing location information, or information for inferring location stored in other nodes. The location technology (e.g., Cell-ID, assisted-GPS, and fingerprint) used by the underlying network depends on its capabilities. Sensors with no geo-locations are identified by their corresponding gateways.

We will use an example of CoAP (Constrained Application Protocol) to illustrate discovery.

Discovery Request: Assume the IP Address of the Management Server is 192.15.10.5. Also assume the Management Server is interested in discovering sensors within 500 m from the location of (37.76724070774898, −122.37890839576721)[2] GPS Coordinates. The management server will send a CoAP GET request to

```
Coap://192.15.10.5:5784/.well-known/core?
& ro=SSN-XG-IRI&sd=yyyyyy=&at30004&lg=-122.37890839576721
&lt=37.76724070774898&md=500&st=2&sr=70
```

Discovery Reply: Upon receiving the request, the CoAP server will start a matching process comparing the request with all stored information in its local data store. Let us assume that the returned set consists of two sensors matching the request. The CoAP server response payload will be

```
</Hts2030HumidSens>;ct=41;      at30004;     lg=-122.37890839576721;
lt=37.76724070774898&md=310;          ro=SSN-XG-IRI;          sd=aaaaaa;
tittle="Humidity-Sensor-2030",
</BitLineAnemomSens>;ct=0; ct=41;at=30004; lg=-122.37890839576721;
lt=37.76724070774898&md=276;          ro=SSN-XG-IRI;          sd=bbbbbb;
tittle="Anemometer-Sensor-111",
```

Table 7.1 summarizes the Registration and Discovery requirements.

---

[2](37.76724070774898, −122.37890839576721) are the GPS Coordinate for a northern California area.

Table 7.1 Summary of IoT Registration and Discovery requirements

Function	Responsibility	Results/outputs
Discovery	Identify IoT sensors, actuators, gateways, and devices via attributes and search protocols	IoT entities, gateways, sensors, and actuators based on filtering criteria
	Identify the location of physical entities	GPS location
	Identify access control policies across management servers and clients (see Sect. 7.5)	Access Control Policy information
	Identify IoT services via attributes and collected data	IoT configured services (outside the scope of this book)
Registration	The process of delivering IoT device information (sensors, actuators, gateways, and IoT entities) to the Management Entity, or to another server, in order for IoT devices to communicate and exchange information	Ability for IoT device (sensors, actuators, gateways, and IoT entities) to register with their associated gateways

Finally, IoT software services may also be discovered by collecting configuration and operational parameters (e.g., using YANG,[3] SNMP MIBs, CLI Outputs). IETF defined a set of requirements for standard-based device (configuration and operational data) management. Key functionalities include:

- Ability to collect configuration and operation data from all IoT devices (e.g., running configuration files) where applicable.
- Ability to extract and then structure/model data from configuration and operation files via an information model.
- Ability to distinguish between configuration data and operational data (i.e., data that describes operational state and statistics).
- Ability for operators to configure the entire network and not just individual devices.
- Ability to check configurations consistency between devices in the network.
- Ability to use text processing tools such as diff and version management tools such as CVS.
- Ability to distinguish between the distribution of configurations and the activation of a certain configuration.

Detailed requirements for discovery of software services are outside the scope of this book.

---

[3] YANG is a tree-structured data modeling language (defined by IETF) used to model configuration and state data [6].

7.4   Communication Manager

The Communication Manager is responsible for providing communications with other platform functions, applications, and devices. This includes supporting the following functionality:

- Ability to provide a global view of the state of the entire underlying platform network. This is needed to address the next requirement.
- Ability to determine the optimal time to establish the communication connection to deliver information between at least two platform entities. Such decision is based on the source delivery request as well as traffic/congestion control optimization techniques within the platform. Data may be stored/buffered for future delivery time per the provisioned Communication Manager policies.
- Ability to deliver required information within the delivery request time.
- Ability to publish its own polices to external systems.
- Ability to provide information to external systems to drive policies describing details of the usage of network resources (i.e., 5% of bandwidth on link X at time T was utilized for service Y).
- Ability to communicate, select paths for a given amount of time, and manage buffers based on communication manager polices.

7.5   Data Management and Repository

Collecting, storing, and exchanging information among various platform entities is one of the key requirements for the IoT Service Platform. Data Storage and Mediation functionalities must include:

- **Data Retrieval**: Data may be retrieved from various sources including IoT devices (e.g., sensors and getaways), IoT network elements (e.g., super-gateways and switches), IoT subscribers or IoT applications. IoT device and network element data is assumed to be collected by collection systems or by collection agents.

- We are using the term "Collection System" to refer to a physical hardware machine (e.g., server, PC) mainly used for data collection. And the term "Collection Agent" refers to a software unit (agent) that resides on a gateway/router blade (or on a computer along with other applications). Hence, Collection System may be the same as Collection Agent (see Problem 30).

- **Data Aggregation**: Data aggregation implies grouping data from similar or diverse sources for further processes. Typically, data from various IoT sources need to be grouped together based on a well-defined data model (e.g., physical locations, device types, subscribers with their assigned devices, etc.). The aggregation syntax should be defined by the data model. Also, data from multiple data collection systems (for the same IoT entity) need to be filtered and aggregated accordingly.

- **Data Parsing**: Data parsing normally implies reading the data, using software, and extracting useful information. Stages of data parsing are hard to define without a concrete use-case but typically include running code to extract specific parameters and writing the extracted data to a database.
- **Data Storing**: The Data Storage and Mediation Function supports taking data from various sources and storing it based on pre-defined policy. Raw data, aggregated data, and parsed data may be stored with different polices (e.g., store raw data for 6 months, store parsed data for 2 years). Associated contextual information is also stored with the data. Examples of contextual information include: data type (e.g., Temperature), data format (e.g., $-100\ °C$ to $+100\ °C$) data source (e.g., Sensor ID and Associated Gateway ID), retrieval time and date (e.g., 03:45:00   PM   EST   on   12/12/2016),   retrieval   location   (e.g., lg $= -122.37890839576721$; lt $= 37.76724070774898$).
- **Access to data based on defined access control policy**: The Data Storage and Mediation needs to have the capability of providing local or remote data access based on a well-defined access control policy. The policy, which is typically defined by the network administrator, needs to capture what types of functions a specific user or application can perform on the data (read-only write-only, read/write). The policy may include temporal access restrictions, and may be role based (e.g., administrator vs. user, etc.).

7.6   Element Manager (Managing IoT Devices and Network Elements)

The element management function is expected to manage IoT sensors, actuators, gateways as well as other devices residing within the platform boundaries. The element management function, as shown in Fig. 7.5, typically utilizes the client-server distributed model where a single management server may manage multiple management clients. In this model, tasks are partitioned between the management server (provider of the service) and the management client (service requester). The management client establishes a connection to the management server over the network to accomplish a particular task (e.g., sending performance results of the last 5 min). Once the management client's task is fulfilled, by the management server, the connection is terminated.

In IoT environment, the management server may be residing in a data center while management client may be residing on the IoT Gateway in an offsite location.

A key function of element management includes:

- Ability for the management client and management server to communicate at any time. Hence, real-time communication is required to send time-sensitive data.
- While it is recommended to use a standardized protocol so that any management server can communicate with any management client, any existing client-server

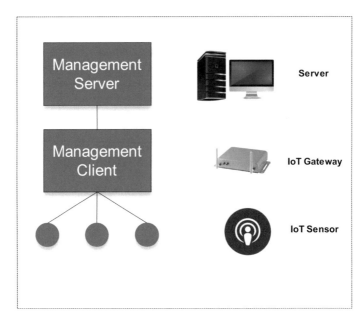

Fig. 7.5 Example of element management function

communication protocol may be utilized. Key examples include TR-069[4] and LWM2M.[5]

- Ability for the management servers (or adaptors) to receive and fully understand (based on an agreed upon protocol) management client requests and/or notifications. For example, air pressure measurements of the oil rig vale.
- Ability for the management clients to receive requests and/or notifications from the management servers (or their adaptors). The management clients may have the ability to fully understand such events and deliver them to targeted sensors, actuators, or device as required. For example, requesting the actuator to shut down a valve.
- Ability for the management server and management clients to address the security requirements as defined later in this chapter and in Chap. 8 including Authorization, Authentication, Access Control, Non-reputation, Data confidentiality, Communication Security, and Data Integrity and Privacy.
- Ability for the super management server to assign different levels of access control privileges when multiple management servers and/or clients exist.

---

[4]TR-069 as a bidirectional SOAP/HTTP-based protocol that was originally for remote management of end-user devices. It was published by the Broadband Forum and entitled CPE WAN Management Protocol (CWMP).

[5]LWM2M (Lightweight Machine-to-Machine) protocol is defined by the Open Mobile Alliance for M2M/IoT, as an application layer communication protocol between a LWM2M Server and a LWM2M Client (located in a LWM2M Device).

- Ability for the super management server to provide read access (with the appropriate access control requirement) to the discovery or other functions to discover access control policy information.
- Ability for the management server to provide read access (with the appropriate access control requirement) to the discovery or other functions to discover managed elements with their latest collected information (e.g., metadata, values) including gateways, sensors, and actuators.
- Ability for the management server to create a new element to be managed (e.g., gateway, sensor), delete an existing element, update any parameters of any existing elements, update the firmware of any element, and to retrieve information of any existing elements.

7.6.1   Configuration (and Provisioning) Management

Configuration management is one of the most important element and network management functions. Configuration management is the process of enabling (or disabling) a service. Before providing the overall requirements for IoT configuration management, it is worthwhile to discuss the main differences between configuration and provisioning management.

The Provisioning function is concerned with the basic process of preparing and equipping an IoT network to provide proper and effective services, while the Configuration function is concerned with the actual enablement or disablement of an IoT service. Provisioning is often equated to initiation of a service or capability, whereas configuration is the final set of touches to deliver the actual service to a particular customer.

Hence, an IoT network is first generically provisioned (e.g., by installing libraries or services on servers) to provide a set of services to any customers. Such provisioning does not imply that a service can simply be launched without additional instructions on which particular server or set of servers to use, which specific set of already provisioned parameter to employ, how to distribute the load when demand increase, etc.

Figure 7.6 shows an example of Device Remote Management/Configuration to address the machine-to-machine (M2M) environment with OMA (Open Mobile Alliance) lightweight M2M protocol, which focuses on constrained cellular and sensor network M2M devices.

Key configuration requirements include:

- Ability to identify IoT devices and their associated management objects and attributes.
- Ability to enable or disable a device capability.
- Ability to update device parameters.
- Ability to roll-back applied changes in the configuration at least to five back versions (tracked by time and date).

Fig. 7.6 Example of configuration management using LW M2M protocol

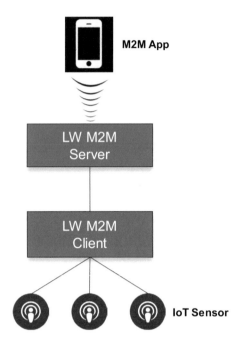

• Ability to reset IoT device parameters to original factory values.

On the IoT network side, an example of network element protocol is the Network Configuration Protocol (NETCONF). It provides mechanisms to install and update the configuration of network elements such as a router or switch using XML to encode the configuration data and the protocol messages.

7.6.2   Fault Management

At the minimum, IoT service providers need to be able to configure new service (turn-on a service for a customer) and then identify any problem or potential problem and have the tools to fix it quickly. No service provider will survive in the market if they do not have the capabilities and processes to discover problems promptly (before they occur in most cases) and take quick action to prevent service interruption or service degradation that could result in Service-Level Agreement (SLA) violation.

Fault management is among the most challenging and important management function of IoT networks. This is due to the fact that large-scale deployment of inexpensive sensors (i.e., with very limited processing capability, storage capacity, and limited energy) means that failures from various defects will not be uncommon. It is also due to the fact that managing IoT devices in remote locations and often

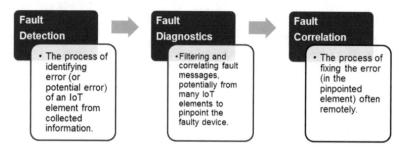

Fig. 7.7 Main stages of fault management function

harsh environments will be demanding, especially when dealing with various IoT topologies and verticals.

Fault Management typically consists of three main functions: fault detection, fault isolation (or diagnostic), and fault correction as shown in Fig. 7.7. In this section, we will first describe these three functions. Then we will introduce fault tolerance and fault or diagnostic signature. Finally we will list the overall fault management requirements for IoT devices and services.

- **Fault Detection** is the process of identifying error (or potential error) of an IoT element typically using collected statistics. The collected data may be time-based (e.g., fault-related data collected from the IoT element by the fault manager function every t seconds) or event-based (e.g., IoT element notifies the fault manager only if pre-defined fault-related conditions are met). When a fault or event occurs in the event-based case, an IoT element will send an alarm or notification to the fault manger (and often notify the network administrator) immediately. An alarm is a persistent indication of a fault that clears only when the triggering condition has been resolved.

- An example of fault-related data is the Simple Network Management Protocol (SNMP) Entity Sensor Management Information Base (MIB) as described by IETF RFC 3433. The Entity Sensor MIB provides generalized access to information related to sensors that are often found in network equipment. The complete list of the MIB information is shown in Table 7.2. One of the key variables of the Entity Sensor MIB is "Entity Sensor Status" with three defined possible values:

  - Entity Sensor Status = 1: indicates that the sensor data value can be obtained (normal operation).
  - Entity Sensor Status = 2: indicates that the sensor data value is unavailable (operational but no data was collected).
  - Entity Sensor Status = 3: indicates that the sensor is broken and cannot collect the sensors data value (failure). Once the failure status is received by the network administrator/operator, S/he needs to investigate the issue further to determine if the failure is due to disconnected wire, out-of-range, violently fluctuating readings, or something else.

Table 7.2 Overview of entity sensor MIB

MIB variable	Description	Examples of potential value
EntitySensorDataType	Entity Sensor measurement data type associated with a physical sensor value	3 = Volts AC 4 = Volts DC 5 = Amperes 6 = Watts 7 = Hertz 8 = Celsius
EntitySensorDataScale	A data scaling factor, represented with an International System of Units prefix	6 = Nano 10 = Kilo 11 = Mega 12 = Giga 13 = Tera 14 = Exa
EntitySensorPrecision	Sensors Precision Range	1 = One decimal place in the fractional part 2 = Two decimal place in the fractional part
EntitySensorValue	Sensor Value	From −999,999,999 To +999,999,999
EntitySensorStatus	Operational Status of Physical Sensor	1 = Ok 2 = Unavailable 3 = Nonoperational
TimeStamp	The time the status and/or value of this sensor was last obtained	10:00:00 AM PST

Fault detection will be triggered if the value of "Entity Sensor Status" variable is 3.

- **Fault Diagnostic and Isolation** (also referred to as Fault root cause analysis) is the process of hierarchal filtering and correlating of fault messages, typically from hundreds of IoT elements or systems, to pinpoint the faulty element to a stage where corrective action can be taken. Such process is often based on artificial intelligence, pattern recognition combined with models of abnormal behavior and/or intelligent rule-based systems.
- Pattern recognition with abnormal behavior models is frequently used in the industry to construct the so-called **Diagnostic Signatures** as a form of accumulated and documented knowledge. Fault Diagnostic and Isolation will then take place at run-time based on matching observed information to the nearest Diagnostic Signature.
- Fault managers may use complex filtering systems to assign alarms to severity levels. Alternatively, they could use the ITU X.733 Alarm Reporting Function's perceived severity field: cleared, indeterminate, critical, major, minor, or warning.
- Fault Isolation (or Fault Diagnostic) in IoT-based network is a challenging problem because of the interactions between different network entities (e.g., wireless sensors, gateways) and protocols.
- **Fault correction** is the process of fixing the error/fault problem, often remotely. A fault manager allows a network administrator to monitor events and perform

actions based on received information. Ideally, the fault manger system should be able to not only correctly identify faults but also to automatically take corrective action, such as to activate the notification system to notify a pre-defined list of administrators (i.e., send e-mail or SMS text to a mobile phone) for intervention when needed, or to launch a program or script to take corrective action.

Critical IoT systems should be designed around the concept of fault tolerance. In principle, they must be able to continue working at least to some acceptable level in the presence of faults. Network element redundancy (e.g., multiple sensors performing identical tasks, dual modular sensing engines in the same sensor, fail-over power supply) is a very common fault tolerance example that is designed to prevent failures due to hardware components.

It should be noted that fault tolerance is not just a property of individual IoT elements; it may also impact IoT communication protocols as discussed in Chap. 5. For example, the Transmission Control Protocol (Chap. 2) was designed as a reliable two-way communication protocol, even in the presence of failed or overloaded communications links. It achieves this by requiring the endpoints of the communication to expect errors such as packet loss, packet reordering, packet duplication and corruption.

The element Diagnostics and Fault Management Function in IoT allows network engineers to troubleshoot sensors and actuators (typically over their gateways) or any other IoT entity remotely. Service troubleshooting (i.e., when devices are working correctly but the service-level parameters are not being meet) is also addressed through this function.

The Diagnostics and Fault Management function supports the following areas:

- Ability to connect and uniquely identify any device in the network including sensors, actuators, gateways, etc. Sensors and actuators are often identified by their corresponding gateways.
- Once the connection is established, Fault Management function requires the ability to retrieve device information that identifies a device, its model and manufacturer. E.g., Device Universal ID, Device Product ID, Device Serial Number, SKU.
- Ability to retrieve device information for the software and firmware installed on the device, e.g., embedded software version.
- Ability to retrieve information related to a battery embedded within the device.
- Ability to retrieve information related to memory in use by a device.
- Ability to reconfigure/change (Write option) device specific parameters to diagnose or fix an identified problem.
- Ability to compare results from main system and backup system (if backup system is deployed and operational) and provide error messages for different results.
- Ability to provide the current list of problems occurring on the network to the fault manager/network management systems/system administrator. Such list is cleared only when the triggering condition has been resolved. Or cleared by the network administrator.
- Ability to retrieve the event logs from any IoT device.

- Ability to allow a system administrator to monitor events from multiple systems/ locations and perform actions.
- Ability to assign alarms to severity levels. E.g., cleared, indeterminate, critical, major, minor or warning.
- Ability to notify administrators of critical and/or other alarms (based on pre-defined rule-based list) via e-mails, text message, call to mobile phones.
- Ability to launch a program or script to take corrective action for critical and/or other alarm types.
- Ability to reboot diagnostic operation.
- Ability to roll-back any changes at any stage.
- Ability to rest IoT device parameters to original factory values.

7.6.3   Performance Management

The Performance Management function can be defined as a mechanism to quantity "how the underlying IoT infrastructure (e.g., IoT network and device layers) is doing?" Is the infrastructure operating under heavy load (e.g., over 90% utilization) and about to run out of bandwidth or is there substantial extra free capacity so a service provider can offer discounted services?

As was mentioned in Chap. 2, IoT is more than just devices at rest; there are also many mobile IoT devices that include wearables, connected vehicles, and even fly-ing drones. A more formal definition of performance management is a set of pro-cesses to measure and monitor the quality and grade of the services that are offered to customers. Quality of Service (QoS) typically refers to performance measures from one element (e.g., delay of one link), whereas Grades of Service (GoS) typi-cally refers to a performance measure of the end-to-end service (e.g., delay of the end-to-end path that a service is taking).[6]

Consequently, a practical description of IoT network performance incorporates three main elements:

- What to measure? Determining what to measure is conceivably the most critical question for IoT management. Smart performance algorithms are useless unless required measurements that drive such algorithms can be collected. In Chap. 3 (Things in IoT), we have identified over a dozen sensor types. Knowing that these sensors are performing correctly is very important. Key sensor perfor-mance measures include: Operating range of input-to-output signals, acceptable noise level produced by sensors, acceptable resolution, and acceptable response time to instantaneous change in input signal.

- Generic measurements for all IoT devices (e.g., gateways, routers) will include device and transport link utilization (based on available bandwidth and capacity),

---

[6] Some researchers use the term QoS to refer to both QoS and GoS as defined above.

end-to-end delay and jitter, packet lost ratios, packet error rates, and any other parameters that impact services carried on the network. These will continue to be important for IoT-based networks.

- Where to measure? Theoretically performance should be measured through the network at all time. Practically, performance should be measured at least between the network end points where the service is delivered. E.g., sensor to gateway, gateway to platform and platform to application.
- How to measure the above parameters and then construct QoS and GoS measures to perform the actual minoring?

Similar to Fault Management, Performance Management supports the following areas for IoT network elements and devices:

- Ability to connect and uniquely identify any device in the network including sensors, actuators, gateways, etc. Sensors and actuators are often identified by their corresponding gateways.
- Once the connection is established, Performance Management function needs to have the ability to ID the device by retrieving device information.
- Ability to retrieve device information for the software and firmware installed on the device, e.g., embedded software version.
- Ability to retrieve information to measure the performance of a device or a module within the device (e.g., battery).
- Ability to measure any performance related parameter including, but not limited to, element utilization, delay, jitter, packet lost, packet arrives with error, amount of memory in use by a device.
- Ability to allow a system administrator to monitor events from multiple systems/locations.
- Ability to notify administrators of critical and/or other performance related activities (based on pre-defined rule-based list) via e-mails, text message, calling mobile phones.

7.6.4   Important Performance Measures for IoT Devices (E.g., Sensors)

The following sensor (and actuators where applicable) performance requirements/characteristics measures are considered important for IoT solutions:

- IoT Sensor's Transfer Function should be plotted (e.g., testing the various ranges of inputs, vendor documentations) to ensure it meets the specific IoT solution requirements. The Transfer Function represents the functional relationship between input signal (physical signal captured by the sensor) and output signal (electrical signal converted by the sensor). Frequently, this relationship is represented by a graph constituting a comprehensive depiction of the sensor characteristics.

- IoT Sensors' Sensitivity should be evaluated and within the minimum acceptable range for the specific IoT solution (e.g., 0.1 variation in temperature sensors may be acceptable for smart homes but not for more critical solutions). The sensitivity is generally the ratio between a small change in electrical output signal to a small change in physical signal. It may be expressed as the derivative of the transfer function with respect to physical signal.
- IoT Sensor's Dynamic Range should be established and documented. Dynamic range is defined as the range of input signals which may be converted to electrical signals by the sensor. Outside of this range, signals cause unsatisfactory accuracy in output.
- IoT Sensor's Accuracy should be established and documented. Accuracy is defined as the maximum expected error between measured (actual) and ideal output signals. Manufacturers often provide the accuracy in the datasheet, e.g., 1% error may be acceptable for some IoT solutions.
- IoT Sensor's Noise Level should be established and documented. As was stated in Chap. 3, all sensors produce some level of noise with their output signals. A sensor's noise is only an issue if it impacts the performance of the IoT system. Smarter sensors must filter out unwanted noise and be programmed to produce alerts on their own when critical limits are reached. Noise is generally distributed across the frequency spectrum. Many common noise sources produce a white noise distribution, which is to say that the spectral noise density is the same at all frequencies.
- IoT Sensor's Resolution should be established and documented. The resolution of a sensor is defined as the smallest detectable signal fluctuation. It is the smallest change in the input that the device can detect. The definition of resolution must include some information about the nature of the measurement being carried out.
- IoT Sensor's Bandwidth (the frequency range) should be established and documented. Some sensors do not operate properly outside their defined bandwidth range.
- IoT Sensor should produce a performance alert and notify its IoT gateway once service issues or interpolation is detected outside its normal operational range (e.g., outside the defined bandwidth, resolution).
- Finally, IoT Sensors should have some ability (depending on the sensors' sophistication level) to work with its IoT gateway to measure the Throughput (actual rate at which the information is transferred), Latency (the delay between the sender and the receiver), Jitter (variation in packet delay at the receiver of the information), and Error Rate (the number of corrupted bits expressed as a percentage or fraction of the total sent) during a specific period of time (e.g., 1 h).

7.6.5   Security Management

Security management is extremely important for IoT. Any security management solution must comprehensively address sensitive data handling, data administration, service subscriptions, data transfer (especially over the Internet), data access control, and identity protection. Given the importance of this area, we have dedicated an entire chapter (Chap. 8) to this critical topic. In this section we will simply list the high level security requirements.

IoT high level security requirements include eight main areas:

- *Data Confidentiality*: ensures that the exchanged messages can be understood only by the intended entities.
- *Data Integrity*: ensures that the exchanged messages were not altered/tampered by a third party.
- *Secure Authentication*: ensures that the entities involved in any operation are who they claim to be. A masquerade attack or an impersonation attack usually targets this requirement where an entity claims to be another entity.
- *Availability*: ensures that the service is not interrupted. Denial of Service attacks target this requirement as they cause service disruption.
- *Secure Authorization*: ensures that entities have the required control permissions to perform the operation they request to perform.
- *Freshness*: ensures that the data is fresh. Replay attacks target this requirement where an old message is replayed in order to return an entity into an old state.
- *Non Repudiation*: ensures that an entity cannot deny an action that it has performed.
- *Forward and Backward Secrecy*: Forward secrecy ensures that when an entity leaves the network, it will not understand the communications that are exchanged after its departure. Backward secrecy ensures that any new entity that joins the network will not be able to understand the communications that were exchanged prior to joining the network.

Detailed discussions of the above areas including existing solutions and gaps will be provided in Chap. 8.

7.7   Firmware Manager

In the past, Firmware Management was not even an issue as older devices rarely required operating system updates. In fact, Firmware is not part of the traditional FCAPS capabilities that we described in Sect. 7.1.

Firmware refers to the device's operating system that controls and operates the device. Firmware is a program written into read-only-memory (ROM), rather than simply being loaded into normal device storage, where it may be easily erased in the event of a crash, and initially added at the time of manufacturing. It is called firmware rather than software to highlight that it is very closely tied to the particular hardware components of a device.

Nowadays, firmware updates are provided by vendors on regular basis, often as a way to fix bugs or introduce new functionality (e.g., Apple's iOS, Cisco's IOS, Samsung's Android).

Key Firmware requirements for IoT solutions include:

- Ability for IoT device to store and maintain multiple firmware images and to manage individual firmware images.
- Ability for IoT management solution to provide a user-friendly device Firmware Management site that provides lifecycle management for firmware associated with a device. This includes

  – Downloadable versions of latest Firmware images.
  – Step by step instructions to download/update images on various supported devices that guarantee full migration of existing settings and applications on an IoT Device.
  – Step by step instructions to remove a Firmware image and roll-back into an older image if needed with full device backup of existing applications and settings.
  – Support for downloading and updating within the same action.
  – Download, update, and removal of Firmware process should be done within a reasonable amount of time (typically less than 10 min) with clear progress bar visible to the user.
  – Q&A and troubleshooting support.

- Ability for IoT management solution to support both wire-line and mobile (the so-called FOTA (Firmware Over-The-Air) firmware upgrade. FTOA is a Mobile Software Management (MSM) technology in which the operating firmware of a mobile device is wirelessly upgraded and updated by its manufacturer. FOTA-capable devices download upgrades directly from the service provider.

7.8 Topology Manager

IoT network topology refers to the arrangement of the various elements (sensors, gateways, switches, links between gateways and switches, etc.). Topology may be physical or logical and is often presented explicitly in a structured graph. Physical Topology is the placement of the actual IoT elements on a graph (e.g., map) as they are connected with physical information (e.g., locations). Logical Topology, on the other hand, displays virtual information such as network virtualization data, data flow on the network.

Key requirements for topology management include:

- Ability to display IoT Physical network that includes all IoT devices (e.g., sensors, actuators) and IoT network elements (gateways, switches, routers). User should have the ability to filter which devices to display.
- Ability to display IoT Virtual network (often on top of a physical view).
- Ability to display specific Element Management parameters (e.g., utilization, devices at faults) based on user selection criteria.
- Ability to filter/configure the topology.
- Ability to retrieve information related to any IoT element.
- Ability to retrieve information related to an IoT protocol.

7.9 Group Manager

Unlike traditional networks, a typical IoT network often contains a large number of IoT devices (e.g., sensors). Hence, it is important to allow network administrators to group IoT elements of the same characteristics into groups instead of managing each element separately.

Group Management is responsible for handling group related requests. The request is sent to manage a group and its membership as well as for any bulk operations, including broadcasting/multicasting, that are supported by the group. Group management security is handled by the element management system.

When facilitating access control using a group, only members with the same access control policy for a resource are included in the same group. Also, only application entities, which have a common role with regard to access control policy, are included in the same group. This is used as a representation of the role when facilitating role based access control.

Group Management Key requirements include:

- Ability to create, retrieve, update, or delete groups. Groups are created by selecting IoT elements of similar characteristics. An IoT element may belong to multiple groups. New members may be added and/or deleted at any time. When new members are added to a group, the group manager should validate if the member complies with the purpose of the group. Requests to create, retrieve, update, or delete are assumed to be initiated by an application.
- Ability to create super group (group of a group). In this case, operations (e.g., Forwarding) are done recursively.
- Ability to initiate and execute a request for the entire members of a group. The request may be a simple notification or read operation (i.e., retrieve information form sensors), or write operation (changing a common parameter).
- Ability to support subscriptions to individual groups.
- Ability to notify group members when they are added to or deleted from a group, or when the group is updated.

7.10   Billing and Accounting

Billing and Accounting management is used to calculate and report the charges based on subscription and/or usage of a service. It supports different charging models including online real-time credit control by interacting with the charging system in the underlying IoT network. Billing polices include the ability to trigger a charge based on specified events and to charge even when the billing system is offline. The system may record information for other purposes such as for event logging. The main charging models include:

- Subscription-based charging (flat rate): Typically a service layer per subscription.
- Event-based charging (per event or task): Charging based on service layer chargeable events. For example, an operation on data (Create, Update, and Retrieve) can be an event.
- Time-based charging: Chargeable events are configurable to initiate information recording. More than one chargeable event can be simultaneously configured and triggered for information recording.
- Usage-based charging: Charge based on bandwidth (or other parameters) consumptions. Users are allowed to change usage level within a task (e.g., high bandwidth for first hour and then switch to lower bandwidth).

Key Billing and Accounting requirements include:

- Ability to bill based on subscription (flat rate), event (per event), time (charge per hour), or usage.
- Ability to allow an application (or network administrator) to develop billing related policies. Further, the Billing and Accounting Module has the ability to start and end the actual billing by applying charging related policies, configurations, and communicating with the charging system in the underlying network.
- Ability to start and end charges based on the defined charges policies. Such charges must be recorded in a billing system/DB.
- Ability to handle offline billing related operations. The offline billing function generates service charging records based on billing polices and recorded information. A service charging record is a formatted collection of information about a chargeable event (e.g., amount of data transferred) for use in billing and accounting.

7.11   Subscription and Notification Manager

Subscription and Notification service provides notifications concerning subscription events. It allows authorized devices and applications to subscribe to a set of notification services, typically from a predetermined list. A notification event may be generic (e.g., a recent security alert) or subscriber-specific (e.g., security alert related to an IoT service and/or device such as end of life date). Subscription and Notification service also provides notifications concerning subscriptions that track event changes on a resource (e.g., deletion of a resource, important change in the resource's events such as a major increase in the temperature reading). The subscription may be provided by the platform itself or by a northbound application communicating with the platform via the API Manager, as shown in Fig. 7.4.

Key requirements for the Subscriptions and Notification Modules include:

- Ability to allow devices and/or applications to subscribe to specific set of services based on right level of authorization. Hence, authorization information may be obtained from the authorization service as we mentioned in Sect. 7.6.5 under Element Management system.[7]
- Ability to allow authorized devices and/or applications to subscribe to a set of notification services from a drop down list.
- Ability to support generic notifications as well as subscriber-specific notifications where notifications are correlated with the subscriber's IoT device or service as mentioned above.
- Ability to support subscription and notification services related to event changes on a resource as mentioned above.
- Ability to provide subscription and notification service in the platform itself and/ or in a northbound application. In the latter case, subscription selection is made in an application that communicates with the platform via the API Manager. Notification may also be sent to such application (if so is selected) via the API Manager.
- Ability to notify devices and/or applications based on subscription and authorization level (e.g., subscribe and notify only for security-related alerts).
- Ability to create and store subscription profile information including device ID, notification address, notification type, notification policies (e.g., notify any time for priority 1 issues, notify from 8 AM to 5 PM for priority 2, etc.).
- Ability to subscribe to a single or multiple resources.
- Ability to store subscription profiles as well as directed notifications along with date, time, and delivery mechanism.

7.12   API Manager

The main function of the API Manager is to manage communication with IoT network and devices, for obtaining network service functions in a common way. It is intended to shield other platform modules from developing their own technology and mechanisms supported by the Underlying Networks.

Key functions of the API Manager include:

---

[7] Alternatively, an Authorization, Authentication and Accounting (AAA) server may be used for device authorization.

- Ability to provide adaptation for different sets of network service functions supported by various Underlying Networks.
- Ability to maintain the necessary connections between the platform entities and the Underlying Network.
- Ability for the API Manager to provide information to the Communication Manager related to the IoT Network so the Communication Manager can include that information determine proper communication handling.

7.13   Commercially Available IoT Platforms

Tens of IoT Platforms exist in the marketplace today. Examples include AWS IoT Platform, Google Cloud IoT Platform, Microsoft Azure IoT Suite Platform, IBM Watson IoT Platform, Salesforce IoT Cloud Platform, Cisco IoT Cloud Connect Platform, Oracle IoT Intelligent Applications Platform, PTC ThingWorx IoT Platform, OpenRemote IoT Platform (open-source focusing on helping engineers creating a range of IoT applications), IRI IoT Voracity Platform (focusing on data discovery, integration, migration, governance, and analytics), Particle Platform, and Altair IoT SmartWorks Platform.

As we mentioned earlier in this chapter, IoT platforms are used to address one or more of the following functions.

1. Rapid and consistent development and deployment of IoT devices and services.
2. Middleware connecting IoT devices and applications to other devices and applications.
3. Streaming data from IoT devices.
4. Profiling customer context data.
5. Device management addressing the FCAPS (Fault, configuration, Accounting, Performance, and Security) functions. See Sect. 7.6 for additional information.
6. Real-time reporting and advanced analytics, e.g., using artificial intelligence algorithms for advanced prediction, service optimization, diagnostics, and trending analysis.
7. Sandbox allowing subject matter experts to test business or technical ideas without (or with limited) programming.
8. Provide API library allowing engineers to import data from other sources (e.g., gateways, Websites, controllers, end application service) and platforms (e.g., using RESTful API).
9. Handle huge data volume from devices, users, applications, websites, and sensors and take actions to give a real-time response.

Selecting the right IoT Platform is challenging and depends greatly on the requirements of the specific solution for hardware, real-time access, custom reports, budget, development skills, and business model.

The purpose of this section is to introduce students and engineers into examples of known IoT platforms and related functionalities. It is not intended to provide recommendations, nor provide feature by feature comparisons. The selected platforms, as shown in Table 7.3, include AWS IoT Platform, Google Cloud IoT Platform, Microsoft Azure IoT Platform, and PTC ThingWorx IoT Platform. The first three are typically considered general-purpose platforms addressing various IoT applications while the last platform (i.e., PTC ThingWorx) is more focused on addressing industrial IoT requirements.

Again, it is important to note that the feature description (Table 7.3) is snapshots at the time of the writing. Such features and capabilities are expected to change over time. Students/engineers are encouraged to log into each platform to understand the latest capabilities.

7.14   Putting All Together

As we mentioned in pervious section, IoT platforms can be divided into two categories: product-centered with a stronger focus on specific products for industrial companies, and general-purpose platform for developers. In many cases, general-purpose platforms are complemented by an accompanying marketplace.

Marketplace is an e-commerce platform owned and operated by a specific vendor (e.g., Amazon). It enables third-party sellers to offer products and/or services online alongside the vendor's regular offerings. This allows the vendor (platform owner) to earn commissions and to create more comprehensive solutions.

Marketplaces have several advantages. First, they bring together offers from multiple suppliers or service providers with minimum investments. Second, they relieve marketplace owners from owning the inventory that their platform sells. Third, they allow platform owners to choose a revenue stream that best fits their market position and business goals. Finally, marketplace owner leaves the more operational side of the business to vendors while focusing on promoting their marketplace brand. Marketplace owners can create a rating and review systems allowing their customers to make informed purchase decisions.

Let us imagine that you are developing an IoT security solution for your own home. In this case, you will need to install and connect your home cameras and sensors to the Internet, select data sources and protocols, and then develop an application for data visualization. You also need to make sure that your data is secure at all time (e.g., data is not altered by third party, your devices are never hacked, and your credentials are always secure) and that your network is reliable and available. You may also chose to combine your data with additional available information (e.g., Weather conditions, Fire and Crime alerts along with Locations) for advanced monitoring especially when you are traveling.

In this case, you will need to subscribe to a platform allowing you to connect your devices, collect data in real-time, and then build (or utilize and existing) interactive dashboards to visualize and track your home data. Such capabilities may be offered by the platform or the associated marketplaces.

Table 7.3 Examples and glimpses of commercially available platforms

	AWS IoT Platform	Google Cloud IoT Platform	Microsoft Azure IoT Platform	PTC ThingWorx IoT Platform
Overview	Almost all IoT platforms allow users to connect their IoT devices and data sources, select supported protocols, build applications, enable security, and define the communication between devices and the Internet.			
Protocols Snapshot	Supports wide variety of communication protocols including custom ones which enable communication b/w devices from different manufacturers, e.g., MQTT. HTTP and WebSockets for asynchronous communication.	Supports wide variety of communication protocols to enable communication between devices from different manufacturers including, e.g., MQTT, HTTP.	Supports wide variety of communication protocols to enable communication between devices from different manufacturers, e.g., AMQP, HTTPS and AMQP. IoT hub Supports SASL and AMQP claim based security in conjunction with AMQP protocol.	Supports wide variety of communication protocols to enable communication between devices from different manufacturers, e.g., MQTT, HTTP, OAuth2, and WebSockets.
Element Management: Fault Management Snapshot	**AWS IoT Device Manager** allows users to troubleshoot device functionality and query the state of IoT devices.	**Google Cloud IoT Core** supports trouble management, e.g., predicting when equipment needs maintenance.	**Microsoft Azure IoT Monitor** provides guidance to reduce the time in diagnosing and troubleshooting.	Supports various functions for troubleshooting including connections to the platform.
Element Management: Configuration Management Snapshot	Users can query the state of device(s) on demand and provide the functionality to apply firmware updated over-the-air.	The device manager allows devices to be configured (in group) through a console or programmatically.	Azure IoT Hub Device Provisioning Service enables zero-touch provisioning to the right IoT Hub.	Includes utilities to provision devices. Allows users to create rule-based Workflows to execute across multiple devices.
Element Management: Accounting and Billing Snapshot	Basic connectivity fee (for platform access) and then usage-based billing (bay for what you use).	Usage-based: Cloud IoT Core is priced according to the data volume.	Basic and standard tier-based billing model. e.g., $0.123 per 1000 operation for device provisioning.	Subscription based with a pay-as-you-go model is supported.

(continued)

Table 7.3   (continued)

	AWS IoT Platform	Google Cloud IoT Platform	Microsoft Azure IoT Platform	PTC ThingWorx IoT Platform
Element Management: Performance Management Snapshot	**AWS IoT Device Manager** monitors, organizes, and provides an interface to manage IoT devices. It provides functionality to register an individual device or in bulk and manage security permissions/ policies.	**Google Cloud IoT Core** provides a solution for collecting, processing, analyzing, and visualizing IoT data in real time. E.g., automatically optimize device performance in real time while predicting downtime.	**Azure Monitor and Resource Health** provides monitoring capabilities with data about the operations of Azure IoT Hub, for instance. Advanced analytics features that can turn connectivity and workflow data into actionable insights.	**ThingWorx Platform** allows user to select data and use it to create specific charts and workflow alerts. Advanced analytics features that can turn connectivity and workflow data into actionable insights.
Element Management: Security Management Snapshot	Data to and from AWS IoT is sent securely over Transport Layer Security (TLS). AWS cloud security mechanisms protect data as it moves between AWS IoT and other AWS services.	Allows users to securely connect, manage, and ingest data using TLS.	Uses TLS based handshake and encryption. Support various security functions including security information and event management, security orchestration, and automation.	Provides transport security, identity management (device and platform), and content & asset management.
	Supports device authentication and authorization (via custom schemes).	Supports device authentication and authorization (via keys and JSON web tokens).	Supports device authentication and authorization (via certificates and keys).	Supports device authentication and authorization.
	Supports various compliance management for security audits.	Supports various compliance management for security audits.	Supports various compliance management for security audits.	Supports various compliance management for security audits.

In general, the following steps are followed:

1. Install your devices.
2. Connect devices to the platform.
3. Select data sources and formats.
4. Add a custom data source via Developer Console (if applicable).

5. Use standard dashboard capabilities (or add a custom dashboard via a developer console) to create your view. Connect dashboard to data source using an offer interface.
6. Continue adding your devices and data sources to enable complete visualization.
7. Customize your dashboard if needed (e.g., Drag and drop widgets into the desired dashboard location, add custom colors).
8. Share dashboard with family members, e.g., adding users with read-only or edit access.
9. Add additional advanced capabilities (if needed).

7.15   Summary

Without a doubt, the IoT Services Platform creates the cornerstone of successful IoT solutions. It is responsible for many of the most challenging and complex tasks of the solution. The Services Platform automates the ability to deploy, configure, troubleshoot, secure, manage, and monitor IoT entities ranging from sensors to applications in terms of firmware installation, patching, debugging, and monitoring just to name a few. The Service Platform also provides the ability for data management and analytics, temporary caching, permanent storage, data normalization, policy-based access control and exposure.

Given the complexity of the services platform in IoT, this chapter grouped the core capabilities into 11 main areas: Platform Manager, Discovery and Registration Manager, Communication (Delivery Handling) Manager, Data Management and Repository, Firmware Manager, Topology Management, Group Management, Billing and Accounting Manager, Cloud Service Integration Function/Manager, API Manager, and Element Manager addressing Configuration Management, Fault Management, Performance Management and Security Management across all IoT entities.

Problems and Exercises

1. This chapter categorized the IoT Services Platform into 11 functions. (a) Name and define each of the 11 functions. (b) List and define the Element Manager functions.
2. What are the traditional FCAPS management functions? Do they also apply to IoT? If so, Are they sufficient?
3. List six reasons why the overall management functions of IoT solutions are more multifaceted than traditional networks.
4. IoT solutions are considered much more complex to manage than traditional networks?

   (a) Why?—List top five factors.
   (b) Why does the Fog Layer introduce new changes for IoT?

5. This chapter mentioned that not all IoT entities will be IP address enabled.

   (a) Why is that? Provide an example of IoT devices that are not IP addresses enabled.
   (b) How do management system track such devices?

6. What is device registration on IoT? Why is it needed?
7. List the key responsibilities of the Discovery Function.
8. It was mentioned in Sect. 5.1 that for non-IP addressed enabled sensors, IoT sensors may be tracked by the combined (a) IP Address of the Gateway and (b) Sensor address. Why both addresses do are needed?
9. Why IoT device self-registration is preferred over the method where a new IoT device have the capability to be identified during the discovery process?
10. The IETF has released NETCONF and YANG which are standards focusing on Configuration management. Name two other older methods that can be used for configuration management? What are their shortcomings?
11. Section 7.7 indicated that Accurate discovery is essential for many management tasks including asset management, network monitoring, network diagnosis and fault analysis, network planning, high availability, and others.

    (a) Provide short definitions of asset management, network monitoring, network diagnosis and fault analysis, network planning and high availability.
    (b) Why is accurate discovery essential for each of the above functions?

12. What are the key differences between Provisioning and Configuration functions? Which one is done first?
13. What are key differences between deployment, Provisioning, and Orchestration?
14. What are the most basic two management functions to provide a new services?
15. Provide an example of Service-Level Diagnostics and Fault Management Function in IoT where all devices are working correctly but the service-level parameters are not being met.
16. Why Fault management is considered by many experts to be most challenging and important management function of IoT-based networks?
17. What are the three main functions of Fault Management? Provide detailed description of each term.
18. What are the concepts of fault tolerance in IoT networks? Give three examples of failures that should be handled by fault tolerance function in IoT-based networks.
19. Fault tolerance is not just a property of individual IoT element; it may also impact the IoT communication protocol. For example, the Transmission Control Protocol (TCP) was design as reliable two-way communication protocol, even in the presence of failed or overloaded communications links. How is this achieved in TCP?
20. There are special software and instrumentation packages designed to detect failures. A good example is a fault masking system. How does Fault Masking system detect failure?
21. What is Diagnostic Signature? Where it used?

22. In priority order, what are the top three IoT management functions that a service provider needs to provide very basic services? Justify your answer.
23. Why Fault management is considered to be very challenging in IoT network? i.e., What are the main differences between managing IoT network and a traditional network?
24. Why IoT management is considered to be most challenging and complex task of the solution?
25. Section 7.1 indicated the need for a complete configuration backups with rollback capabilities as a key requirement for the IoT Platform Manager. What is configuration roll-back? Why is it needed? Provide an example?
26. What are the definitions of Sensitivity and Dynamic Range? What are the typical units of Sensitivity and Dynamic Range?
27. What is Hysteresis? What is a typical unit of Hysteresis?
28. What is a Firmware? What does it do? Why is it called so?
29. Why Firmware Images are loaded into ROM and not the device storage?
30. How come Firmware Management was not part of the tradition FCAPS?
31. Data may be retrieved from various IoT sources including IoT devices and network elements (e.g., sensors, gateways, switches), IoT subscribers, and IoT applications. IoT device and network element data is assumed to be collected by collection systems or by collection agents.

    (a) What are the key differences between a collection system and a collection agent?
    (b) What is IoT subscriber data? How is the data collected?
    (c) What is an IoT application data? How is the application data collected?

32. In a table list three Subscription and Notification requirements along with examples of a subscriber and notification message.

References

1. IoT – Converging Technologies for Smart Environments and Integrated Ecosystems, Reviewer Publishers, Online: http://www.internet-of-things-research.eu/pdf/Converging_Technologies_for_Smart_Environments_and_Integrated_Ecosystems_IERC_Book_Open_Access_2013.pdf
2. Internet of Things, Evolving the Manufacturing Industry, Online: http://www.cisco.com/c/en/us/solutions/internet-of-things/iot-products/services.html
3. The Internet of Things: Between the Revolution of the Internet and the Metamorphosis of Objects, Gerald Santucci, Online: http://cordis.europa.eu/fp7/ict/enet/documents/publications/iot-between-the-internet-revolution.pdf
4. From the Internet of Computers to the Internet of Things, Friedemann Mattern and Christian Floerkemeier, Distributed Systems Group, Institute for Pervasive Computing, Online: http://www.vs.inf.ethz.ch/publ/papers/Internet-of-things.pdf
5. Reaping the Benefits of the Internet of Things, Cognizant Reports, May 2014, http://www.cognizant.com/InsightsWhitepapers/Reaping-the-Benefits-of-the-Internet-of-Things.pdf
6. Philip N. Howard (8 June 2015). "How big is the Internet of Things and how big will it get?". *The Brookings Institution*. Retrieved 26 June 2015.

7. Stefan Wallina and Claes Wiksrom, "Automating Network and Service Configuration Using NETCONF and YANG": Online: http://www.tail-f.com/wordpress/wp-content/uploads/2013/03/Tail-f-NETCONF-YANG-Service-Automation-LISA-Usenix-2011.pdf

8. BJORKLUND, M. YANG - A Data Modeling Language for the Network Configuration Protocol (NETCONF). RFC 6020, Oct. 2010.

9. Broadband Forum Technical Report TR-069, CPE WAN Management Protocol, Issue 1 Amendment 5, Version 1.4, Nov 2013.

10. Open Mobile Alliance M2M Device Management Specifications, Online: http://openmobilealliance.hs-sites.com/lightweight-m2m-specification-from-oma

11. Open Mobile Alliance LightweightM2M Version 1.0, Online: http://technical.openmobilealliance.org/Technical/technical-information/release-program/current-releases/oma-lightweightm2m-v1-0

12. Sokullu, R. and Karaca, O., "Fault Management for Smart Wireless Sensor Networks," Ubiquitous Intelligence & Computing and 9th International Conference on Autonomic & Trusted Computing (UIC/ATC), Sept 4, 2012.

13. G. Stanley and Associate, White Paper, "A Guide to Fault Detection and Diagnosis", Online: http://gregstanleyandassociates.com/whitepapers/FaultDiagnosis/faultdiagnosis.htm

14. IETF Network Working Group, Request for Comments (RFC) 3433, Entity Sensor Management Information Base, Online: https://tools.ietf.org/html/rfc3433

15. G. Huston, "Measuring IP Network Performance", the Internet Protocol Journal, Vol 6, Number 1, Online: http://www.cisco.com/web/about/ac123/ac147/archived_issues/ipj_6-1/measuring_ip.html

16. H Hui-Ping, X Shi-De, M Xiang-Yin, "Applying SNMP Technology to Manager Sensors in IoT", The Open cybernetics & Systemic Journal, 2015, pp. 1019-1024, Online: http://benthamopen.com/contents/pdf/TOCSJ/TOCSJ-9-1019.pdf

17. L. Adaro, Monitoring 101 eBook, Nov 2015, Online: https://thwack.solarwinds.com/docs/DOC-187523

18. Stanford Sensor Course, Online: http://web.stanford.edu/class/me220/data/lectures/lect02/lect_2.html

19. B. Hedstrom, A. Watwe, S. Sakthidharan "Protocol Efficiencies of NETCONF versus SNMP for Configuration Management Functions", University of Colorado, May 2011, Online: http://morse.colorado.edu/~tlen5710/11s/11NETCONFvsSNMP.pdf

20. OMA LightweightM2M v.10, Open Mobile Alliance, Online: http://technical.openmobilealliance.org/Technical/technical-information/release-program/current-releases/oma-lightweightm2m-v1-0

21. S. Duquet, Smart Sensors, Enabling Detection and Ranging for IoT and Beyond, Ladder Technology Magazine Elektronik Praxis, April 2015, Online: http://leddartech.com/smart-sensors

22. 50 Sensors Applications for Smarter World, Libelium, Online: http://www.libelium.com/top_50_iot_sensor_applications_ranking/

23. P. Seneviratne, Internet Connected Smart Water Sensors, September 2015, Online: https://www.packtpub.com/books/content/internet-connected-smart-water-meter

24. P. Jain, Pressure Sensors, Prototype PCB from $10, Online: http://www.engineersgarage.com/articles/t

25. D. Merrill, J. Kalanithi, P. Maes, "Siftables: Towards Sensor Network User Interfaces", Online: http://alumni.media.mit.edu/~dmerrill/publications/dmerrill_siftables.pdf

26. Whatis.com, Online: http://whatis.techtarget.com/definition/firmware

27. Mobileburn, Online: http://www.mobileburn.com/definition.jsp?term=firmware

Chapter 8
Internet of Things Security and Privacy

8.1 Introduction

The Internet of Things (IoT) promises to make our lives more convenient by turning each physical object in our surrounding environment into a smart object that can sense the environment, communicate with the remaining smart objects, perform reasoning, and respond properly to changes in the surrounding environment. However, the conveniences that the IoT brings are also associated with new security risks and privacy issues that must be addressed properly. Ignoring these security and privacy issues will have serious effects on the different aspects of our lives including the homes we live in, the cars we ride to work, and even the effects that will reach our own bodies.

If your home does not already have a smart meter, it will soon have multiple of those meters that are dedicated to monitor and control the power consumption, the heating, and the lighting of your house. This is not to mention the smart gadgets that will be found all over your house such as the smart camera that notifies your smartphone during business hours when movement is detected, the smart door that opens remotely, and the smart fridge that notifies you when you are short of milk. Imagine now the level of control that an attacker can gain by hacking those smart meters and gadgets if the security of those devices was overlooked. In fact, the damage caused by cyberattacks in the IoT era will have a direct impact on all the physical objects that you use in your daily life. The same applies to your smart car as the number of integrated sensors continues to grow rapidly and as the wireless control capabilities increase significantly over time, giving an attacker who hacks the car the ability to control the windshield wipers, the radio, the door lock, and even the brakes and the steering wheel of your car. Our bodies will not also be safe from cyberattacks. In fact, researchers have shown that an attacker can control remotely the implantable

© The Author(s), under exclusive license to Springer Nature Switzerland AG 2022
A. Rayes, S. Salam, *Internet of Things from Hype to Reality*,
https://doi.org/10.1007/978-3-030-90158-5_8

and wearable health devices (e.g., insulin pumps and heart pacemakers) by hacking the communication link that connects them to the control and monitoring system. This gives the attacker, for example, the ability to tune the injected insulin dose causing serious health problems that may even cause death to patients wearing those smart health devices. In fact, such concerns have made doctors disable the wireless capability of the heart pacemaker of Dick Cheney, the former US vice president, in order to protect him from such malicious attacks.

The security risks are also extremely serious when IoT devices are used in business enterprises. If an attacker hacks any of those smart objects that are used in a big enterprise, then the sensing capabilities that those smart objects have can be used by the attacker to spy on the enterprise. Such cyberattacks can also be used to steal sensitive information such as the company earnings report and credit card information. In fact, these stealing attacks are common in big enterprises such as the largest financial hacking case in the US history, which took place in 2013, where a group of five hackers stole $160 million from credit cards and over hundreds of millions in criminal loot.

Maintaining users' privacy in IoT is also crucial as there is an enormous amount of information that an outsider can learn about people's life by eavesdropping on the sensed data that their smart house appliances and wearable devices report. In fact, people will be living in a "Big Brother" world where smart things record our daily activities anytime and everywhere. The advances in the fields of facial, speech, and human activity recognition amplify the amount of information that the sensed data can reveal if it falls in the wrong hands. Even if your IoT objects are merely reporting metadata, you would be surprised by the amount of information that an outsider can learn about your personal life when aggregating the metadata collected from multiple hacked objects that surround you over time. It is thus essential to find solutions to preserve people's privacy in the IoT era.

The objective of this chapter is to shed the light on some of the security and privacy issues that the IoT paradigm is exposed to. We also survey the techniques that were proposed to address these issues. Some of the discussed techniques prevent security breaches from taking place, while others try to detect malicious behavior and trigger an appropriate mitigating countermeasure. The rest of the chapter is organized as follows. Section 8.2 identifies the new security challenges that are encountered in the IoT paradigm. Section 8.3 identifies the IoT security requirements. Section 8.4 briefly describes the three domains in the IoT architecture. Sections 8.5–8.7 survey the security attacks and countermeasures at the cloud domain, the fog domain, and the sensing domain, respectively. Section 8.8 discusses approaches for securing IoT Devices. The section starts by providing several examples of IoT devices used in security attacks, and then discusses solutions including MUD and DICE. Finally, Sect. 8.9 summarizes the chapter and provides directions for future work related to the area of IoT security.

8.2 IoT Security Challenges

IoT has unique characteristics and constraints when it comes to designing efficient defensive mechanisms against cybersecurity threats that can be summarized by:

1. *Multiple Technologies*: IoT combines multiple technologies such as radio-frequency identification (RFID), wireless sensor networks, cloud computing, virtualization, etc. Each of these technologies has its own vulnerabilities. The problem with the IoT paradigm is that one must secure the chain of all of those technologies as the security resistance of an IoT application will be judged based on its weakest point which is usually referred to by Achilles' heel.
2. *Multiple Verticals*: The IoT paradigm will have numerous applications (also called verticals) that span eHealth, industrial, smart home gadgets, smart cities, etc. The security requirements of each vertical are quite different from the remaining verticals.
3. *Scalability*: According to Cisco, 26.3 billion smart devices will be connected to the Internet by 2020. This huge number makes scalability an important issue when it comes to developing efficient defensive mechanisms. None of the previously proposed centralized defensive frameworks can work anymore with the IoT paradigm, where the focus must be switched to finding practical decentralized defensive security mechanisms. An IoT solution needs to scale cost-effectively, potentially to hundreds of thousands or even millions of endpoints.
4. *Availability:* Availability refers to characteristic of a system or subsystem that is continuously operational for a desirably long period of time. It is typically measured relative to "100% operational" or "never failing." A widely held but difficult-to-achieve standard of availability for a system or product is known as "five 9 s" (available 99.999% of the time in a given year) availability. Security plays a major rule in high availability as network administrators often hesitate to use needed threat-response technology functions (e.g., network discovery as illustrated in Chap. 7) for fear that such functions will take down critical systems. Even a simple port scan causes some IoT devices to stop working, and the cost of downtime can far exceed the cost of remediating all but the most severe incidents. In some instances, network administrators would rather have no cybersecurity protection rather than risk an outage due to a false positive. This leaves them blind to threats within their control networks. Companies often add redundancy to their systems so that failure of a component does not impact the entire system.
5. *Big Data*: Not only the number of smart objects will be huge, but also the data generated by each object will be enormous as each smart object is expected to be supplied by numerous sensors, where each sensor generates huge streams of data over time. This makes it essential to come up with efficient defensive mechanisms that can secure these large streams of data.
6. *Resource Limitations*: The majority of IoT end devices have limited resource capabilities such as CPU, memory, storage, battery, and transmission range. This makes those devices a low-hanging-fruit for denial of service (DoS) attacks

where the attacker can easily overwhelm the limited resource capabilities of those devices causing a service disruption. In addition to that, the resource limitations of those devices raise new challenges when it comes to developing security protocols especially with the fact that the traditional and mature cryptography techniques are known to be computationally expensive.

7. *Remote Locations*: In many IoT verticals (e.g., smart grid, railways, roadsides), IoT devices, epically sensors, will be installed in unmanned locations that are difficult to reach. Attackers can interfere with these devices without being seen. Cyber and physical security monitoring systems must be installed in safeguarded location, operate in extreme environmental conditions, fit in small spaces, and operate remotely for routine updates and maintenance avoiding delayed and expensive visits by network technicians.

8. *Mobility*: Smart objects are expected to change their location often in the IoT paradigm. This adds extra difficulties when developing efficient defensive mechanisms in such dynamic environments.

9. *Delay-Sensitive Service*: The majority of IoT applications are expected to be delay-sensitive, and thus one should protect the different IoT components from any attack that may degrade their service time or may cause a service disruption.

8.3   IoT Security Requirements

We summarize in this section the security requirements for IoT. These requirements include:

- *Confidentiality*: ensures that the exchanged messages can be understood only by the intended entities.
- *Integrity*: ensures that the exchanged messages were not altered/tampered by a third party.
- *Authentication*: ensures that the entities involved in any operation are who they claim to be. A masquerade attack or an impersonation attack usually targets this requirement where an entity claims to be another identity.
- *Availability*: ensures that the service is not interrupted. Denial of service attacks target this requirement as they cause service disruption.
- *Authorization*: ensures that entities have the required control permissions to perform the operation they request to perform.
- *Freshness*: ensures that the data is fresh. Replay attacks target this requirement where an old message is replayed in order to return an entity into an old state.
- *Non-repudiation*: ensures that an entity cannot deny an action that it has performed.
- *Forward Secrecy*: ensures that when an object leaves the network, it will not understand the communications that are exchanged after its departure.
- *Backward Secrecy*: ensures that any new object that joins the network will not be able to understand the communications that were exchanged prior to joining the network.

8.4 IoT Three-Domain Architecture

Before introducing IoT security issues, we briefly describe in this section the three-domain architecture that we consider in our security analysis.

As illustrated in Figs. 8.1 and 8.2, the architecture is made up of the following three domains:

1. *IoT Sensing Domain*: This domain is made up of all the smart objects that have the capability to sense the surrounding environment and report the sensed data to one of the devices in the fog domain. The smart objects in the sensing domain are expected to change their location over time.

Fig. 8.1 Mapping of IoT domains

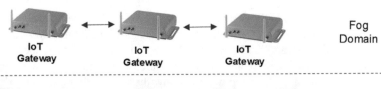

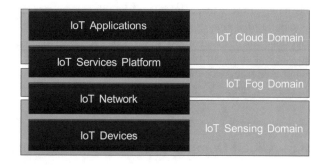

Cloud Domain

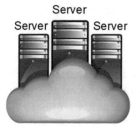

Server
Server   Server

Fog Domain

IoT Gateway   IoT Gateway   IoT Gateway

Sensing Domain

Fig. 8.2 The IoT domains

2. *Fog Domain*: This domain consists of a set of fog devices that are located in areas that are highly populated by many smart objects. Each fog device is allocated a set of smart objects where the allocated objects report their sensed data to the fog device. The fog device performs operations on the collected data including aggregation, preprocessing, and storage. Fog devices are also connected with each other in order to manage the communication among the smart objects and in order to coordinate which fog device will be responsible for handling which object as objects change their location over time. Each fog device is also connected to one or multiple servers in the cloud domain.

3. *Cloud Domain*: This domain is composed of a large number of servers that host the applications that are responsible for performing the heavy-computational processing operations on the data reported from the fog devices.

We analyze in the following sections the security attacks and countermeasures at each one of those three domains. We follow a top-down order where we describe the attacks and countermeasures that are encountered at the cloud domain, the fog domain, and the sensing domain. For each one of those domains, we identify the most popular security attacks and then describe how these attacks are launched, what vulnerabilities they exploit, and what countermeasure techniques can be used to prevent, detect, or mitigate those attacks.

8.5   Cloud Domain Attacks and Countermeasures

As mentioned earlier, the cloud domain holds the IoT applications that are performing different operations on the data collected by the IoT objects. Each IoT application is dedicated one or multiple virtual machines (VMs) where each VM is assigned to one of the servers in the cloud data center and gets allocated certain amount of CPU and memory resources in order to perform certain computing tasks. The cloud data center is made up of thousands of servers where each server has certain CPU, memory, and storage capacities, and thus each server has a limit on the number of VMs that it can accommodate. The servers in the cloud data center are virtualized which allows multiple VMs to be assigned to the same server as long as the server has enough resource capacity to support the resource requirements of each hosted VM. Figure 8.3 shows an illustration of how multiple VMs can be assigned to the same server, thanks to virtualization (more details on virtualization were discussed in Chap. 6). Each IoT application is hosted on a VM that has its own operating system (OS). The hypervisor (sometimes also called the virtual machine manager) monitors those running VMs and manages how these VMs share the server's hardware. The hypervisor also provides the logical separation among the VMs and also separates each VM from the underlying hardware. The hypervisor has also a migration module that manages how to move a VM that is currently hosted on the server to another server. The migration module also manages the reception of a VM that is moved from other servers.

Fig. 8.3 Illustration of how multiple IoT applications can be hosted on the same server, thanks to virtualization

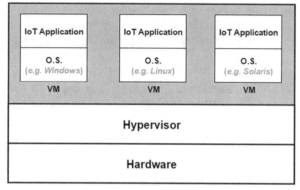

Cloud Server

Cloud computing is considered a high-risk environment for many businesses and consumers as they feel its perimeter cannot be defined nor controlled. In addition, many government agencies must comply with regulatory statutes, such as the Health Insurance Portability and Accountability Act (HIPAA), the Sarbanes-Oxley Act of 2002 (SOX), and the Federal Information Security Management Act (FISMA). The IoT applications running in the cloud domain are susceptible to numerous security attacks. We summarize next the most popular ones:

1. *Hidden-Channel Attacks:* Although there is a logical separation among the VMs running on the same server, there are still some hardware components that are shared among those VMs such as the cache. This opens opportunities for data leakage across the VMs that reside on the same server. Three steps are followed by the attacker in order to leak information from a target VM. These three steps are explained next:

   (a) *Step1: Mapping Target VM*: The first step toward launching an attack against a VM in a cloud data center is to locate where the target VM resides. A cloud data center is typically divided into multiple management units called clusters, where each cluster is located in a certain geographical location and is made up of thousands of servers. Each cluster is divided into multiple zones (sometimes called "pods") where each zone consists of a large number of servers. Although clients have the choice to specify in which cluster their VM resides, they do not have control on selecting the zone or the server within the zone where their VM will reside as this decision is made based on the cloud provider's scheduling algorithm which is not released publicly. In order to know where a target VM resides, the attacker needs only to know the external IP address of that VM where each VM hosted on the cloud has usually two IP addresses: an external address used to communicate with any entity that is located outside the cloud cluster and an internal address used only within the cloud cluster and is only visible within the cloud cluster. The attacker can infer based on the VM's external

IP address on what cluster the VM resides, as cloud clusters are usually placed in different geographical locations and have different IP addresses. Now in order to identify in what zone within the cluster the target VM resides, the attacker needs to know the target VM's internal IP address as the internal IP addresses for all VMs within the same zone have the same network prefix. In order to identify the VM's internal IP address, the attacker rents a VM in the same cluster as the one where the VM resides. The rented VM is then used to query the DNS server of the cloud cluster where the internal IP address of the target VM can be fetched. By observing the internal IP address of the target VM in the DNS query, the attacker can tell what zone within the cloud cluster the VM is hosted in.

(b) **Step2**: **Malicious VM Placement**: having identified on what cluster and on what zone the target VM resides, the next step toward launching an attack against the target VM is to place a malicious VM on the same server where the target VM resides. In order to do that, the attacker rents a VM in the same cluster as the target VM. The cloud provider's scheduling algorithm places the rented VM on one of the servers within one of the cluster's zones. The attacker performs a traceroute from the rented VM to the target VM where the routing path that separates the rented VM and the target VM is identified. If the identified routing path shows multiple hops that separate the target VM and the rented VM, then the attacker knows that the rented VM was not placed on the same server as the target VM. The attacker then releases the rented VM and requests a new one. The cloud provider's scheduling algorithm selects a server to host the requested VM. The attacker performs a traceroute from the new rented VM to the target VM in order to know whether or not the target VM and the new rented VM reside on the same server. The attacker continues releasing then renting new VMs and performing a traceroute until he/she identifies that the cloud provider's scheduling algorithm has placed the rented VM on the same server as the target VM.

(c) **Step3**: **Cross-VM Data Leakage**: Having placed a malicious VM on the same server as the target VM, the attacker now tries to learn some information about the target VM by exploiting the fact that although VMs are separated logically, thanks to virtualization, they still share certain parts of the server's hardware such as the instruction cache and the data cache. The attacker can now, for example, learn what lines of cache (data or instruction) the target VM has accessed recently. This can be done as follows. When the shared cache is assigned to the malicious VM that is under the control of the attacker, the attacker fills the whole shared cache by dummy data. The malicious VM then yields the shared cache to the target VM which performs some data access operations. The malicious VM sends an interrupt after a short time from yielding the cache to the target VM asking to assess the cache so that the target VM yields the cache for the malicious VM. Now the malicious VM probes the different lines of the cache asking to fetch the dummy data that were previously filled in the cache. By observing the time it takes to access each chunk of the dummy data, the malicious VM can tell which chunks of the dummy data were fetched from the cache and which chunks were fetched from memory as they were replaced by data that was accessed by the target VM. This gives information to the malicious VM about what addresses the target VM has accessed recently. Knowing what addresses the target VM accesses over time can help the malicious VM recover parts of the security keys that the target VM is using.

(d) Different countermeasures can be taken to prevent hidden-channel attacks from taking place. The first twos steps needed to launch this attack (mapping the target VM and placing a malicious VM on the same server as the target VM) can be prevented by not allowing the VMs hosted in the cloud data center to send probing packets such as traceroute packets. Preventing data from being leaked across VMs that are hosted on the same server can be achieved by one of the following techniques:

- *Hard Isolation*: The basic idea behind this preventive technique is to maintain high levels of isolation among the VMs. One way to do this is to separate the cache dedicated for each VM through hardware or software. Another way to achieve hard isolation is by assigning only one VM to each server. Although this completely prevents data leakages across VMs, it is not a practical solution as it leaves the servers within the cloud data center underutilized. A better way to achieve hard isolation is by letting each cloud client specify a list of trusted cloud users called the *white list*. The cloud client is fine with sharing the server with only the VMs belonging to the *white list users*. New scheduling algorithms are needed in that case in order to decide on what server each VM should be placed such that the security constraints of each VM that are specified by the white and black lists are met. A key limitation of this technique is that each VM must have a list of identified untrusted VMs.

- *Cache Flushing*: This technique flushes the shared cache every time the allocation of the cache is switched from a VM to another. The downside of this countermeasure is that the VMs running on the server will experience frequent performance degradation as the shared cache will be emptied every time a switch from a VM to another occurs, which increases the time needed to access and fetch data.

- *Noisy Data Access Time*: This technique adds random noise to the amount of time needed to fetch data, which makes it hard to tell whether or not the data was fetched from the cache or from the memory. By doing this, it becomes harder for a malicious VM to identify what segments of the cache were populated by another VM that shares the same server. Of course this has a price as the fetched data gets delayed a little bit due to the noise (variable time delay) that is added to the time needed to fetch the data.

- *Limiting Cache Switching Rate*: A mitigation technique to limit the amount of data that can be leaked across VMs can be achieved by limiting how often the cache is switched from a VM to another. The idea here is that if the cache is not switched from a VM to another too soon, then the content of the cache will be modified a lot by the VM that possess the cache. This makes it hard for another VM to attain fine-grained knowledge of what data the previous VM has accessed when probing the cache.

2. *VM Migration Attacks*: The virtualization technology supports live *VM migration*, which allows moving a VM transparently from a server to another. The term *live* refers here to the fact that the application running on the VM is disrupted for a very short duration due to this migration where the disruption is as low as hundreds of milliseconds. Before delving into the security issues that VM migration brings, we explain briefly the mechanism for performing VM migration and the scenarios where VM migration is usually performed.

   The mechanism of moving a VM from a source server to a destination server is done by copying the VM's memory content. The VM's hard disk content does not need to be copied as it is usually stored on a network-attached storage (NAS) device and can be accessed from any location within the cloud cluster. If the destination server where the VM will be moved to lies on the same local network as the source server, then the VM keeps the same IP address even after migration in order to avoid the need for communication redirection. Maintaining the same IP address even after moving to another server is done after copying the memory content of the VM by sending a gratuitous ARP reply packet that informs the routing devices within the cloud about the VM's new physical address, so that any packet destined to the VM's IP address gets routed to the VM's new location on the destination server. Each server has a dedicated module in the hypervisor called the VM migration module that is responsible for sending the VM content for the source server or receiving the VM's memory content for the destination server.

   VM migration is very useful in multiple scenarios. Consider, for example, the case when a server that is hosting some VMs needs to be taken offline for maintenance or for patch installation. VM migration can be used in this case to move all the VMs currently running on the server into other servers so that the server can be taken down for maintenance without terminating the running VMs that are hosted on that server. VM migration is also a very useful tool for managing the servers in the cloud data center where it can be used to balance the workload among the servers or to consolidate the scheduled VMs on fewer number of powered servers so that a larger number of servers can be powered down to save energy. However, the conveniences that VM migration brings raise new security threats. The attacks that exploit VM migration can be divided into two subcategories based on the target plane:

   (a) *Control Plane Attacks*: These attacks target the module that is responsible for handling the migration process on a server which is called the migration module that is found in the hypervisor. By exploiting a bug in the migration module software, the attacker can hack the server and take full control over the migration module. This gives the attacker the ability to launch malicious activities including:

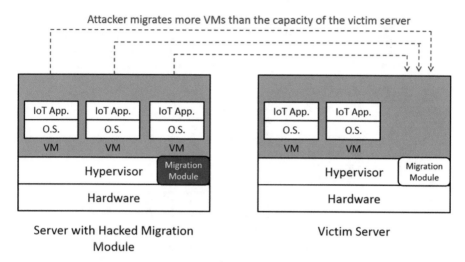

Fig. 8.4 Illustration of the migration flooding attack

- *Migration Flooding*: This attack is illustrated in Fig. 8.4 where the attacker moves all the VMs that are hosted on the hacked server to a victim server that does not have enough resource capacity to host all the moved VMs. This causes a denial of service of the applications running in the VMs of the victim server as there will not be enough resources to satisfy the demands of all the hosted VMs leading into VM performance degradation and VM crashes.

- *False Resource Advertising*: The hacked server claims that it has a large resource slack (a large amount of free resources). This attracts other servers to off-load some of their VMs to the hacked server so that the cloud workload gets distributed over the cloud servers. After moving VMs from other servers to the hacked server, the attacker can exploit other vulnerabilities to break into the offloaded VMs as now these VMs are placed on a server that is under the control of the attacker.

(b) *Data Plane Attacks*: These constitute the second type of VM migration attacks, and those attacks target the network links over which the VM is moved from a server to another. Examples of data plane attacks include:

- *Sniffing Attack*: where an attacker sniffs the packets that are exchanged between the source and destination and reads the migrated memory pages.
- *Man-in-the-Middle Attack*: the attacker fabricates a gratuitous ARP reply packet similar to the one that is usually sent when a VM moves from a server to another. This fabricated ARP packet informs the routing devices that the physical address where the victim VM resides was changed to become the physical address of the attacker's malicious VM. Now the incoming packets that are destined to the victim get routed to the new physical address where the attacker resides. The attacker can then pas-

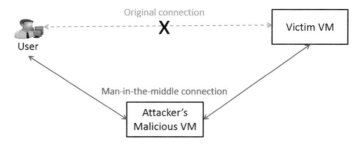

Fig. 8.5 Man-in-the-middle attack

sively monitor the received packets while continuing to forward them to the actual physical address where the victim VM resides so that the victim does not detect that any malicious activity is going on. The attacker can also modify the content of the received packets if the integrity of the packets is not protected by any security mechanism. An illustration of the man-in-the-middle attack is shown in Fig. 8.5.

- Having explained the VM migration attacks, we now discuss the possible countermeasures. Unfortunately, little attention was given to secure VM migration where the focus was more on how to optimize the performance degradation or the energy overhead associated with those migrations. In order to secure VM migration, mutual authentication should be performed between the server initiating the migration and the server that will be hosting the migrated VM. The control messages that are exchanged between the servers to manage the migration should also be encrypted and signed by the entity that is generating those control messages in order to avoid altering the content of those control messages and in order to prevent other entities from fabricating fake control messages. Sequence numbers or timestamps should also be included in the exchanged control messages in order to prevent a malicious entity from replaying an old control message that was sent earlier. Also, gratuitous ARP Reply packets that update the physical address of the VM should be accepted only after authentication in order to prevent man-in-the-middle attacks. The reader interested in learning more about VM migration attacks and countermeasures is referred to [19] for further information on this topic.

3. *Theft-of-Service Attack*: In this attack a malicious VM misbehaves in a way that makes the hypervisor assigns to it more resources than the share it is supposed to obtain. This extra allocation of resources for the malicious VM comes at the expense of the other VMs that share the same server as the malicious VM, where these victim VMs get allocated less share of resources than what they should actually obtain, which in turn degrades their performance.

   Xen is a well-known hypervisor that is susceptible to this attack. One of the main roles of Xen hypervisor is to decide to which VM among the ones running

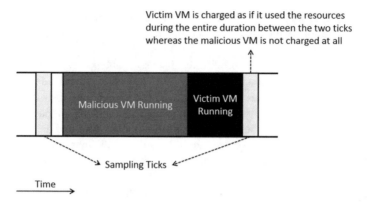

Fig. 8.6   Illustration of the theft-of-service attack

on the server each physical core should be assigned to over time. In order to do that, Xen samples every 10 ms to check the VMs that are utilizing the cores. Xen then assumes that the VM that is detected to be using one of the cores at the sampling time has been using the server's core during the entire 10 ms. The hypervisor then calculates how much time each VM has been assigned the cores. VMs that utilized the cores less than the remaining VMs are given higher priority to utilize the server's core in the future in order to guarantee a fair allocation of the shared resources.

The fact that Xen performs periodic sampling can be exploited by a malicious VM by using one of the cores at times other than the sampling time. As illustrated in Fig. 8.6, the malicious VM can yield the acquired core to another VM shortly before the sampling tick. The hypervisor then assumes that the other VM that has yielded the core has been using the core during the entire 10 ms. The malicious VM does not get logged as using the core and thus keeps having high priority to use the cores in the future.

Two countermeasures were proposed to handle this attack. The first countermeasure is to log more accurately the start and end time when each VM was utilizing the cores using accurate clocks. Another solution is to randomize the sampling times.

4. *VM Escape Attack*: Virtual machines are designed in a way that isolate each VM from the other VMs running on the same server, which prevents VMs from accessing data that belongs to other VMs that reside on the same server. However, in reality software bugs can be exploited to break this isolation. If a VM escapes the hypervisor layer and reaches the server's hardware, then the malicious VM can gain root access to the whole server where it resides. This gives the VM full control on all the VMs hosted on the hacked server. Different techniques were proposed to prevent a malicious VM from bypassing the hypervisor layer and obtaining the root privileges. An example of such techniques is CloudVisor which basically adds an extra isolation layer between the hardware and the hypervisor through nested virtualization that prevents the malicious VM from

obtaining the root privileges even if it bypasses the hypervisor layer. Other architecture solutions were also proposed to avoid VM escape attacks and could be found in [28].

5. *Insider Attacks*: In all the previously discussed attacks, we were treating the administrators of the cloud data center as trusted entities, and we were focusing only on the attacks that are originating from other malicious VMs that are hosted in the cloud data center. However, some sensitive applications may have serious concerns about hosting their collected information on the cloud data center in the first place as the cloud data center administrators will in that case have the ability to access and modify the collected data. Different techniques were proposed to protect the data from these insider attacks. Homomorphic encryption is a form of encryption that can be used to prevent such attacks as it allows the cloud servers to perform certain computing operations on encrypted input data to generate an encrypted result. This encrypted result when decrypted matches the result of performing the computational operation on the unencrypted input data. Applying homomorphic encryption in the IoT paradigm allows cloud servers to perform the necessary processing operations on the encrypted data that is collected from the smart devices without giving the cloud servers the ability to interpret neither the input data nor the result as they are both encrypted using a secret key that is not shared with the cloud. Only the smart objects and the user running the IoT application can interpret these data as they have the key needed for decryption. Another form of protection against insider attacks is to chop the data collected by the smart object into multiple chunks and then to use a secret key to perform certain permutations on those chunks before sending the data to the cloud servers. This allows storing the data on the cloud servers in an uninterpretable form for the cloud administrators. Only authorized entities that have the secret key can return the stored data to an interpretable form by performing the correct permutations.

For convenience, Table 8.1 summarizes all the cloud domain attacks that were discussed in this section. The second, third, and fourth columns of Table 8.1 describe, respectively, the vulnerability that causes this attack, what security requirement each attack violates, and what are the countermeasures that can be used to prevent or detect and mitigate each attack.

8.6   Fog Domain Attacks and Countermeasures

Recall that the fog domain is made up of a set of fog devices where each fog device collects the sensing data that is reported from a set of smart objects. The fog device performs different operations on the collected data which include data aggregation, data preprocessing, and data storage. The fog device may also perform some reasoning operations on the collected data. After processing and aggregating the collected data, the fog device forwards these data to the cloud domain. It is worth

Table 8.1 Summary of the security attacks in the cloud domain

Attack	Vulnerability reason	Security violation	Countermeasures
Hidden-channel attack	Shared hardware components (e.g., cache) among the server's VMs	Confidentiality	Hard isolation Cache flushing Noisy data access time Limiting cache switching rate
VM migration attacks	VM migration software bugs VM migration is performed without authentication Memory pages copied in clear	Confidentiality Integrity Availability	Server authentication Encrypting migrated memory pages
Theft-of-service attack	Periodic sampling of VMs' used resources	Availability Non-repudiation	Fine-grain sampling using high precision clocks Random sampling
VM escape attack	Hypervisor software bugs	Confidentiality Availability Integrity	Add an isolation domain between the hypervisor and hardware
Insider attacks	Lack of trust in cloud administrators	Confidentiality Integrity	Homomorphic encryption Secret storage through data chopping and permutation based on a secret key

mentioning that not only fog devices are connected with the cloud domain, but also fog devices are usually connected with each other in order to allow the fog devices connecting different smart objects to communicate directly with each other and in order to coordinate assigning objects to fog devices as their location changes. Fog devices can be independent components or could be built on top of existing gateways. Each fog device provides computing resources to be used by the IoT smart objects that are located close to the fog device. These computing resources are virtualized in order to allow the connected objects to share the computing resources that are offered by the fog device where each object or set of connected objects are allocated a virtual machine that performs the necessary data processing operations.

One can see that the computing capabilities provided by fog devices are very similar to the computing services provided by the servers in the cloud as they are both virtualized environments. The high similarities between the fog domain and the cloud domain make the fog domain susceptible to all the cloud domain attacks that were described in Sect. 8.5.

Although the fog domain is highly similar to the cloud domain, there are three key differences that distinguish fog devices from cloud servers:

1. *Location*: Unlike cloud servers which are usually located far from smart objects, fog devices are placed in areas with high popular access and thus are placed close to the smart objects. This placement plays an important role in giving the fog devices the ability to respond quickly to changes in the reported data. This also gives the fog devices the ability to provide location-aware services as smart objects connect to the closest fog device, and thus each fog device knows the location of the objects connected to it.

2. *Mobility*: Since the location of the smart object may change over time, then the VMs created to handle those objects at the fog domain must be moved from a fog device into another, in order to keep the processing that is performed in the fog device close to the object that is generating data.
3. *Lower Computing Capacity*: The fog devices that are installed in a certain location are expected to have a lower computing capacity when compared to capacities offered by cloud data centers as the latter are made of thousands of servers.

These characteristics raise new security threats that are specific to the fog domain and that distinguish it from the cloud domain. The security threats that are specific to the fog domain are the following:

- *Authentication and Trust Issues*: The fact that fog devices do not require a large facility space or a high number of servers compared to cloud data centers will encourage many small and less-known companies to install virtualized fog devices in dense areas and to offer these computing resources to be rented by the smart objects that are near the installed fog devices. Unlike cloud data centers which are offered by well-known companies, fog devices are expected to be owned by multiple and less-known entities. An important security concern that needs then to be taken into account when assigning a smart object to a fog device is to authenticate first the identity of the owner of the fog device. Authentication is not enough, as the smart object also needs to decide whether or not the owner of the fog device can be trusted. Trust is an important aspect as a smart object will be assigned to different fog devices belonging to different entities as their location may change over time. Reputation systems such as those that were proposed in peer-to-peer networks in or to rank cloud providers in can be used to select a trustworthy fog device among the available ones in the area surrounding each smart object.
- *Higher Migration Security Risks*: Although VM migration is common in both the cloud and the fog domains, there is an important difference between the migration in the cloud domain and that in the fog domain. While the migrated VMs in the cloud domain are carried over the cloud data center's internal network, the migrations from a fog device into another are carried over the Internet. Thus there is a higher probability that the migrated VMs get exposed to compromised network links or network routers when moving a VM from a fog device into another. This makes it vital to encrypt the migrated VM and to authenticate the VM migration messages that are exchanged among the fog devices.
- *Higher Vulnerability to DoS Attacks*: Since fog devices have lower computing capacities, this makes them a low-hanging-fruit for denial of service (DoS) attacks where attackers can easily overwhelm fog devices when compared to the cloud data centers, where a huge number of servers that have high computing capacity are available.
- *Additional Security Threats Due to Container Usage*: In order to provide the computing needs for a larger number of connected objects, the fog device may use containers rather than VMs to allocate the resource demands for each connected object. The main difference between a container-based virtualization and

full virtualization is the fact that containers share not only the same hardware but also the same operating system with the other containers that are hosted on the same fog device (refer to Chap. 6). This is unlike the full virtualization (which was illustrated in Fig. 8.3) where only the hardware is shared among multiple VMs and each VM has its own operating system. The low overhead of containers allows larger number of objects to be served by the fog device. However, sharing the same operating system among the containers dedicated for objects that belong to different users raises serious security concerns as the opportunities for data leakage and for hijacking the fog device increase significantly. The industry needs to address these gaps in container security to enable IoT applications at scale.

- *Privacy Issues*: We mentioned before that each smart object will be connected to one of the fog devices that are close to it. This means that the fog device can infer the location of all the connected smart objects. This allows the fog device to track users or to know their commuting habits which may break the privacy of the users carrying those objects. New mechanisms should be developed in order to make it harder for fog devices to track the location of the smart objects over time. Furthermore, the advancement in wireless signal processing has made it possible now to identify the presence of humans and track their location, their lip movement, and their heartbeats by capturing and analyzing the wireless signals that are exchanged between the sensing objects and the fog domain. This advancement makes it possible for any entity to install a reception device close to your home that analyzes the wireless signals that are emitted from your home in order to spy on your daily activities. The work in [47] is among the first papers that identified these risks where the authors in that paper propose a device called an obfuscator that prevents leaking such information by emitting signals that make it hard for an unauthorized receiver to infer the amplitude, the frequency, and the time shift of the originally exchanged signals. The obfuscator does not only prevent such leakages but also acts as a relay that rebroadcasts some of the sent messages which increases the transmission rate between the sensing objects and the fog domain.

8.7 Sensing Domain Attacks and Countermeasures

The sensing domain contains all the smart objects, where each object is equipped with a number of sensors that allow the object to perceive the world. The smart object is also supplied with a communication interface that allows it to communicate with the outer world. The smart object reports the sensed data to one of the fog devices in the fog domain. This is done by either creating a direct connection with the fog device if the smart object is directly connected by wires or has the wireless transmission capability to reach that fog device or in a multi-hop fashion where the smart object relies on other smart objects that lie along the path to the fog device to deliver the sensed data (as illustrated in Fig. 8.7).

Fig. 8.7 Multi-hop versus
direct connection between
the smart object and the
fog device

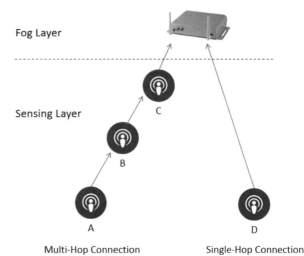

The sensing domain is susceptible to multiple attacks. We summarize next some
of the most well-known ones:

1. *Jamming Attack*: This attack causes a service disruption and takes one of
   two forms:

   (a) *Jamming the Receiver*: This attack targets the physical domain in the OSI
       stack of the receiver (where the receiver is the fog device in the case of a
       direct connection or another object in the case of a multi-hop connection)
       where a malicious user (called the jammer) emits a signal (called the jam-
       ming signal) that interferes with the legitimate signals that are received at
       the receiver side. The interference degrades the quality of the received signal
       causing many errors. As a result, the receiving end does not acknowledge the
       reception of these damaged packets and waits for the sender to retransmit
       those packets.

   (b) *Jamming the Sender*: Unlike the previous attack, this type targets the data
       link layer at the OSI layer of the sending object where the jammer in this
       attack sends a jamming signal that prevents the neighboring objects from
       transmitting their packets as they sense the wireless channel to be busy and
       back off waiting for the channel to become idle.

       There are different jamming strategies that a jammer may follow to
       launch a jamming attack. The most well-known ones are summarized next:

- *Constant Jamming*: The attacker continuously transmits a random jamming signal all the time. The main limitation of this attack is that it can be detected easily
  by observing random bits that do not follow the pattern dictated by the MAC
  protocol. Another main limitation is the fact that it requires the jamming device
  to be connected to a source of power as it requires lots of energy.

- Deceptive Jamming: This is similar to the constant jamming with the exception that the jammer conceals its malicious behavior by transmitting legitimate packets that follow the structure of the MAC protocol rather than sending random bits.
- Reactive Jamming: This is a strategy for jamming the receiver that is suitable for the case when the jamming device has a limited power budget. The jammer in that case listens to the medium and transmits a jamming signal only after it senses that a legitimate signal is being transmitted in the medium. This is more power efficient than continuously transmitting signals as listening to the channel consumes less power than transmitting signals.
- *Random Jamming*: The jammer alternates between sending a jamming signal and remaining idle for random periods of time in order to hide the malicious activity.
- More sophisticated jamming attacks have also emerged that intend to increase the service disruption time, reduce the probability of detection, increase the abilities to recover from the countermeasure that the victim node may take, while also reducing the power that the jamming device requires. An example of a power efficient advanced jamming attack would be to jam only the acknowledgment packets that nodes exchange rather than jamming the whole transmitted data packets as the former are shorter than the latter and thus require less power to jam while causing the same damage.
- Different preventive and detective techniques were proposed to address jamming attacks. We summarize next the most popular ones:
- *Frequency Hopping*: This is a preventive technique where the sender and receiver switch from a frequency to another in order to escape from any possible jamming signal (IEEE 802.15.4 TSCH discussed in Chap. 5 is an example of a wireless technology that employs this technique). Switching from a frequency to another is based on a generated random sequence that is known only for the sender and receiver. If the jammer is aware of the use of this preventive strategy, then the jammer has to switch from a frequency to another trying to collide with the frequency used by the sender and receiver. The interaction between the hopping strategies of the legitimate nodes and that of the jammer in that case can be modelled as a two-player game, where game theory can be used to come up with a hopping strategy that reduces the chances of colliding with the frequency sequence of the jammer.
- *Spread Spectrum*: This technique uses a hopping sequence that converts the narrow band signal into a signal with a very wide band, which makes it harder for malicious users to detect or jam the resulting signal. This technique is also very efficient when the transmitted data are protected by an error-correction technique as it allows the reconstruction of the original signal even if few bits of the transmitted data were jammed by the attacker.
- *Directional Antennas*: The use of directional antennas can mitigate jamming attacks from being successful as the sender and receiver antennas will have less sensitivity to the noise coming from the random directions that are different from the direction that connects the sender and the receiver.
- *Jamming Detection*: Different detective techniques were proposed in the literature to detect jamming attacks. The receiver can detect that it is a victim of a

jamming attack by collecting features such as the received signal strength (RSS) and the ratio of corrupted received packets. Advanced machine learning technique can then be used to differentiate jamming attacks from the degradation caused by the poor quality of the channel due to normal changes in the wireless link. We point the reader to the survey in [2] for further information about jamming intrusion detection systems.

2. *Vampire Attack*: This attack exploits the fact that the majority of IoT objects have a limited battery lifetime where a malicious user misbehaves in a way that makes devices consume extra amounts of power so that they run out of battery earlier thereby causing a service disruption. The damage caused by this attack is usually measured by the amount of extra energy that objects consume compared to the normal case when no malicious behavior exists.

We identify four types of vampire attacks based on the strategy used to drain power:

(a) *Denial of Sleep*: Different data link layer protocols were proposed to reduce the power consumption of smart objects by switching them into sleep whenever they are not needed. Examples of these protocols include S-MAC and T-MAC protocols. The idea behind these protocols is to agree on a duty-cycle schedule where objects exchange control messages in order to synchronize their schedules so that they agree on transmitting signals at certain cycles while remaining asleep for the rest of the time. An adversary can now launch a denial of sleep attack which prevents objects from switching to sleep by simply sending control signals that change their duty-cycles keeping them active for longer durations. The adversary can still succeed in launching this attack even if the control messages that synchronize the duty-cycles of the objects are encrypted. When the control messages are encrypted, the adversary can capture one of those encrypted control messages and replay it (resend it) at a later point of time causing the nodes to change their synchronization and their schedules. The adversary needs in that case to use traffic analysis techniques that rely, for example, on the length of the packets and the rate at which packets are exchanged in order to distinguish the control messages from the data messages that the nodes exchange since the content that packets carry is hidden by encryption.

(b) *Flooding Attack*: The adversary can flood the neighboring nodes with dummy packets and request them to deliver those packets to the fog device, where devices waste energy receiving and transmitting those dummy packets.

(c) *Carrousel Attack*: This attack targets the network layer in the OSI stack and can be launched if the routing protocol supports *source routing*, where the object generating the packets can specify the whole routing path of the packets it wishes to send to the fog device. The adversary in that case specifies routing paths that include loops where the same packet gets routed back and fourth among the other objects wasting their power. Figure 8.8 illustrates this attack.

Fig. 8.8 Illustration of the carrousel attack where the numbered arrows show the path specified by the malicious objects that the packets generated by the malicious object follow

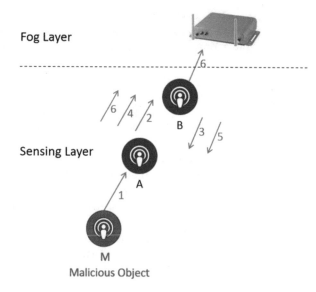

Fig. 8.9 Illustration of the stretch attack

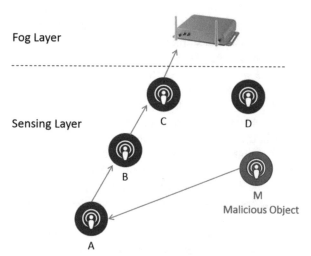

(d) *Stretch Attack*: This attack also targets the network layer in the OSI stack. If the routing protocol supports source routing, then a malicious object can send the packets that it is supposed to report to the fog device through very long paths rather than the direct and short ones as illustrated in Fig. 8.8. Even if source routing is not supported, the attacker can select a next hop that does not have the shortest path to the fog device in order to increase the power consumption of the objects that will be responsible to deliver those packets (Fig. 8.9).

The adversary can further amplify the amount of wasted energy by combining flooding attack with carrousel attack and stretch attack. The adversary in that case floods the neighboring objects with a large number of generated packets and specifies long paths with loops that the packet should follow in order to increase the amount of wasted power.

Denial of sleep attacks can be mitigated by encrypting the control message that arranges the schedules of the node while including a timestamp or a sequence number in the encrypted control message. This prevents the adversary from succeeding, in replaying an old control message, by checking the encrypted timestamp or the encrypted sequence number that the replayed control message is not a new message but an old one that someone replayed to cause disruption. Flooding attacks can be mitigated by limiting the rate of the packets that each object may generate. Carrousel attacks can be mitigated by making each object that is requested to forward a packet based on a route specified by the source check the specified path where packets with loops within their paths are dropped as they are most likely originating from malicious users. Finally, stretch attacks can be mitigated by disabling source routing or by making sure that the forwarded packets are making progress toward their destination and are not following long paths.

3. *Selective-Forwarding Attack*: This attack takes place in the case when the object cannot send its generated packets directly to the fog device but must rely on other objects that lie along the path toward the fog device to deliver those packets. A malicious object in this attack does not forward a portion of the packets that it receives from the neighboring objects. A special case of this attack is the *black-hole attack* where the attacker drops the entire set of packets that it receives from the neighboring objects. The best way to prevent packet drops from taking place for sensitive IoT applications is to increase the transmission capability of the objects so that they can reach the fog device directly without the need for help from intermediate objects. Unfortunately not all IoT objects are expected to have high transmission range to reach the fog device and thus will be relying on other objects to deliver their packets, which makes them susceptible to this attack. Different solutions were proposed to mitigate the number of dropped packets. *Path redundancy* is one of those solutions, where each object forwards each generated packet to multiple neighboring objects, where multiple copies of the same packet get delivered to the fog device through different paths. This decreases the chances of not having at least a copy of each generated packet delivered to the fog device. The main limitation of this mitigation technique is that it has a high energy overhead as it increases significantly the traffic. Rather than mitigating the damage caused by those attacks, the approach in [6, 8] tries to detect malicious objects that are dropping the sent packets so that packets can be routed through different paths that avoid those objects. Detecting the presence of objects that are dropping packets along certain paths can be done by selecting certain trusted objects as checkpoints. Each time a checkpoint receives a packet, it sends an acknowledgment to the object that generated that packet. The acknowledgment includes a unique identifier for the packet that was received along with a signed hash for the acknowledgment's content. This guarantees that no other entity fabricates fake acknowledgment packets and that no other entity

can alter the content of these acknowledgments. The interested reader may refer to [7] for a complete overview on the countermeasures that can be used against selective-forwarding attacks.

4. *Sinkhole Attack*: A malicious object claims that it has the shortest path to the fog device which attracts all neighboring objects that do not have the transmission capability to reach the fog device to forward their packets to that malicious object and count on that object to deliver their packets. Now all the packets that are originating from the neighboring nodes pass by this malicious node. This gives the malicious node the ability to look at the content of all the forwarded packets if data is sent with no encryption. Furthermore, the malicious object can drop some or all of the received packets as we explained previously in the selective-forwarding attack. Figure 8.10 illustrates how the network topology changes before and after this attack. Techniques to detect and isolate the malicious objects were proposed and are based on the idea of collecting information from the different objects where each object reports the neighboring objects along with the distance to reach those objects. A centralized intrusion detection system is then used to rely on the reported information to identify objects that are potentially providing misleading information. Detecting such attack becomes harder when multiple malicious nodes collude to hide each other.

Finally, Table 8.2 summarizes all security attacks in the sensing domain that were discussed in this section. The second column of the table shows what layer in the OSI stacks the attack targets, whereas the third, fourth, and fifth columns describe, respectively, the vulnerability reason, the security requirement that the attack breaks, and the defensive countermeasures against each attack.

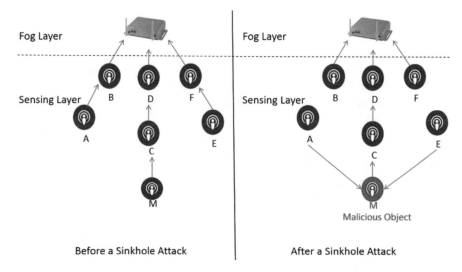

Fig. 8.10 Network topology before and after a sinkhole attack. The malicious object M claims that it has a shorter route to reach the fog device which attracts the neighboring objects A and E to rely on M to deliver their packets

Table 8.2 Summary of the security attacks targeting the sensing domain

Attack	Target OSI layer	Vulnerability reason	Security violation	Countermeasures
Jamming attack	Physical Data link	Shared wireless channel	Availability	Frequency hopping Spread spectrum Directional antennas Jamming detection techniques
Vampire attack	Data link Network	Limited battery lifetime	Availability Freshness	Rate limitation Drop packets with a source route that contains a loop Monitor whether or not the forwarded packets are making progress toward their destination
Selective-forwarding attack	Network	Limited transmission capability	Availability	Increase transmission range Path redundancy Choose certain intermediate objects as checkpoints to acknowledge received packets
Sinkhole attack	Network	Limited transmission capability	Confidentiality Availability	Analyze the collected routing information from multiple objects

8.8   Securing IoT Devices

In this section we will provide several examples of IoT devices being used to launch security attacks (Sect. 8.8.1), in addition to two solutions that attempt to secure IoT devices, namely MUD (Sect. 8.8.2) and DICE (Sect. 8.8.3).

8.8.1   IoT Devices Gone Rogue

With the increase of practical deployments, IoT devices have proven to be easy targets for hackers who turn compromised devices into active actors to carry out their attacks on networked IT infrastructure. This is especially true in the context of distributed denial of service (DDoS) attacks. Insecure IoT devices represent a growing pool of compute and communications resources that is open to misuse. These devices can be hijacked to spread malware, recruited to form botnets that may attack other Internet users, and even can be used to attack critical national infrastructure, or the structural functions of the Internet itself.

There are multiple recent examples of IoT devices being used as attack vectors. We will highlight some of them next.

8.8.1.1   Botnets

A botnet is a typically large collection of networked computers (bots) that are under remote control from some malicious third party over the Internet. Usually, these computers would have been compromised by an outside attacker who controls aspects of their functionality without the owners' consent or knowledge.

Because there are many such computers in a typical botnet, the attacker has access to a quasi supercomputer that can be employed for malicious purposes. Furthermore, since the bots are distributed geographically and organizationally over the Internet, the quasi supercomputer can be difficult to deter. The first botnet was developed in 2001 to send spam, and that is still a common use. Another common use for botnets is for DDoS attacks, in which a target server is constantly bombarded with network traffic until it is overwhelmed beyond its capacity and forced to go offline.

In 2016, a DDoS attack rendered much of the Internet inaccessible on the US East coast, and the attack was perpetrated by the Mirai botnet. Mirai took advantage of insecure IoT devices in a simple but clever way: It scanned large blocks of the Internet for open Telnet ports, then attempted to log in using username/password combinations that are frequently used defaults for these devices and never changed. With this simple approach, it was able to recruit an army of compromised closed-circuit TV cameras and routers, ready for launching a DDoS attack.

The reason why the botnet was so effective was due to the fact that it leveraged a large number of IoT devices which often include an embedded stripped-down Linux operating system. These devices had no built-in ability to be patched remotely and were in physically remote or inaccessible locations.

8.8.1.2 Webcams

Webcams are often marketed as consumer products for baby monitoring or as security devices. In one instance, a webcam manufacturer had faulty software on their products that allowed anyone with knowledge of the webcam's IP address to view the camera's video feed, and sometimes listen in through the embedded microphones. Another manufacturer's product was susceptible to remote code-injection attack, which allowed a malicious user to get administrative access to the camera, thereby placing the user at a risk of being spied upon. The remote execution flaw not only allows an attacker to set their own custom password to access the device, but also to add new users with administrative access to the interface, download malicious firmware or reconfigure the product as they please.

8.8.1.3 Casino Fish Tank

Security firm Darktrace published a report where it revealed that an unnamed casino in North America was hacked through an Internet-connected fish tank. That connection allowed the tank to be remotely monitored, automatically adjust temperature and salinity, and automate feedings. In this incident, the vulnerable smart tank was used as an easy backdoor into the casino's network. Once the attackers gained access to the tank, they scanned the casino's network for other vulnerabilities and moved laterally to other places in the network where they were able to steal 10 gigabytes of private data from the casino. The tank's communication patterns with the casino's network appeared normal enough. However, the data that it was pumping through to the Internet was highly suspect. It was the only tank system that transmitted data to a remote server in Finland, which it was in communications with. It also did so by employing protocols that are normally used for streaming audio or video.

8.8.1.4   Cardiac Devices

Cardiac devices, such as pacemakers and defibrillators, are used to monitor and control patients' heart functions and prevent heart attacks. In 2017, the FDA announced that St Jude's Medical implantable cardiac devices had security vulnerabilities that would enable an attacker to access these devices, where they could deplete the battery or administer incorrect pacing or shocks. The vulnerabilities were in the transmitter that reads the device's data and remotely shares it with physicians.

8.8.1.5   Vehicles

In 2015, Charlie Miller and Chris Valasek, two security researchers, exposed the security vulnerabilities in automobiles by hacking into cars remotely, controlling the cars' various functions from the radio volume to the brakes. They did so by leveraging day-zero exploits that give attackers wireless access to the car via the Internet. This was done by sending commands through the vehicle's entertainment system to its dashboard functions, steering, brakes, and transmission, all remotely from their laptops. The entertainment system served as an excellent entry point, because automakers are increasingly enabling the linking of these systems to the Internet. From that entry point, Miller and Valasek's attack pivots to an adjacent chip in the car's head unit (the hardware for its entertainment system), silently rewriting the chip's firmware to plant their code. That rewritten firmware is capable of sending commands through the car's internal computer network, known as a CAN bus, to its physical components like the engine and wheels.

Proper identification of connected devices is the first step when securing any network. With IoT, the asset inventory problem is compounded due to the sheer scale of "things," and there is a key requirement to efficiently and unambiguously identify connected devices for onboarding and ongoing management. With the ongoing rapid growth in the number of IoT devices, malicious actors view these devices as a soft attack surface from where to launch their attacks onto any other target in the network. As such, it is critical to provide mechanisms and capabilities for securing these devices. Two such mechanisms are MUD and DICE, which will be covered in detail next.

8.8.2   MUD

Manufacturer Usage Descriptor (MUD) is an embedded software standard defined by the IETF (RFC 8520) to help reduce the vulnerability surface of IoT devices by employing network policy (whitelisting approach). It aims to reduce the scope of malware injection and hijacking of over-the-air firmware updates. It also addresses the scenario of devices that are no longer being actively maintained by their original manufacturer.

MUD enables IoT device manufacturers to advertise formal device specifications, including the intended communication patterns for a given device when connected to the network. The network can then leverage this advertised intent, or profile, to formulate a tailored and context-specific access control policy, to guarantee that the device communicates only within the specified parameters. This way the network behavior of the device, in any operating environment, can be locked down and verified rigorously. In this context, MUD becomes the delegated identifier and authoritative enforcer of policy for IoT devices on the network. MUD works by enabling networks to automatically permit each IoT device to send and receive only the traffic it requires to perform as intended while blocking unauthorized communication with the device.

The MUD solution consists of three key components, as shown in Fig. 8.11.

- A unique identifier, in the form of a Universal Resource Locator (URL), that an IoT device advertises when it connects to the network.
- An Internet hosted profile file that this URL points to. This file contains an abstracted policy that describes the level of communication access which the IoT device needs to perform its intended functionality.
- A core process that receives the URL from the IoT Device, retrieves the profile file from the MUD File Server, and establishes the appropriate access control policies in the network to restrict the communication patterns for that IoT device.

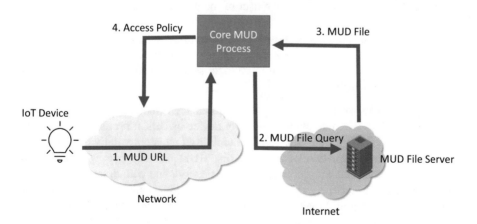

Fig. 8.11 MUD architecture

MUD leverages mechanisms that have existed in network infrastructure, including routers and switches, for over a decade. In what follows, we will describe the MUD workflow and associated mechanisms that can be used in more details.

1. The IoT device informs the network of the MUD URL using any one of the following existing protocols: DHCP, LLDP, or via a certificate in an IEEE 802.1X exchange. Once the device has communicated the URL to the network, its task in supporting MUD is done. The goal is to keep the device prerequisites as simple as possible for IoT device manufacturers.
2. The URL is received from the network by a Core MUD Process. This module may reside in one of many potential systems, depending on the nature of the network infrastructure. For instance, in an enterprise network, it may be part of the Policy (e.g., AAA) server. In a home network, it may be provided by the Internet service provider (ISP) or by the customer premise equipment (CPE) vendor. In a mobile service provider network, it might be part of an operational support system (OSS).
3. The Core MUD Process resolves the MUD URL and retrieves the profile file from the MUD File Server. This file is a declaration of intent that specifies what access the device is intended to have in the form of an abstract policy. The rationale being that an IoT device may be designed to communicate with a single or small number of controllers or with similar Things, or that for a given service, it should or should not have access to the local network.
4. The Core MUD Process translates these abstract intent definitions into a context-specific access control policy that the local network infrastructure can consume. How that translation occurs will vary depending on the network deployment. Some networks may use Access Control Lists (ACLs). Other networks may rely on segmentation using VLANs or VNIs, while others may use service groups or some other access control mechanism.
5. An administrator may then approve, reject, or modify the policy, based on deployment specifics. This policy may be merged with other policies, for instance, to take into account the user of the device or the device's deployment location.
6. The Core MUD Process pushes the merged policy to the associated systems of the network infrastructure (for example, switches, routers, etc.). This can be achieved using some configuration protocol such as NETCONF, Radius, or any alternative mechanism.

MUD provides a clear value proposition to device owners, network administrators and IoT device manufacturers. First, for device owners, it limits the impact and extent of exploitation of any security vulnerability that is potentially discovered in their IoT devices. For network administrators, MUD provides them with better visibility of the types of Things connected to the network and with the type of policies that they require. This helps them with better inventory management, risk assessment, and remediation. Finally, for device manufacturers MUD alleviates any support, financial liability, or brand damage that may arise due to compromised devices.

8.8.3   DICE

Device Identifier Composition Engine (DICE) is a collection of hardware and soft-ware mechanisms for cryptographic IoT device identity, attestation, and data encryption. DICE is an industry standard created by the Trusted Computing Group (TCG).

IoT devices that perform encryption use a private key called a Unique Device Secret (UDS) in order to secure their operation. It is possible for an attacker to leak this key by compromising the code running on the chip. Having access to the private key can enable the attacker to impersonate the device and even to replace its firm-ware. Therefore, it is paramount to prevent the disclosure of the UDS. The key to DICE is its ability to break up the boot process for any device into layers and to combine unique secrets and a measure of integrity for each of these layers. This way, if malware is present at any stage of the boot process, the device is automati-cally re-keyed and secrets protected by the legitimate keys remain safe.

DICE implements three measures to secure the UDS:

- **Power-on Latch**: The power-on latch locks read access to the UDS before early boot-code transfers control to subsequent execution layers.
- **Cryptographic One-way Functions**: A cryptographic one-way function com-putes a hash of the UDS to store in RAM so that in the event of RAM disclosure by compromised code, the original UDS is safe.
- **Tying Key Derivation to Software Identity**: To prevent compromise of the device by attempts to modify the early boot-code, the cryptographic one-way function uses a measurement of the boot code as input together with the UDS. The function outputs a key called the Compound Device Identifier (CDI) taking both the UDS and early boot code hash as input (optionally taking the hardware state and configuration as input as well). This process ensures that modification of early boot code generates a new key so that the UDS is secure.

The reason for tying the CDI derivation to the code that is booting on the device is to guarantee that a firmware update automatically results in the device being re-keyed. This behavior is desirable to address two security problems, specifically:

1. If an attacker changes the code that boots on the device with the intent of stealing keys, the attacking program (with a different hash) ends up obtaining a different key than the original authorized program.
2. If authorized code contains a security vulnerability that leads to CDI compro-mise, then the device must be re-keyed after patching. The CDI derivation func-tion ensures that patching the vulnerable firmware automatically results in a new CDI being computed.

DICE introduces a simple security approach that does not increase the silicon requirements for IoT devices. It targets constrained devices where traditional Trusted Platform Modules (TPM) may be unfeasible due to limitations related to cost, power, physical space, etc. As such, it is possible to implement it in the tiniest microcontrollers.

DICE is predicated upon a hardware root of trust for measurement. It works by organizing the boot into layers and creating secrets unique to each layer and configuration based on UDS (refer to Fig. 8.12). If a different code or configuration is loaded, at any point in the boot chain, the secrets will be different. Each software layer keeps the secret it receives completely confidential. If a vulnerability exists and a secret is leaked, it patches the code automatically and creates a new secret, effectively re-keying the device. In other words, when malware is present, the device is automatically re-keyed and secrets are protected.

DICE provides strong device identity, attestation of device firmware and security policy, and safe deployment and verification of software updates. The latter are often a source of malware and other attacks. Another key benefit for device manufacturers is that they are no longer required to maintain databases of unique secrets.

8.9   Summary and Future Directions

This chapter analyzed IoT from a security and privacy perspectives. Ignoring security and privacy will limit the applicability of IoT and will have serious results on the different aspects of our lives given that all the physical objects in our surrounding will be connected to the network. In this chapter, the IoT security challenges and IoT security requirements were identified. A three-domain IoT architecture was considered in our analysis where we analyzed the attacks targeting the cloud domain, the fog domain, and the sensing domain. Our analysis describes how the different attacks at each domain work and what defensive countermeasures can be applied to prevent, detect, or mitigate those attacks. We hope that the research and industry communities will pay attention to the discussed security threats and will apply appropriate countermeasures to address those issues. We also hope that security and privacy will be considered at the early design stage of IoT in order to avoid the common pitfall of considering security as an afterthought.

We end this chapter by providing some future directions for IoT security and privacy:

- *Fog Domain Security*: The fog domain is a new domain that was introduced to bring the computing capabilities to the edge of the network. We believe that further attention should be paid to this domain as it has not received enough

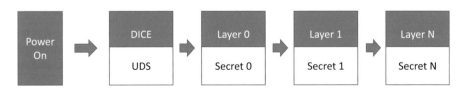

Fig. 8.12 DICE architecture

attention from the academia and the industry. The focus should be on identifying threat models related to the fog domain and also on finding efficient solutions that can run on the fog devices that are available in the market.

- *Collaborative Defense*: We identified while surveying the related work that what the literature on IoT security lacks is a collaborative solution where the different domains (cloud, fog, and sensing) interact with each other to stop or mitigate a certain attack. We believe that an interdomain-defensive solution will be way more effective than applying countermeasures at each domain separately, where the different domains can interact and collaborate in order to stop any ongoing malicious activity.
- *Lightweight Cryptography*: This is a highly important topic that has gained a significant attention recently and is anticipated to be very important for the future of IoT where the objective is to find efficient cryptographic techniques that can replace the traditional computationally expensive ones while achieving an acceptable level of security.
- *Lightweight Network Security Protocols*: Not only the cryptographic computations must have lower overhead but also the network security protocols that are used for communication. Many efforts are being paid by the research and industry communities to find cross-domain-optimized security protocols that achieve the necessary security protection while maintaining a low overhead.
- *Digital Forensics*: Although tracking the location of smart objects is considered a privacy violation, it also has some useful cases. Consider, for example, the case where police rely on tracking the smart objects that are carried by a missing person in order to identify the missing person's location. Digital forensics in the IoT era will play an important role in solving the different forensic cases as they will all become technology related. This area is also expected to receive further attention in the future where different techniques can be used to extract knowledge from the smart objects.

Problems and Exercises

1. The authors have broken IoT security challenges into seven areas. Name them. Why big data is an issue for IoT?
2. What techniques can be applied to prevent cross-VM data leakage? Explain how the hard isolation technique can be achieved.
3. What are some of the typical uses of VM migration in cloud data centers? What are the two types of attacks that are related to VM migration?
4. Who is the entity that initiates insider attacks, and how can homomorphic encryption be used to prevent such attacks?
5. What are the three key differences that distinguish fog devices from cloud servers? Provide a brief explanation of each difference.
6. Which provides more protection against security attacks: container-based virtualization or full virtualization? Why?
7. What are the two connection approaches that the smart objects may use to communicate with the fog device? Which approach is more secure and can this approach always be used?

8. What are the four strategies that a jammer may follow in order to launch a jamming attack? Which strategy is suitable when the jammer have limited energy budget?

9. What are vampire attacks? Name their types.

10. What is network high availability? What is network redundancy? How are they related?

11. Chapter 3 discusses three different ways to obtain information for IoT "things": sensors, RFID, and video tracking. In a table, compare the security for the three technologies.

12. What is limiting cache switching rate? How can it be accomplished? Explain how it works.

References

1. D. Willis, A. Dasgupta, S. Banerjee, Paradrop: a multi-tenant platform for dynamically installed third party services on home gateways, in *SIGCOMM workshop on distributed cloud computing*, (ACM, New York, NY, 2014)
2. W. Xu et al., Jamming sensor networks: attack and defense strategies. Network IEEE **20**(3), 41–47 (2006)
3. W. Ye, J. Heidemann, D. Estrin, Medium access control with coordinated adaptive sleeping for wireless sensor networks. Networking, IEEE/ACM Transactions **12**(3), 493–506 (2004)
4. T. Van Dam, and K. Langendoen, An adaptive energy-efficient MAC protocol for wireless sensor networks. in *Proceedings of the 1st international conference on Embedded networked sensor systems,* ACM, 2003
5. K.P. Dyer, et al., Peek-a-boo, i still see you: Why efficient traffic analysis countermeasures fail. in *Security and Privacy (SP), 2012 IEEE Symposium,* IEEE, 2012
6. J. Park, et al., An Energy-Efficient Selective Forwarding Attack Detection Scheme Using Lazy Detection in Wireless Sensor Networks. in *Ubiquitous Information Technologies and Applications*, (Springer, The Netherlands, 2013), pp. 157–164
7. L.K. Bysani, and A.K. Turuk, A survey on selective forwarding attack in wireless sensor networks. in *Devices and Communications (ICDeCom), 2011 International Conference,* IEEE, 2011
8. B. Xiao, B. Yu, C. Gao, CHEMAS: Identify suspect nodes in selective forwarding attacks. J. Parallel Distrib. Comput. **67**(11), 1218–1230 (2007)
9. P. Thulasiraman, S. Ramasubramanian, and M. Krunz, Disjoint multipath routing to two distinct drains in a multi-drain sensor network. in *INFOCOM 2007. 26th IEEE International Conference on Computer Communications*, IEEE, 2007
10. H.-M. Sun, C.-M. Chen, and Y.-C. Hsiao, An efficient countermeasure to the selective forwarding attack in wireless sensor networks. in *TENCON 2007–2007 IEEE Region 10 Conference*, IEEE, 2007
11. A. Grau, Can you trust your fridge? Spectrum, IEEE **52**(3), 50–56 (2015)
12. C. Li, A. Raghunathan, and N. K. Jha, Hijacking an insulin pump: Security attacks and defenses for a diabetes therapy system. in *e-Health Networking Applications and Services (Healthcom), 2011 13th IEEE International Conference,* IEEE, 2011
13. D. Evans, *The internet of things how the next evolution of the internet is changing everything.* Technical report, CISCO IBSG, 2011

14. R. Thomas, et al., Hey, you, get off of my cloud: exploring information leakage in third-party compute clouds. in *Proceedings of the 16th ACM conference on Computer and communications security*, ACM, 2009

15. M. Dabbagh, B. Hamdaoui, M. Guizai and A. Rayes, Release-time aware VM placement. in *Globecom Workshops (GC Wkshps)*, (2014), pp. 122–126

16. M. Dabbagh, B. Hamdaoui, M. Guizani, A. Rayes, Toward energy-efficient cloud computing: Prediction, consolidation, and overcommitment. Network, IEEE **29**(2), 56–61 (2015)

17. M. Dabbagh, B. Hamdaoui, M. Guizani, A. Rayes, Efficient datacenter resource utilization through cloud resource overcommitment, in *IEEE Conference on Computer Communications Workshops (INFOCOM WKSHPS)*, 2015, pp. 330–335

18. R. Boutaba, Q. Zhang, and M. Zhani, Virtual Machine Migration in Cloud Computing Environments: Benefits, Challenges, and Approaches. in *Communication Infrastructures for Cloud Computing*, ed. by H. Mouftah and B. Kantarci (IGI-Global, Hershey PA, 2013), pp. 383–408

19. D. Perez-Botero, *A Brief Tutorial on Live Virtual Machine Migration from a Security Perspective*, University of Princeton, Princeton, 2011

20. W. Zhang, et al., Performance degradation-aware virtual machine live migration in virtualized servers. in *International Conference on Parallel and Distributed Computing, Applications and Technologies (PDCAT)*, 2012

21. V. Venkatanathan, T. Ristenpart, and M. Swift, *Scheduler-based defenses against cross-VM side-channels*. Usenix Security, (2014)

22. T. Kim, M. Peinado, and G. Mainar-Ruiz, Stealthmem: System-level protection against cache-based side channel attacks in the cloud. in *Proceedings of USENIX Conference on Security Symposium, Security'12. USENIX Association*, 2012

23. H. Raj, R. Nathuji, A. Singh, and P. England, Resource management for isolation enhanced cloud services. in *Proceedings of the 2009 ACM workshop on Cloud computing security*, ACM, 2009, pp. 77–84

24. Y. Zhang and M. K. Reiter, Duppel: Retrofitting commodity operating systems to mitigate cache side channels in the cloud. in *Proceedings of the 2013 ACM SIGSAC Conference on Computer; Communications Security*, CCS '13. ACM, 2013

25. P. Li, D. Gao, and M. K. Reiter, Mitigating access driven timing channels in clouds using stopwatch. in *IEEE/IFIP International Conference on Dependable Systems and Networks (DSN)*, 2013, pp. 1–12

26. R. Martin, J. Demme, and S. Sethumadhavan, Timewarp: Rethinking timekeeping and performance monitoring mechanisms to mitigate sidechannel attacks, in *Proceedings of the 39th Annual International Symposium on Computer Architecture*, 2012

27. F. Zhou et al., Scheduler vulnerabilities and coordinated attacks in cloud computing. in *10th IEEE International Symposium on Network Computing and Applications (NCA)*, 2011

28. K. Panagiotis, and M. Bora, Cloud security tactics: Virtualization and the VMM. in *Application of information and communication technologies (AICT), 2012 6th International Conference.* IEEE, 2012

29. F. Zhang et al., CloudVisor: retrofitting protection of virtual machines in multi-tenant cloud with nested virtualization. in *Proceedings of the Twenty-Third ACM Symposium on Operating Systems Principles,* ACM, 2011

30. T. Taleb, A. Ksentini, Follow me cloud: interworking federated clouds and distributed mobile networks. IEEE Network **27**, 12 (2013)

31. E. Damiani et al., A reputation-based approach for choosing reliable resources in peer-to-peer networks. in *Proceedings of the 9th ACM conference on computer and communications security.* ACM, 2002

32. W. Itani et al., Reputation as a Service: A System for Ranking Service Providers in Cloud Systems. in *Security, Privacy and Trust in Cloud Systems.* (Springer, Berlin Heidelberg, 2014). pp. 375–406

33. J. Sahoo, M. Subasish, and L. Radha, Virtualization: A survey on concepts, taxonomy and associated security issues. in *Second International Conference on Computer and Network Technology (ICCNT)*, 2010

34. S.Yi, Q. Zhengrui, and L. Qun, Security and privacy issues of fog computing: A survey. in *Wireless Algorithms, Systems, and Applications*, (Springer International Publishing, 2015), pp. 685–695

35. E. Oriwoh, J. David, E. Gregory, and S. Paul, Internet of things forensics: Challenges and approaches. in *9th International Conference on Collaborative Computing: Networking, Applications and Worksharing (Collaboratecom)*, IEEE, 2013, pp. 608–615

36. Z. Brakerski, V. Vinod, Efficient fully homomorphic encryption from (standard) LWE. SIAM J. Comput. **43**(2), 831–871 (2014)

37. E. Lauter, Practical applications of homomorphic encryption. in *Proceedings of the 2012 ACM Workshop on Cloud computing security workshop*, ACM, 2012

38. C. Hennebert, D. Jessye, Security protocols and privacy issues into 6lowpan stack: A synthesis. Internet of Things Journal IEEE **1**(5), 384–398 (2014)

39. Daily Tech Blogs On Line, http://www.dailytech.com/Five+Charged+in+Largest+Financial+Hacking+Case+in+US+History/article32050.htm

40. M. Miller, Car hacking' just got real: In experiment, hackers disable SUV on busy highway (The Washington Post, 2015), online: http://www.washingtonpost.com/news/morning-mix/wp/2015/07/22/car-hacking-just-got-real-hackers-disable-suv-on-busy-highway/

41. 2015 Data Breach Investigation Report, Verizon Incorporation (2015)

42. M. Dabbagh et al., Fast dynamic internet mapping. Futur. Gener. Comput. Syst. **39**, 55–66 (2014)

43. Forrester, Security: The Vital Element of the Internet of Things, 2015, online: http://www.cisco.com/web/solutions/trends/iot/vital-element.pdf

44. F. Adib and D. Katabi, See through walls with WiFi!, vol. 43. (ACM, 2013)

45. S. Kumar, S. Gil, D. Katabi, and D. Rus, Accurate indoor localization with zero start-up cost, in *Proceedings of the 20th Annual International Conference on Mobile Computing and Networking*, ACM, 2014, pp. 483–494

46. G. Wang, Y. Zou, Z. Zhou, K. Wu, and L. Ni, We can hear you with Wi-Fi!, in *Proceedings of the 20th Annual International Conference on Mobile Computing and Networking*, ACM, 2014, pp. 593–604

47. Y. Qiao, O. Zhang, W. Zhou, K. Srinivasan, and A. Arora, PhyCloak: Obfuscating sensing from communication signals, in *Proceedings of the 13th USENIX Symposium on Networked Systems Design and Implementation (NSDI)*, 2016

48. T. Yu, et al., Handling a trillion (unfixable) flaws on a billion devices: Rethinking network security for the internet-of-things, *Proceedings of the 14th ACM Workshop on Hot Topics in Networks*, 2015

49. M. Dabbagh, B. Hamdaoui, M. Guizani, A. Rayes, Software-defined networking security: pros and cons. IEEE Commun. Mag. **53**, 73 (2015)

Chapter 9
IoT Vertical Markets and Connected Ecosystems

The Internet of Things is expected to connect over 20 billion "things" to the Internet by 2020, covering a broad range of markets and applications. As IoT becomes more cost effective and easier to deploy, new contenders and industry players are expected to enter the market. Hence, existing companies will be forced to disrupt or be disrupted. For the leaders of any of these companies, this begs two main questions: Firstly, what new business models to employ in order to deliver better and cheaper service? And secondly, who to partner with to bring services to market quicker and at a lower cost?

In this chapter, we will first introduce, in Sect. 9.1, the key IoT application domains, which are often referred to in the literature as IoT verticals. Alphabetically, key verticals include Agriculture and Farming, Energy, Enterprise, Finance, Healthcare, Industrial, Retail, and Transportations.

These verticals will include data sources (e.g., sensors, RFIDs, video cameras, etc.) producing wealth of new information about the status, location, behaviors, usage, service configuration, and/or performance of systems, products, or devices. In Sect. 9.2, we will present the new business model which is mainly driven by the availability of new information, thereby offering extraordinary business benefits to the companies that manufacture, support, and service those systems, products, or devices, especially in terms of customer relationships. In Sect. 9.3, we will present the top requirements to deliver "Anything as a Service" in IoT followed by a specific use case.

Finally, the manifold IoT verticals in combination with the new business model will undeniably introduce opportunities for innovative partnerships. No single vendor will be able to address all business requirements. We will describe the requirements for such model in the last section.

© The Author(s), under exclusive license to Springer Nature Switzerland AG 2022   247
A. Rayes, S. Salam, *Internet of Things from Hype to Reality*,
https://doi.org/10.1007/978-3-030-90158-5_9

9.1   IoT Verticals

There is no agreement across the industry on the number of IoT verticals. The number ranges from a few to over a dozen across various standards and marketing collaterals. The oneM2M and ETSI standard bodies have identified ten IoT verticals: Agricultural and Farming, Energy, Finance, Healthcare, Industrial, Public Services, Residential, Retail, and Transportation. Other companies have used a slightly different categorization to include Energy, Transportation, Education, Healthcare, Commerce, Travel and Tourism, Finance, IT, and Environment.

As we mentioned in pervious chapters, the objective is not to divide IoT into verticals and silos. On the contrary, the real impact of IoT will only occur when data from the silos is combined to create completely new types of applications. In other words, an IoT application should be able to manage IoT elements from many verticals with common parameters, open data models, and APIs. The collected data from IoT elements, combined with the new knowledge emerging in the area of "big data," will create the framework for many new types of applications. This progress will drive the growth of IoT.

In this chapter, we will describe IoT use cases using a modified version of the oneM2M and ETSI categorizations, as shown in Fig. 9.1. The IoT verticals include: Agriculture and Farming, Energy, Oil and Gas, Enterprise, Finance, Healthcare, Industrial, Retail, and Transportations.

It is important to note that some IoT standard bodies have used the term "Energy" as a comprehensive label to include "Energy Consumption" in smart buildings/cities as well as "Oil and Gas" in the petroleum industry (e.g., to monitor oilrigs, pipelines, and emission). We believe IoT Energy and IoT Oil and Gas are two separate verticals. Energy comes from Oil and Gas as well as other sources such as solar and winds. In addition, energy is about managing smart meters, smart buildings, and smart cities, while Oil and Gas is more about process and asset management in the petroleum industry. More information will be provided in Sects. 9.1.2 and 9.1.3.

Fig. 9.1 IoT verticals

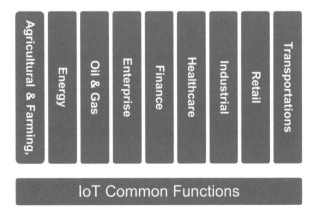

9.1.1   IoT Agriculture and Farming

According to the World Agriculture reports, global food consumption is expected to grow by 70% by 2050. IoT is well positioned to transform the agriculture industry and enable farmers to increase the quantity and quality of their crops at reasonable costs. IoT farming techniques are already increasing crops productivity and creating economies of scale for farmers. This is critical especially with the recent environmental challenges farmers are facing, such as increased water shortage in many regions of the World and the diminishing availability of farmland.

IoT sensor-based agriculture solutions are used to monitor soil moisture, crop growth, livestock feed levels, and irrigation equipment. The solutions utilize analytics to analyze operational data combined with weather and other information to improve decision-making.

Top IoT Agriculture and Farming use cases include:

- **Advanced Yield Monitoring**: Farming companies have introduced solutions to monitor and control various types of crops to deliver better results. For instance, wine quality is being monitored by installing sensors to monitor soil moisture and trunk diameter in vineries to optimize the amount of sugar in grapes. Similar techniques are used for water management by sensing the soil and determining the optimal amount of water required as part of green initiatives.
- **Optimal Seeding**: Based on soil analysis and historical weather data, IoT enabled solutions determine the best kind of seeds and optimal row spacing as well as seeding depth. They also produce soil fertilization recommendations that include type and amount.
- **Optimal Water Usage**: Monitoring and controlling surrounding environmental conditions to determine water usage to capitalize on the production of fruits and vegetables. This includes utilizing weather forecast information to prevent damage due to ice formation, heavy rain, drought, snow, or strong wind. The humidity levels are also monitored in crops such as hay and alfalfa to avoid fungus and other bacterial contaminants.
- **Livestock Monitoring**: Monitoring, tracking, and controlling farm animals (cows, goats, chickens, etc.) in open grasslands or indoor locations such as cages or stables. IoT is also used to monitor animal toxic gas levels, study ventilation, and warn on air quality to protect farm animals from harmful gases from excrements.
- **Farming as a Service**: see Sect. 9.2.

9.1.2   IoT Energy Solutions

IoT Energy covers smart buildings offering dynamic monitoring of overall energy consumption, thereby allowing their occupants or tenants to see when they are consuming power during peak hours at abnormally high rates. This allows the tenants to optimize energy usage while maintaining comfort. It also covers smart cities offering automatic dynamic optimization of global energy consumption on the streets, highways, and public facilities.

IoT Energy use cases include:

- **IoT Smart Meters**: IoT smart meters record electrical power consumption on regular basis (e.g., hourly, every 15 min) and send collected information to the power company for monitoring and billing.

- IoT smart meters benefit power companies as well consumers. Power companies use the collected information to construct usage patterns and trend analyses to predict future energy usage especially during peak hours. They plan for such peaks with additional supply and by offering very attractive offers to customers to conserve energy. Customers use the information to view, typically on the portal of the power company, hourly electric and daily gas energy usage data. Consumers use the detailed hourly, daily, weekly, or monthly information to make smarter energy choices (e.g., use washing machine after 7 PM for cheaper rate).

- **Smart Homes (Connected Home)**: Connected home is defined as any home with at least one connected device (e.g., connected appliance, home security system, and door or motion sensor). Connected devices can learn usage patterns and enable remote operation to reduce energy consumption (e.g., water heaters, air conditioning, and lighting.)

- Connected devices send information to service provider systems, which in turn, quickly analyze the data and notify homeowners if needed, or directly send alerts to homeowners. The first model is often a subscription-based service in which a homeowner subscribes to a service (e.g., home security company) while the second model is non-subscription model (e.g., home security camera installed by homeowner and connected over the home Wi-Fi gateway). Can you name an example of model 2? (see Problem 8).

- **Other Cases**: IoT is also used to monitor and optimize solar energy plants performance. How? (see Problem 10).

- To meet the IoT key promise of making human lives better, all connected home devices should come together into a single connected IoT system or connected service provider system offering the homeowner full and simple access and control.

9.1.3   IoT Oil and Gas Solutions

Ever since the explosion and sinking of the Deepwater Horizon oil rig in the Gulf of Mexico in April 2010, which was recognized as the worst oil spill in U.S. history, combined with the increase in strict government regulations, IoT has been at the core of the oil and gas industry transformation. It is not only enabling full real-time monitoring of oilrigs but also allowing contingent workforce to run near real-time maintenance of critical assets.

IoT Oil and Gas is used for predictive maintenance, pipeline monitoring, emission control, and location intelligence. It is also used for near real-time alert and trending analysis using sensors, installed on various equipment and augmented with ERP (enterprise resource planning) data to trigger maintenance workflows for asset management and fleet operations monitoring.

- **Connected Oil and Gas fields**: IoT sensors are being installed to monitor and control oil wellheads, pipelines, and equipment, to enhance the overall oil field remote operations, to enable predictive maintenance, and provide comprehensive facility operations at reasonable costs. Hence achieving better reliability and productivity from the fields.

- Also Connected Oil and Gas fields reduce the need for site visits (e.g., site visits to unmanned offshore platforms), hence reducing the associated hazards and improving personnel safety.
- **Downstream Applications**: IoT Oil and Gas also can play a role in downstream operations such as Oil and Gas storage, transportation, refineries, and distribution (e.g., petrol station fuel tanks can be monitored by distribution companies to dispatch tank trucks).

9.1.3.1   Oil and Gas Exercise

Chemical injection stations (Fig. 9.2) are used to dose corrosion inhibiting and biocide chemicals into oil pipelines. This eliminates the growth of organisms and reduces the corrosion rate of the pipelines in order to prolong their operational life.

One chemical station is required to dose at a rate of 0.4 gpm (gallons per minute) of chemicals per 10,000 bpd (barrels per day) of oil in the pipeline. In an existing plant, the station is set to dose at a constant 0.4 gpm. Considering the following pipeline flowrate profile during a day, calculate the quantity of chemicals saved per day by applying IoT to control the chemical injection station.

Answer
We only need to examine the part of the timeline where the flow within the pipeline drops below the 10,000 bpd threshold, as that is where the IoT solution will yield savings over the constant/static solution.

The flow within the pipeline drops to 8000 bpd for 12 h. During this time, the variable dosage supplied by the IoT solution drops to $8000/10{,}000 \times 0.4$ gpm $= 0.32$ gpm.

The amount of chemical dispensed by the IoT solution for those 12 h is $= 0.32$ gallons/min $\times 12$ h $\times 60$ min/h $= 230.4$ gallons.

The non-IoT solution would have dispensed during the same time $= 0.4 \times 12 \times 60 = 288$ gallons.

The savings $= 288 - 230.4 = 57.6$ gallons.

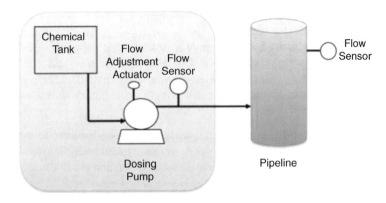

Chemical Injection Station

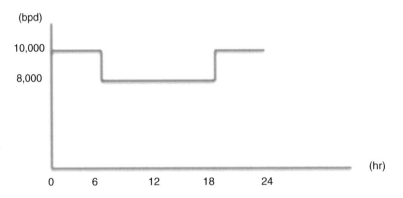

Fig. 9.2 Oil and Gas Exercise

9.1.4   IoT Smart Building Solutions

As with smart homes (under smart energy), smart buildings utilize sensors and controllers to monitor and automatically trigger services to save valuable time in cases of emergency (e.g., fire, intrusion, or gas leak). With the smart building system, services like video monitoring, light control, air-condition control, and power supply control are often managed from the same control center. In this section, we will focus on Smart buildings as an enterprise solution, as specified in the oneM2M standards.

- **Safety Monitoring and Alerting**: Examples include **noise level monitoring** in urban zones and sounding alarms in real time, **electromagnetic field levels monitoring** by measuring the energy radiated by cell stations and other devices, **chemical leakage detection** in rivers by detecting leakages and wastes of factories in rivers, **air pollution** and control of CO_2 emission factors, pollution emitted by cars and toxic gases generated in farms, as well as **earthquake early detection**.

- **Smart Lighting**: In smart lighting, IoT is used to minimize energy consumption, provide weather adaptive lighting in street-lights and to automate maintenance.
- **Flooding, Water Leakage, and Pollution Monitoring**: Monitoring of safe water levels in rivers, lakes, dams, and reservoirs. Detection of the presence of toxic chemical. Monitoring of tanks, pipes, and pressure variations. Real-time control of leakages and waste in the sea.
- **Detection of Hazardous Gases and Radiation Levels**: Detection of gas levels and leakages in and around industrial buildings and chemical factories. Monitoring of ozone levels during the meat drying process in food factories. Distributed measurement of radiation levels in the surroundings of nuclear power stations to generate leakage alerts.
- **Other use cases include** detection of garbage levels in containers to optimize the trash collection routes, preemptive monitoring of burning gasses and fire conditions to define alert zones, snow level measurement to know in real time the quality of ski tracks and alert avalanche prevention security corps, monitoring vibrations and earth density to detect dangerous patterns in land conditions, and monitoring of vibrations and structural conditions in buildings and bridges.

9.1.5 IoT Finance

While IoT Financial solutions are not as obvious as other IoT verticals, the financial industry has indeed benefited greatly from IoT. For many financial services businesses, the reality is that their business model is based on the flow of information, rather than on actual sensors and physical objects. As we mentioned in Chap. 1, some financial companies (e.g., Square, Intuit) have introduced IoT platform-based solutions connecting customers instantly with financial institutes and services. Such process used to be tedious and required time that often resulted in losing prospective deals to competitors. Banks are using IoT-based facial recognition solutions to identify important customers when they walk into the bank so they can be offered first class treatment.

Auto insurance companies are working with technology companies and communication service providers to install sensors-based IoT telematics solutions in automobiles, to track driver behaviors in order to improve underwriting and pricing of policies. Other use cases include:

- **IoT Usage-based auto insurance**: Sensors are installed in vehicles to track actual mileage, car location, and driving areas. In addition, IoT-based claim filing system is utilized allowing drivers to file claims using their smart phones eliminating the need for expensive agents and paper work.

- **IoT Solution to reduce Fraud and Liability**: In highly delicate work environments (e.g., chemical or nuclear plants, physical activities), smart sensors may be embedded in employees' uniforms. This allows the IoT solution to monitor employee whereabouts in high-risk areas, warn them in real time of any potential danger, and prevent them from entering restricted areas. This should result in safer work environments for the employees and reduce fraudulent workplace related claims for the employer.
- **IoT Safety Solutions**: Sensors embedded in commercial infrastructure can monitor safety breaches such as smoke, mold, or toxic fumes, allowing for adjustments to the environment to head off or at least mitigate a potentially hazardous event.
- Other use cases include IoT-based commercial real estate building-management systems to speed up the overall building management processes, location-based near-field communication (NFC) Payment Processing, paperless mortgage applications including home inspection and the approval process.

The progression of financial IoT is not without its challenges. Most driving consumers and corporations are uncomfortable with the notion of being "watched" at all time. Many have asked for limits on the collection and use of sensor-based data. This is a critical area for the industry to address by introducing balanced solutions that allow the collection of adequately limited data while protecting the interests of clients and markets. Full disclosure of collected data (what are you collecting about me?) as well as the secure handling and use of personal information (who has access to my data and how is it being used?) are already being demanded by consumers and corporations.

9.1.6   IoT Healthcare

Healthcare is considered as one of the most important verticals for IoT. Healthcare providers as well as patients are in great position to benefit from IoT. Intelligent IoT wearable devices in combination with mobile apps are allowing patients to capture their health data easily and send medical information for up to the minute analysis. Hospitals are using IoT for real-time tracking of important medical devices, personnel, and patients.

Examples of IoT Healthcare use cases include:

- **Fall Detection**: Fall detection is considered a main public health concern among senior citizens. The number of wearable medical devices, systems, and companies offering services intended at detecting falling have increased radically over recent years. Fall detection alert systems, typically worn around the waist or neck, include intelligent accelerometers that differentiae normal activities from actual falls. Fall detection solutions are already improving the quality of life of many elderly or disabled people living independently. It should be noted that smart phones also use accelerometers to determine vertical and horizontal display based on orientation.

- **Tracking of medical devices**: Accurate tracking of expensive medical devices is very essential for hospitals especially in crowded emergency rooms with large medical staff. IoT solutions are being used to identify the exact location of such devices, identify last user, and then auto adjust the device setting, if applicable, based on the fingerprint of the current user.
- **Medical fridges for hospitals**: Sensors are being embedded in medical fridges for hospitals and medical offices to dynamically control temperatures inside mobile and stationary freezers filled with vaccines, medicines, and organic elements.
- Other use cases include measuring ultraviolet radiation and warning people of the hazard of sun exposure especially during certain hours.

As is the case with IoT financial, IoT healthcare has its own share of challenges. The security of IoT data and devices as well as government regulations are considered by many as the most important concerns for patients and healthcare providers. Patients are concerned about employers gaining access to their medical records, especially when they register their BYOD mobile devices. Some physicians and healthcare IT departments are still adjusting to using and securing mobile devices in their operations. Finally the lack of standards and communication protocols around IoT put the development of solutions at risk.

9.1.7   IoT Industrial

Industrial equipment and machines used in the overall manufacturing process, for instance, are becoming more digitized with capabilities to connect to the Internet. At the same time, manufacturers are looking at ways to advance operational efficiency such as supply chain and quality control, by utilizing such equipment to gather important data for their business to remain competitive and provide services at reasonable costs.

IoT is used to establish networks between machines, humans, and the Internet, thereby creating new ecosystems. It is also used to identify business gaps and opportunities, as we will cover in Sect. 10.3. Examples of Industrial use cases include:

- **Predictive maintenance**: Predictive maintenance covers all connected assets in industrial plants (e.g., water treatment site). By utilizing real-time data collected from sensors and cameras, combined with advanced analytics it is possible for companies to anticipate equipment failures and respond faster to critical situations. Advanced analytics is a hot research area that includes artificial intelligence and machine learning. With machine learning, computers can develop algorithms on their own by analyzing data overtime. These algorithms can then be used to make predictions.

- **Connected Factory**: as the name indicates, connected factory means connecting the entire factory network to the Internet with full monitoring and controlling solution. Connected factory typically includes mobile operation center for comprehensive and secure management.
- **Connected Mine**: In connected mines, all mining vehicles, mining operation, mining asset tracking and personal safety equipment are connected.
- **Supply Chain Control**: Monitoring of storage conditions along with the supply chain and product tracking for traceability purposes.

9.1.8   IoT Retail

According to a survey by Infosys, more than 80% of consumers are willing to pay up to 25% more for a better experience. This translates to a huge opportunity to be gained with IoT by collecting and analyzing information about products and customer interests and then gaining actionable insights from this information. Input sources include point of sale (PoS), supply chain sensors, RFID as well as video cameras in the store.

- **Full Tracking of Products in Stores**: With IoT, retailers have full visibility into products and merchandise with digital supply chains. This makes it possible for retailers to emphasize on top selling products by offering more personal choices to fulfill and enhance the overall customer experience. It also makes it possible to determine under-selling products as well as overstocked and low stock products.
- **Full Automation of Product Delivery**: Range of delivery options may be offered to the customers including pick up in-store, home or car delivery, or retrieval from another location such as smart lockers from local 24-h stores. In the latter case, smart lockers are equipped with sensors that send automatic messages to customers reminding them to pick up.
- On the business side, some retailers have capitalized on IoT to redesign their distribution system to leverage larger stores as distribution centers. In this case, larger stores are used to offer a larger range of products to smaller stores for collection the same day, thereby extending customers' choice of delivery and collection options.
- **Flexible Shopping and Loyalty Programs**: Retailers are already using Web technologies such as cookies, Wi-Fi, and video cameras to track customers shopping behavior to enhance customer experience and send special offers based on buying patterns or even online browsing and search trends. For instance, retailers are using Bluetooth beacons in combination with shopping apps on customers' smart phones to generate heat maps that show how consumers move around stores (why would customer download retailer apps?—see Problem 11). For customers who are not willing to download retailers apps, Wi-Fi triangulation is alternatively used to generate detailed heat maps.

- **Customer Engagement Suite**: As we mentioned in Chap. 1, some companies have introduced customer engagement tools that include email marketing services. These tools allow businesses to target specific customer segments with customized promotions based on actual purchase history. Square also introduced Square Payroll tool for small business owners to process payroll for their employees.
- **Interactive consumer engagement and operations**: Using real-time video cameras, in-store programmable devices and in-store display screens retailers can deliver smarter messaging based on what customers are looking at. This allows them to influence buying decisions, including up-sells.

9.1.9   IoT Transportation

As industry regulations force transportation and logistics organizations to do more with less, many companies have already discovered the benefits of using IoT to offer new services, improve efficiency and security, significantly gain real-time visibility of their operations, and save on fuel just to name a few advantages.

Top use cases include:

- **Smart and Connected Parking**: Smart parking addresses one of the causes of pollution in urban areas. We all have been in situations where we drive back and forth looking for a parking spot. Smart and Connected Parking has addressed this problem very effectively. With smart parking service, drivers can easily find available parking spaces, pay parking fees and even make advance reservations. Making parking reservations may be available for limited people such as VIPs or the disabled, since ordinary parking service needs to satisfy first-come-first-served rule.
- **Smart Roads and Traffic Congestion**: Smart roads include Intelligent Highways with warning messages and diversions based on sensors capturing climate conditions and traffic events like accidents and traffic jams. Traffic congestion solutions monitor traffic as well as pedestrian levels to optimize driving and walking routes.
- **Connected Rail**: Connected rail solutions are used to connect trains, tracksides, stations, and passengers. For instance, IoT is used to automatically alert passengers of scheduling and safety issues on their smart devices as well as offering onboard entertainment. IoT is also used to implement solutions to meet governmental and industrial safety compliance requirements at a minimum cost.
- Other use cases include continuous quality of shipment monitoring, which encompasses observing vibrations, location, temperature, strokes, container openings, and storage incompatibility detection. For instance, emissions warning on containers storing flammable goods close to others containing explosive material. Control of routes followed for delicate goods like medical drugs, jewels, or dangerous merchandise are also included.

9.2   IoT Service Model: Anything as a Service

IoT enabled devices and products will provide a wealth of information about their status, location, behaviors, usage, service configuration, and performance. This information, if leveraged correctly, offers extraordinary business benefits to the companies that manufacture, support, and service those products, especially in terms of customer satisfaction.

With the availability of such data combined with cost effective Internet-based communications, many companies are starting to ponder why would they stop at selling a product and forgo very essential feedback information, when they can also sell a service with the right to monitor the actual usage and behavior of the product in the deployed environment. Usage information are not only used to service a product/device and prevent service deterioration by verifying contract Service Level Agreements (SLAs) but also to learn about the product in the field and determine the most essential set of future enhancements. Feedback information may be categorized by market segments but generally include common set of specific information such as which features are used the most, which features are used the least, which features are never used and feature usage patterns (feature A is used with feature B).

IoT is bending the traditional linear value chain by allowing companies to economically connect to products and collect essential data. The data is then analyzed and correlated with business intelligence (BI) and Intellectual Capital (IC), and used to provide a proactive, predictive, and preemptive service experience. This is made possible with the creation of a "feedback loop" through which the heartbeats of manufactured objects continually flow back though the complex business systems that create, distribute, and service those products. Adopters of this new IoT service model are in a great position to deliver extraordinary business performance and break away from their competition.

With this model, many companies are already offering at least a form of their products (or main features of such products) as a service with an always-on connection to fully monitor actual usage and behavior in the deployed environment. Next we will present a few key examples.

9.2.1   Thrust as a Service

Aircraft engine manufacturers are moving from the business of selling engines to the business of selling thrust as a service. In fact, Rolls-Royce has been offering such services for the last several years. It sends jet engine telemetry data to data centers for full analysis and diagnostics. An inspection can be scheduled at the correct time or spare parts can be directed to the right destination even before the pilots or the airline know that one of their engines has a problem.

Today most of Rolls-Royce engines are not sold, but rented out on an hourly basis under their TotalCare® program, and a center is monitoring maybe hundreds

Fig. 9.3 Connected Jet Engine

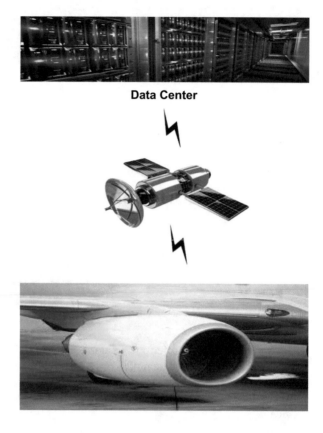

Data Center

or even thousands of engines at the same time. This model allows Rolls-Royce to accumulate a wealth of engine operational data and enables it to consult airlines on best practices. This makes it difficult for third parties to take maintenance business away from Rolls-Royce. Figure 9.3 illustrates the framework of "Thrust as a Service."

Other aircraft engine manufacturers have similar programs. Airlines do not pay for the engines, but for the time they are flying. With this model, engine manufacturers have a strong incentive to improve the reliability of their engines and drive out third-party maintenance providers.

9.2.2 Imaging as a Service

Hospitals and large medical facilities worldwide are being challenged with high cost of medical equipment and increased government regulations. Vendors of medical imaging machines (e.g., Magnetic Resonance Imaging (MRI) machines, Computed Tomography (CT) scanners, and X-ray machines) are taking advantage

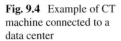

Fig. 9.4 Example of CT
machine connected to a
data center

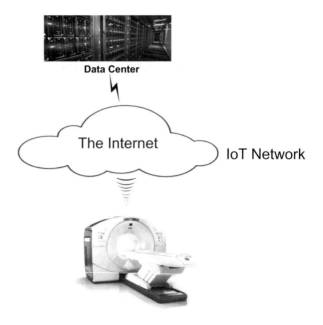

of such challenges and offering "Imaging as Service" provisions. The new connected "as a service" business model is not only reducing imaging equipment operational costs, but also offering equipment manufacturers, service providers, and
hospitals new revenue streams. Figure 9.4 depicts an example of imaging as a
service.

9.2.3   Farming as a Service

Agriculture machinery and chemical companies are also realizing the value of the
new IoT service model. Tractors and many farming machines are being equipped
with sensors and actuators. Agriculture machinery and chemical companies are
partnering together to offer Farming as a Service (FaaS) where the farming machines
are brought to a farm during seeding seasons. The machines analyze the soil square
feet by square feet, send the data back to the agriculture machinery company data
centers, where the data is analyzed in real time, and the result is sent to actuators to
release into the soil the best matching kinds of seeds and the right amount of
fertilizers.

Farming machines (e.g., tractors) may be connected over cellular (e.g., 4G) networks or drones as shown in Fig. 9.5. In the latter case, drones are deployed by
agriculture machinery companies just for the duration of seeding. Drones are typically used when the cellular signal is weak. What is another method of connecting
agriculture machinery to the network? (see Problem 7).

Data Center

Fig. 9.5 Farming as a service (FaaS)

9.2.4 IT as a Service

Another and perhaps less obvious example is the IoT network provider itself. Virtually all modern businesses/enterprises are powered by technologies, and visibility into the underlying infrastructure is mission critical. In the past, businesses relied on IT to deliver mission-critical business functions (e.g., customer portals, finical applications, email, supply chain systems, and a myriad of other crucial services that need to work flawlessly to prevent any impact on services and customers).

Today, businesses can no longer afford waiting for IT to provide all infrastructure capabilities.

As IT infrastructure continues to grow and become more complex, especially with the proliferation of hardware, software, applications, VMs, cloud services, and mobile devices, providing visibility into that infrastructure is a constantly moving target.

Vendors of IoT hardware and software solutions (e.g., sensors, gateways, routers, switches, platforms) are also offering "Feature as a Service." For instance, a network vendor may own IoT getaways (or IoT routers and switches) and simply offers connection services with guaranteed SLAs (service level agreements). As with pervious examples, the networking vendor can only do so by enabling its IoT elements (e.g., gateways, routers, switches) to collect and send data to the vendor's data centers for service monitoring, analysis, and diagnostic. Such model also allows the

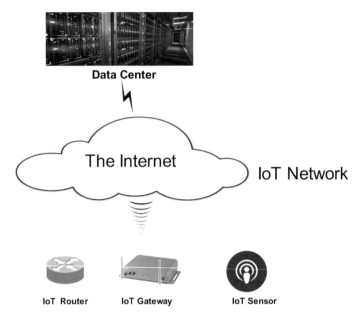

Fig. 9.6 IT as a service

vendors to gather a wealth of operational data and enables them to offer consultation to other enterprises on best practices (Fig. 9.6).

It should be noted that in all of the above examples:

- Any device or system (e.g., jet engine, medical imaging equipment, IoT gateways) downtime represents a loss of revenue or time, none of which airlines, hospitals, or IoT service providers are willing to lose. With IoT "as service" model, jet engines, medical imaging equipment as well as IoT network elements are covered via service contracts with the original equipment manufacturers. Through remote predictive monitoring and maintenance, service contract providers can fix problems before the service is even impacted.
- The ability for manufacturers to connect and pull intelligence from their systems (e.g., jet engine, medical imaging equipment, IoT gateways) has been available for some time now, primarily as an outgrowth from their own support and maintenance service offers. With IoT, a new "as a service" model is being realized. Services on top of connectivity are improving equipment ROI and competitiveness for equipment vendors and stakeholders (e.g., hospitals, OEMs, and service providers). Also, in existing solutions, connectivity may not be realized over the Internet, rather over dedicated links and proprietary networks. However, many vendors are indeed building IoT platforms to transition from propriety rigid and expensive solutions into open economical IoT-based solutions.

9.3 Enabling "Anything as a Service"

In this section, we will describe the requirements for end-to-end intelligent service automation. This includes the basic requirements for specific instrumentation and telemetry data to be provided by the product, embedded management capabilities as well as vertical-specific intellectual capital to provide a proactive, predictive, and preemptive service experience addressing the operations and health of the product.

Regardless of IoT verticals or underlying technologies, "Anything as a Service" can only be realized with several key capabilities. In this section, we will list these capabilities in ten main areas. Once the capabilities are enabled across the IoT layers, systems (e.g., IoT Platform as we specified in Chap. 6) are required to automate the end-to-end functionalities.

Given the difficulties with providing generic answers across IoT verticals, we will use the Thrust as a service as the guiding example for illustrations.

1. Which data to collect and from which entities? E.g., for the Thrust as a service example, the data includes: jet engine operational parameters including engine RPM (Revolutions Per Minute), fuel consumption, temperature, pressure, aircraft aerodynamic and mechanical operational parameters such as wind speed, ground speed, positions of flaps, positions of slats, positions of spoilers, positions of ailerons, positions of rudders, positions of elevators, positions of horizontal stabilizers, fuel level, etc.
2. How to collect (or sense) such data? E.g., using embedded pressure, temperature, or speed sensors, or by tapping into aircraft control bus messages, etc.
3. Once the data is collected and while it is in the Fog layer, what type of local analysis (e.g., by the collection agent itself) is required? E.g., an hour of flight generates terra-bytes worth of data. It makes sense to compress this data by filtering out and compacting duplicate sensor readings before transporting the data over expensive satellite links.
4. How to transmit the collected (or locally analyzed) data from the device to backend data centers securely and with minimum impact on the network? E.g., utilize satellite links for critical data that needs to be delivered in real-time, and airport Wi-Fi while the aircraft is docked at the gate for non-critical data.
5. How to entitle, validate parse, and analyze the collected data once it is received by the backend system? Hence, entitlement, data validation, data parsing, and data analysis require interactions with the supplier/partner backend systems and databases including intellectual capital information. E.g., matching the data with the correct models based on the jet-engine model and aircraft type. Segregating one airline's flight data records from those of another airline, etc.
6. Which service based performance (e.g., end-to-end delay), diagnostic and security compliance measures should be calculated at the backend and by which algorithms? E.g., fuel economy can be a function of the engine RPM, wind speed and direction (head vs. tail), flaps/slats positions, etc. Complex algorithms come into play for that single performance metric.

7. Which thresholds (e.g., Quality of Service, Grade of Service) should Step #6 estimated measures be evaluated against?
8. If Step #6 estimated measures are above the threshold, what type of real-time and none-real-time actions should be taken in the impacted device and/or the network? Which algorithms? E.g., suggest alternate flaps/slats settings on take-off or landing to minimize fuel consumption.
9. If action is needed, which secure protocol should be used access the device/network from the backend system and take action? E.g., using Secure Socket Layers (SSL) to encrypt communication between the aircraft and data centers.
10. Finally, which trending algorithms should be used to predict future measures?

Determining the required feature data (Question 1) is considered to be the most critical and difficult question especially for new technology. Feature data can only be defined if the performance measures and trending algorithms are well defined and understood.

9.3.1   Example: IoT IT Services

We will use the example of IT infrastructure as a service. Specifically, we will assume an IT infrastructure (e.g., IoT Gateways and network switches) is deployed by an IT company to provide "IT Service" to a transportation company.

IoT-based IT Service requires identifying every managed entity with an IP address, collecting data from these managed entities and performing event correlation based on vendor best practices and intellectual capital. Such information is used to proactively predict network and service performance and to provide information about future trends and threats to enable proactive remediation. This way, network planners/administrators can take action before a problem occurs thereby preventing risk-inducing conditions from occurring at all.

The most essential input for an IT service is well defined standardized embedded measurements to be collected from the network devices. This includes data subscribing to the standardized YANG (Yet Another Next Generation) data modeling language for the Network Configuration Protocol (NETCONF) or Simple Network Management Protocol (SNMP) MIBs. NETCONF and the older SNMP are network management protocols developed and standardized by the Internet Engineering Task Force (IETF).

NETCONF and SNMP are essential for FACPS (Fault, Accounting, Configuration, Performance, and Security) management. When NETCONF and SNMP data is not sufficient, "syslog" and the output of Command Line Interface (CLI) commands are also utilized. In fact, many network devices are configured to send syslog messages to an event collector, such as a syslog server, in response to specific events. The syslog protocol separates the content of a message from the transport of the message. In other words, the device sending the syslog message does not require any communication from the devices transporting or logging the message. This enables devices, which would otherwise be unable to communicate, to notify network administrators of problems. The syslog standard is documented in Request for Comments (RFC) 3164 and RFC 5424 of the IETF.

It should be noted that unlike the jet engine and medical machine examples (Sect. 9.2), which mainly employ mechanical or external sensors, IT services rely on embedded software to sense and collect data from the device. Other embedded measurements include IP SLA and Netflow as mentioned in Chap. 1.

The collected statistics are then consumed by various algorithms, utilizing the Intellectual Capital (IC) information[1] to calculate management and contract renewal related measures as outlined in steps 3–6 above. IC is another critical input for IP based smart services.

Figure 9.7 shows an overview of IoT IT services. A service becomes proactive by adding advanced software analytics algorithms to the collected data, and then delivering this results in a actionable way that provides critical value for the customers. IoT Services provide a proactive, predictive, and preemptive service experience that is automated and intelligence based to address the operations, health, performance, and security of the network. It securely automates the collection of device, network, and operations information from the network. The collected information is analyzed and correlated with the vendor's vast repository of proprietary intellectual capital turning it into actionable intelligence to aid network planners/administrators increase IT value, simplify IT infrastructure, reduce cost and streamline processes.

Fig. 9.7 Overview of IoT IT services

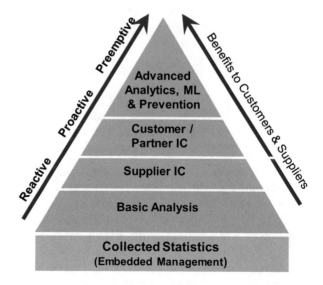

---

[1] IC information is typically captured by analyzing collected data overtime against the supplier intelligence and data bases (e.g., Microsoft collects and analyzes data from its Windows customers over the Internet).

IoT IT services enable network vendors and technology service providers to provide solutions through machine-to-machine[2] interactions that automatically provide real-time visibility and issue resolution. Such intelligence enables people-to-people interactions and enhanced social media collaboration. The interactions enable vendors and service providers to continue growing their critical intellectual capital.

Another essential requirements for IoT IT services is the smart agent with automated two-ways always-on connectivity between the device (or the network) and service management backend systems that typically reside in the network operation center (NOC), at the network supplier, or managing partner. This connection is used to (a) send uninterrupted near-real-time device/network intelligence from the device/network to the service management system(s) and to (b) allow network management system(s) to connect to the device/network to take action to prevent service outage or service deterioration.

Thus, one of the key differences between traditional network management and IoT IT service is the fact that IoT IT services utilize uninterrupted, persistent machine-to-machine or machine-to-person diagnostics, fortified with intellectual capital and best practices, in a blend designed to give network administrators deep visibility into the network. Network management solutions themselves may be connected to backend services.

With IoT IT services, network administrators have direct view and intelligence at the device, network, operations, and application layer providing automated reports and recommendations. This end-to-end approach results in network intelligence that enables network vendors (typically responsible for network and service warranty), customers/clients (network owners), and partners (typically responsible for operating, monitoring, and maintaining the network by working with vendors and customers) to deliver proactive services including regular monitoring, proactive notification, and remote remediation to enhance the customers' network availability and performance.

9.4   Connected Ecosystems

As was mentioned in quite a few chapters in this book, the number of devices connected to the Internet is already in billions and expected to reach over 20 billion in just a few years. Each of these devices is in a position to create a set of new automated services that are essential to business as well as the advancement of the world economy. Today's businesses are already requiring manufacturers to supplement their products with intelligence and connectivity. With such capabilities, IoT layers and domains will be drivers for major software development as well as services support in devices, infrastructure, platforms, and applications. No single vendor will be

---

[2] The term "Machine" refers to managed entity with an IP address such as router, switch, router interface.

able to handle a complete IoT vertical, let alone offering an end-to-end solution. IoT go to market will be driven by complex partnerships that includes a combination of Original Equipment Manufacturers (OEM), Value-added Resellers (VAR), Systems Integrators (SI), and Independent Software Vendors (ISV). IoT products, hardware and software, as well as end-to-end solutions will be developed in multi-dimensional partnerships, meaning that they are developed to integrate into IOT devices, networks, platforms, applications, and/or service. They will also be utilized to extend an IoT enabled service portfolio.

On the device and network side, for instance, suppliers have been exploiting the device embedded intelligence and connectivity capabilities to offer IoT-based services changing the traditional maintenance and support from reactive to proactive approach. These services are typically offered as part of remote management of network equipment and assets, which provides proactive network monitoring, health checkups, diagnostics, and software repairs in addition to technical support.

Suppliers are also realizing that connected devices continue to generate information value not just for services but over their lifespans. They now know the current location of the device, when it was first installed, important specifications, diagnostics, availability of spares, replacement alternatives, repair instructions, support status, and so on. This information can then be used by manufacturers and their partners for sales and marketing efforts, product development, and new customer services.

Analysts believe that manufactures who have been exposed to the values driven by connected device have a superior advantage. Their businesses will be shaped by new, significant revenue opportunities emerging from the availability of the information provided by these newly connected devices.

In the reaming of this chapter, we will describe the new IoT ecosystem-based business model, using IT use cases for illustration, and then describe the key gaps to allowing OEM, VAR, SI, and ISV to form partnership to develop end-to-end IoT solutions.

9.4.1 IoT Services Terminologies

As we just mentioned in Sect. 9.2, suppliers have been able to connect their devices (e.g., Jet Engines) to send information to their data centers for some time even before IoT fully materialized. However, proprietary communication protocols and algorithms were often utilized. The proprietary algorithms were used by tools to sense, collect, store, analyze, and transport the data. Proprietary systems are rigid in nature, developed to support a single solution and are prohibitory expensive to support and maintain (e.g., over satellites).

IoT promises to provide an open and efficient solution that can be utilized across multiple environments and technologies. The Internet Protocol itself has been shown to present a proficient and open approach to support "as a service" model as illustrated in Chap. 3.

Before we introduce IoT Ecosystem solutions, however, we will define the key terminologies to be used in the rest of this chapter.

- Product, device, or machine refers to an "entity to be managed" such as IoT Gateway, router, switch, card on the switch, platform, or application, network management system. Such entity is expected to have a unique identifier (i.e., IP Address).
- Supplier (or vendor) refers to the company that manufactures, sells, and/or leases the device/machine. e.g., Cisco is a supplier of networking devices, Rolls-Royce or GE is a supplier of jet engines, Caterpillar is a supplier of heavy machinery.
- Enterprise (or network owner) refers to a business/company that has purchased services and purchased or leased the required devices/products that are required to run the services. e.g., AT&T is a customer of Cisco and Owner of AT&T network. An end subscriber to AT&T services is a customer of AT&T and an owner of a device managed by AT&T.
- Partner refers to the third party company that partners with a vendor to service a customer network. The partner may be an OEM, VAR, SI, ISV, or business partner on the service level, e.g., IBM is a partner of Cisco that may be hired by AT&T to manage/service AT&T network.

9.4.2 IoT Connected Ecosystems Models

In this section, we will describe multiple flavors of ecosystem models that have resulted from the IOT models with connectivity and device intelligence. But first, we will describe the traditional model. Historically, vendors have sold their products to an enterprise, The enterprise fully manages the products on their own, as shown in Fig. 9.8, or the enterprise outsources the management of such products to a single or multiple partners, as shown in Fig. 9.9.

In IoT, the support paradigm is expected to be a combination of the above two models. We will refer to this model as a Full Ecosystem Model which has been empowered by Virtualization and Cloud Computing. Figure 9.10 shows a flavor of such model with Customer–Partner–Supplier Relationships. In this model, network vendors and/or their partners are often contracted by the network owners to manage the network as well as the services that are offered on the networks.

The depth of such contracts varies between companies and typically depends on the structure, resources, and expertise of the client. It can range from a limited device warranty service where vendors are responsible for the health of their devices by providing TAC (Technical Assistance Center) support and RMA (Return Material Authorization) to full Managed Service where the network vendor and/or its partner is responsible for the comprehensive management functions as well as the end-to-end services offered by the network owner to end customers. In this case, the enterprise may own some aspect of the service management (e.g., in charge of monitoring and fixing level 0 and level 1 problems). The partner owns more complex aspects of

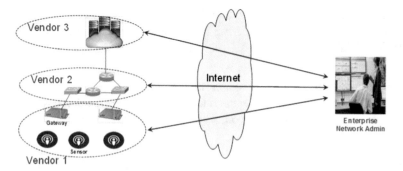

Fig. 9.8 Traditional support model—limited to vendors and enterprises

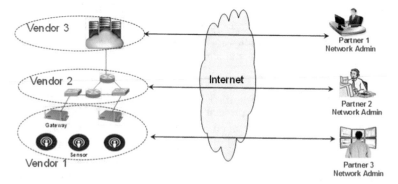

Fig. 9.9 Traditional support model—limited to vendors and partners

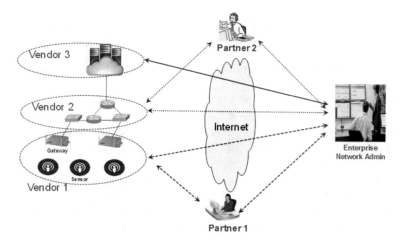

Fig. 9.10 Full ecosystem model with Customer–Partner–Supplier relationships

service management (e.g., level 2) and the vendor is responsible for levels 3 and 4 which may include fixing defects by subject matter experts as well as, RMAs and firmware update support.

It should be noted that:

- Level 0 typically means self-support by searching support documentations such as FAQs and information from the Internet. It allows users to access and resolve issues on their own without contacting a local Helpdesk or Service Desk for resolution.
- Level 1: is the initial support level responsible for basic customer issues.
- Level 2: is a more in-depth technical support level than Level 1 with more experienced technicians with knowledgeable on a particular product or service.
- Level 3: is the highest level of support in a three-tiered technical support model responsible for handling the most difficult or advanced problems.
- Level 4: While not universally used, a fourth level often represents an escalation point beyond the organization, e.g., The Research & Development organization that have developed the code and algorithms.

Other flavors of the Full Ecosystem Model include multiple partners and even vendors for the same IoT layer (e.g., sensors from multiple vendors). In this case, data integrity is very essential to prevent partner 1, for example, from accessing data managed by partner 2 especially when partner 1 and 2 are competitors.

In all of these three cases (Figs. 9.8, 9.9, and 9.10), the value of an IT product has been limited to the product itself and a traditional maintenance and support contract. With IoT, these support and "break-fix" contracts provide a valuable augmentation to the product for customers and have a potential to grow to a considerable scale.

9.4.3   IoT Connected Ecosystems Models Key Capabilities

The IoT Ecosystem model cannot work properly without addressing data privacy, standardization, and security.

Data privacy is vital to prevent data from being exposed to hackers and competitors. Data privacy is very delegate in IoT connected ecosystem model: Data must be shared but only with the appropriate vendors and/or partners to speed up the discovery of any potential issue. With multiple partners managing, the three way ecosystem model that includes vendor-partner-enterprise (Fig. 9.9) required a full proof secure system guarantees sensitive data does not fall into the wrong hands.

Security is important for every player including the enterprises, vendors, partners, and of course the end customers. Ecosystem players are not willing to risk investments unless standard technologies and methodologies are first established.

Standardization is essential to deliver scalable and flexible solutions to the market at reasonable price. It makes it possible for individual stakeholders to partner and work with IoT hardware (e.g., sensors, getaways) and software (e.g., IoT platform and applications) vendors, application developers, solution integrators, data content owners, and connectivity providers.

Outsourcing the management and operation of the network is gaining significant attractiveness in recent years. It benefits the enterprises in so many ways. Examples of such benefits include:

1. Allowing enterprises to concentrate on their own business and leave IT related functions to the experts. This is especially important for Small or Medium Business (e.g., Small Banks, Retailers) with limited IT resources.
2. Allowing network owners to introduce and deploy new technologies quickly. Network owners do not need to hire or train subject matter experts every time a new service/technology is introduced.
3. Allowing enterprises to take more intelligent risks (e.g., trying multiple technologies at the same time) by taking advantage of Cloud Computing to lease required infrastructures only for the duration of service.
4. Allowing network vendors and partners to manage the full lifecycle of the products and use the collected information to develop smarter products customized for the customer. For example, a farming equipment company may offer embedding soil analysis system that analyzes farm soil in real-time and determines the best type and amount of fertilizer, in addition to the business of selling farming traditional equipment.
5. Allowing network vendors and partners to compare the network health and KPI (Key Performance Indictors) with other networks of the same type and provide reports to the customers to repair and/or improve the network and service performance.

Key capabilities to enable connected ecosystem models include:

1. Ability to acquire essential data from managed devices or products in timely fashion. Depending on the specific IoT vertical, such capability requires agreements on the data to be collected, APIs and embedded storage via smart agents for instance. Smart agent may be defined as capability that resides on the device or product to collect the required data on regular basis or on demand. It should also have the ability to notify northbound applications based on programmed conditions (e.g., notify northbound application when the temperature change is more than 1 degree).
2. Ability for Supplier or Partner to analyze the data in timely fashion with a service platform as we mentioned in Chap. 7.
3. Ability for suppliers or partners to correlate collected data against Intellectual Capital (IC) and business intelligence rules and other databases to produce actionable results.
4. Two-ways connectivity: Connectivity allowing devices and products to send data securely to the supplier and/or partner service platform systems. It also allows the service platform system to access the device or product secularly to take action when required.

5. Secure entitlement and data transfer capability to register and entitle customer networks and communicate securely (via encryption and security keys) with service providers or network vendors as we mentioned in detailed in Chap. 8.

With the above capabilities, services will be transitioned from being reactive to being proactive and predictive.

9.5 Summary

This chapter introduced key IoT verticals that included Agriculture and Farming, Energy, Oil and Gas, Enterprise, Finance, Healthcare, Industrial, Retail, and Transportations.

Some standard bodies have used the term "Energy" to include energy consumption in smart cities as well as "Oil and Gas" in the petroleum industry. We believe "IoT Energy" and "IoT Oil and Gas" should be treated as two separate verticals. This is due to the fact that energy is produced from many other sources (e.g., Winds, solar) with focus on energy consumption. However, Oil and Gas focuses more on process and asset management for the petroleum industry.

The chapter then presented a new IoT business model, driven by the availability of new information, and offering key business benefits to the companies that manufacture, support, and service those systems, products, or devices.

Next the chapter presented the top requirements to deliver "Anything as a Service" that includes: ability to determine: which data is needed? How to capture the data? What type of local analysis is needed? How to transmit the data? How to entitle, validate parse, and analyze the collected data once it is received by the backend system? Which service based performance? Which QoS and GoS thresholds? What type of real-time and non-real-time actions should be taken in the impacted device and/or the network and which algorithms? Which secure protocol should be used access the device/network from the backend system and take action? And which trending algorithms should be used to predict future measures?

Multiple IoT verticals in combination with the new ecosystem business model were also introduced. The chapter clearly showed that no single vendor would be able to address all business requirements. Finally the chapter listed the key benefits of the proposed IoT Ecosystem partner and the capabilities to enable connected ecosystem models to function properly.

Problems and Exercises

1. What are the top ten IoT verticals As defined by oneM2M and ETSI standard bodies?
2. This chapter stated that the real impact of IoT will only occur when data from the silos is combined to create completely new types of applications. What does this mean? Why is it important?
3. What are the top two challenges to the farming industry? Why does IoT address these challenges?

4. Some Companies identified the Six-Pillar for IoT to include Connectivity, Fog Computing, Security, Data analytics, Management and automation and Application Engagement Platform. What is meant by each area? Why each of these areas is essential?

5. Complete the following Tables

IoT solution	Definition	IoT vertical
Smart and Connected Parking		
Structural health		
Noise Urban Maps		
Smartphone Detection		
Electromagnetic Field Levels detection		
Traffic Congestion		
Smart Lighting		

6. Three main use cases were listed for IoT Agriculture and Farming. List another use case.

7. What is the definition of a connected home? Provide an example.

8. Devices in connected homes can send information to service providers or directly to homeowners. List example for each case.

9. In the Farming as a Service (FaaS), Agriculture machinery companies are utilizing drones when the cellular coverage is not available.

   (a) Beside drones, what other technology may be used?
   (b) How does drone technology work?
   (c) Compare pros and cons of Drones vs. The other Technology in part (a).

10. Describe how IoT is used to monitor and optimize solar energy plants?

11. Experts believe that the lack of IoT standards and communication protocols is putting development in risk especially in Healthcare and Finical. Why is that?

12. Define the top requirements and framework to introduce "Heat as a Service" under smart building?

13. In IoT Retails use cases, retails use customer smart phones to generate heat maps that show how consumers move around stores. Why would a customer download retailer apps? What can the retailer do if customers are not willing to download the app?

14. In a table format, compare the transport, end device and place of analytics for Thrust as a Service, Imaging as a Service, Farming as a Service, and IT as a Service.

15. In Sect. 9.3 mentioned that the IT infrastructure for business is growing and becoming moving target with complexity. How is the infrastructure becoming more complex? Provide examples.

16. Describe the operational model of IT as a Service (ITaaS)? Which organization is delivering the service? Which organization is receiving the service? How is the service deliver?

17. With the availability of IoT data combined with cost effective Internet-based communications, many companies starting to contemplate why they would they

stop at selling a product and forgo very essential feedback information, when they can also sell a service with the right to monitor the actual usage and behavior of the product in the deployed environment. Usage information are not only used to service a product/device and prevent service deterioration by verifying contract level Service Level Agreements but also to learn about the product in the field and determine the most essential set of future enhancements. Provide an example.

18. With IoT, who do service providers determine "which features, of a particular product, are used the most?"
19. What is IoT-based IT Service? What are to tow top requirements for IoT-based IT and why are they needed?
20. What are the key differences between traditional network management and IoT IT service?
21. (a) Why businesses are requiring manufacturers to supplement their products with intelligence and connectivity? (b) Why is it difficult for single vendor to provide a complete IoT solution? (c) List three typical partnerships that vendors needs to establish to provide complete IoT solutions.
22. Some IoT standard bodies have combined "IoT Energy" and "IoT Oil and Gas" into one vertical, called "Energy". However, the authors have decided to keep "IoT Energy" and "IoT Oil and Gas" as two separate verticals. What was their arguments based upon?
23. What is the 80–20 Business Rule? Which IoT Businesses does it apply to?
24. Why many suppliers are utilizing IoT connectivity to generate information value not just for services but over their lifespans? Provide examples of such information.
25. What Level 0–4 support in Technical Services? Is there a Level 0? If so, what is it?
26. What is IoT Full Ecosystem Model? Which major technology has made make such model feasible?
27. What are the top three requirements that are required for the IoT Connected Ecosystems Model to work? Provide a brief summary of each requirement?
28. Why outsourcing the management and operation of an IoT network is gaining significant attractiveness in recent years?
29. What are the top five capabilities to enable connected ecosystem model for IT-based service?

References

1. OneM2M White Paper, "The Inter operability enabler for the entire M2M and IoT Ecosystem", January 2015, Online: http://www.onem2m.org/images/files/oneM2M-whitepaper-January-2015.pdf
2. OneM2M Technical Report: OneM2M-TR-0001-UseCase, September 23, 2013.
3. TM Forum Vertical Markets and connected Ecosystems, Online: https://www.tmforum.org/vertical-markets-connected-ecosystems/

4. Internet of Things World Forum Cisco Day 1 Presentation, October 14, 2014, Online: http://www.slideshare.net/BessieWang/iot-world-forum-press-conference-10142014
5. A. Slaughter, G. Bean, & A. Mittal, "Connected barrels: Transforming oil and gas strategies with the Internet of Things", Deloitte University Press, August 12, 2015. Online: http://dupress.com/articles/internet-of-things-iot-in-oil-and-gas-industry/
6. SAP, "The CEO Perspective: Internet of Things for Oil and Gas Top Priorities to Build a Successful Strategy", 2014, Online: https://www.sap.com/bin/sapcom/en_us/downloadasset.2014-10-oct-30-20.the-ceo-perspective-internet-of-things-for-oil-and-gas-pdf.html
7. 5Liblium, 50 Sensor Applications for a Smarter World, Online: http://www.libelium.com/top_50_iot_sensor_applications_ranking/
8. "Rolls-Royce Totalcare: Meeting the needs of Key Customers", Executive Briefing #6, March 2013, Online: http://www.som.cranfield.ac.uk/som/dinamic-content/media/Executive%20Briefing%206%20-%20RR%20Totalcare%20-%20Mtg%20the%20Needs%20of%20Key%20Customers%20-%208%20Mar%2010%20v9.pdf
9. "Driven by Increased Demands on Healthcare Suppliers, Connected Medical Imaging Equipment to Grow at a 17% CAGR", ABI Research, Mach 20, 2015, Online: https://www.abiresearch.com/press/driven-by-increased-demands-on-healthcare-supplier/
10. "IoT Services for Medical Imaging Equipment Market: MRI, Xray, CT Scanners, and Tomography", ABI Research, Online: https://www.abiresearch.com/market-research/product/1021023-iot-services-for-medical-imaging-equipment/
11. Federal energy Regulatory Commission Assessment of Demand Response and Advanced Meeting Staff Report, December 2008, Online: http://www.ferc.gov/legal/staff-reports/12-08-demand-response.pdf
12. J. Eckenrode, "The Internet of Things and financial services: Too much—or not enough—of a good thing?", Deloitte Quick Look Blog, October 7, 2015, online: https://quicklookblog.com/2015/10/07/the-internet-of-things-and-financial-services-too-much-or-not-enough-of-a-good-thing/
13. Technology Target, A guide to healthcare IoT possibilities and obstacles, Online: http://searchhealthit.techtarget.com/essentialguide/A-guide-to-healthcare-IoT-possibilities-and-obstacles
14. "The IoT, What the IoT Means for the Public Sector", Hitachi Data Systems, Online: http://www.isaca.org/Groups/Professional-English/cybersecurity/GroupDocuments/IoT%20in%20the%20Public%20Sector.pdf
15. A. Rayes, Book Chapter: "IP-Based Smart Services", Network Embedded Management and Applications, Springer, 2012, http://www.springer.com/engineering/signals/book/978-1-4419-6768-8
16. C. Chen, L. Yuan, A. Greenberg, C. Chuah and P. Mohapatra, "Routing-as-a-Service (RaaS): A Framework For Tenant-Directed Route Control in Data Center", IEEE/ACM Transactions on Networking, Vol. 22, No.5, Oct 2014.
17. K. Lakshminarayanan, I. Stoica, S. Shenker, J. Rexford, "Routing as a Service", Online: http://www.icsi.berkeley.edu/pubs/networking/routingservice04.pdf
18. "Real-World IoT Case Studies, Best Practice and IoT Methodology", entrsise-iot.org, Online: http://enterprise-iot.org/
19. Internet of Things: http://www.mckinseyquarterly.com/The_Internet_of_Things_2538
20. "The Internet of Things: How the Next Evolution of the Internet Is Changing," Cisco IBSG white Paper, Dave Evans, April 2011.
21. P. Marshall, "The 80/20 Rule of Time Management: Stop Wasting Your Time," Online: https://www.entrepreneur.com/article/229813
22. GE Services: http://www.geaviation.com/services/index.html
23. Boeing commercial Aviation Services: http://www.boeing.com/commercial/aviationservices/index.html
24. Boeing Gold Care Solution: http://www.boeing.com/commercial/goldcare/index.html
25. Caterpillar Services: http://www.cat.com/parts-and-service

26. John Deere Services and John Deere Farm Sight: http://www.deere.com/wps/dcom/en_US/
    services_and_support/services_support.page,   http://www.deere.com/en_US/CCE_promo/
    farmsight/index.html
27. Go Green Intuitive, June 2016, Online: https://gogreeninitiative.org/wp/
28. Computer Networking A Top-Down Approach, 5th Edition, J. F. Kurose and K. W. Ross,
    Addison Wesley.
29. Glen Allmendinger and Ralph Lombreglia, *Four Strategies for the Age of Smart Service*: http://
    hbr.org/product/four-strategies-for-the-age-of-smart-services/an/R0510J-PDF-ENG
30. J. Case, et al., "Simple network management protocol (SNMP)," RFC 1157, 1990.
31. R. Presuhn, et al., "Management Information Base (MIB) for the Simple Network Management
    Protocol (SNMP)," ed: STD 62, RFC 3418, 2002.
32. L. Steinberg, "Troubleshooting with SNMP and Analyzing MIBS," ed: McGraw-Hill
    Companies, 2000.
33. Alfa Images Online: http://alfa-img.com/show/farm-planters.html
34. The Internet of Everything – Vision and Strategy, Cisco Systems, March 2013, Online: http://
    www.slideshare.net/Cisco/2-internet-of-everything-vision-and-strategy-rob-lloyd-final
35. Integrated Framing Systems, Online: http://www.monsanto.com/investors/documents/whis-
    tle%20stop%20tour%20vi%20-%20aug%202012/wst-ifs_posters.pdf
36. How the Internet of Things is Reinventing Retail, Position Paper by ComQi, July 2015, Online:
    http://www.comqi.com/internct-things-reinventing-retail/
37. Infosys Study "Rethinking Retail Insights from consumers and Retail into an Omi-Channel
    Shopping Experience", 2014, Online: https://www.infosys.com/newsroom/press-releases/
    Documents/genome-research-report.pdf
38. "How the Internet of Things Is Improving Transportation and Logistics", Supply
    Chain 247, September 9, 2015, Online: http://www.supplychain247.com/article/
    how_the_internet_of_things_is_improving_transportation_and_logistics/zebra_technologies

Chapter 10
The Blockchain in IoT

10.1 Introduction

The role of centralized governance over networks and entities has allowed for the mass control of digital media and private life. As the Internet has evolved, researchers and developers have looked for new ways to distribute control and trust. Blockchain technology was first introduced in 2008 with the famous Bitcoin whitepaper by pseudonym Satoshi Nakamoto. Since then, we have seen a global wave of interest and investments into the world of cryptocurrencies and digital assets. While some are just trying to invest into cryptocurrencies, others believe more in the underlying technology behind it—blockchain.

Through the use of blockchain technology, one can decentralize an entire network—never relying on a central entity—and can place trust across all users instead of one central node. By distributing the data throughout the network, any one person or computer can contact their closest node to retrieve information residing on a common ledger.

Many expect that blockchain technology has the potential to transform a range of different industries. Because of this, blockchain is already being used and researched by many of the leading companies in technology. While many efforts are still in their infancy, and there are many challenges to solve, it is expected that blockchain has the power to propel significant transformations in the IoT sector.

Cisco estimates that there will be roughly 26 billion devices connected to the Internet by 2020. Server-client models will struggle to scale to such demand. Centralized models mean high maintenance costs for the manufacturer, and limited consumer trust in devices that are always connected to the Internet [3]. Blockchains facilitate the sharing of services and assets like never before. These types of possibilities have led companies like IBM, Cisco, and Intel to contribute to blockchain in IoT efforts.

© The Author(s), under exclusive license to Springer Nature Switzerland AG 2022
A. Rayes, S. Salam, *Internet of Things from Hype to Reality*,
https://doi.org/10.1007/978-3-030-90158-5_10

There are countless digital currencies and innovative applications being developed on top of blockchain. The impact of these efforts will be hard to predict. In IoT, blockchains can facilitate things like M2M transactions, automated firmware updates, or even the tracking of food quality and control. Imagine cars automatically negotiating rates for parking spaces, or drones automatically reserving and paying for a landing pad. These are just a few possibilities, and in this chapter, we explore further how the blockchain can impact the IoT domain.

The chapter is organized in the following way. Section 10.2 defines the blockchain. We describe the difference between Bitcoin and blockchain, and provide an overview of how blockchain has evolved over time. In Sect. 10.3 we dive into how blockchains work, and review the features that make the technology important. Section 10.4 introduces how the blockchain may impact notable use cases in IoT, and reviews the advantages and disadvantages of blockchain technology. Lastly in Sect. 10.5, we go over security considerations within blockchain and IoT.

10.2   What Is the Blockchain?

Before learning what a blockchain is, we should first understand why Bitcoin and the blockchain were introduced together in the original Bitcoin whitepaper. Bitcoin was presented as the *peer-to-peer electronic payment system*, and blockchain was the proposed *mechanism* that allowed it to work. A peer-to-peer digital currency needs a mechanism that allows its users to trust each other without the need for a central authority (like a bank). It is in the Bitcoin whitepaper that Satoshi Nakamoto proposes such a mechanism. More specifically, Nakamoto proposes the blockchain as the solution to the double spending problem—how to tell if a user, or device, has spent the same digital coin more than once. Double spending is particularly hard to detect in a distributed system like Bitcoin, because there is no central authority tracking balances. This means that without a solution like the blockchain ledger, it is easy for a user to send the *same* coin to different users before anyone in the network learns of the fraudulent transactions. Blockchain is therefore what allows Bitcoin to be a trustless system, and is the key innovation responsible for the success of Bitcoin and other cryptocurrencies that later emerged.

> What is needed is an electronic payment system based on cryptographic proof instead of trust, allowing any two willing parties to transact directly with each other without the need for a trusted third party…In this paper, we propose a solution to the double-spending problem using a peer-to-peer distributed timestamp server…—Nakamoto, 2008

10.2.1   *Bitcoin and Blockchain*

It is important to make a clear distinction between Bitcoin and the blockchain. As mentioned earlier, the blockchain is the *mechanism* that allows Bitcoin to work. Thus, Bitcoin can be considered to be an *application* that uses blockchain—but

Fig. 10.1 Bitcoin vs.
Blockchain

Table 10.1 Categories of Blockchain

Categories	Description
Blockchain 1.0	Blockchains used for currencies
Blockchain 2.0	Use of smart contracts within blockchains
Blockchain 3.0	Applications beyond currency and financial markets

blockchain can be used on its own. It can be used to enable other cryptocurrencies, or as we will see in the next section, blockchain can also enable an array of different applications beyond Bitcoin and other cryptocurrencies (Fig. 10.1).

A simple analogy we can use is that of the *car* and the *combustion engine*. A car uses a combustion engine to function, but the combustion engine can be used to power other systems such as buses, trucks, boats, electrical generators, etc. Thus, we can think of the blockchain as the combustion engine and Bitcoin as the car. Bitcoin is just the first example of many possible applications of blockchain technology.

10.2.2   Evolution of Blockchain

Since its introduction in 2008, the blockchain has evolved as it has been adapted in a wide range of applications and industries. In Table 10.1, we break down the different categories of blockchain as proposed by Melanie Swan in the book *"Blockchain, a Blueprint for a new economy."*

Blockchain 1.0: Blockchain 1.0 consists of the use of blockchain in digital *currency* applications for the decentralization of money or payment systems. This includes Bitcoin, other cryptocurrencies, and payment systems. In the beginning, these were the first applications to employ blockchain as a technology.

Blockchain 2.0: The next major innovation in blockchain, considered Blockchain 2.0, is a technology known as *"contracts."* Beyond peer-to-peer payment systems, Blockchain 2.0 includes the transfers of other property such as stocks, bonds, and smart property. It also includes *"smart contracts,"* which are described later in this section.

Blockchain 3.0: Blockchain 3.0 consists of all applications beyond currency and markets. This includes the use of blockchain in areas like healthcare, governments, and commercial settings. In Sect. 10.5 of this chapter, we cover a couple of these segments, and the potential use cases of blockchain in IoT.

10.2.3   Defining Blockchain

A blockchain is composed of a distributed digital ledger that is immutable—cannot be edited—and is shared among all participants in a blockchain network. More specifically, a blockchain is a data structure composed of timestamped and cryptographically linked blocks. Each block has a cryptographic hash, a list of validated transactions and a reference to the previous block's hash. Through this mechanism, nodes can verify that a participant owns an asset without the need for a central governing authority. The key characteristics behind the success of blockchain are as follows:

1. Decentralized architecture.
2. A "trustless" system.
3. Consensus mechanism.
4. History of transactions.
5. Ensured immutability.

We consider these as the key factors that have made the technology transformational. The blockchain allows for participants to engage in trustless peer-to-peer transactions. In short, it is said that decentralized, trustless transactions are the key innovation of the blockchain [1].

10.3   How Blockchains Work

A blockchain is just what the name implies, a group of blocks linked, or chained, together cryptographically. It also keeps record of all transactions that have ever been executed by nodes on the network. In this section, we provide an overview of how blockchains work by using Bitcoin as an example. We examine how transactions are created, how they are broadcasted, how they are recorded into blocks, and how they are accepted into the distributed network of nodes.

Important Definitions

Nodes: Any computer or device connected to a blockchain network.
Ledger: A shared and distributed history of all transactions and balances.
Mining/Miners: In Bitcoin, mining is the process of generating a new legitimate block by applying proof-of-work. There are people that dedicate their nodes to "mine" new blocks. These nodes are considered "miners."

Consensus: A consensus algorithm is the mechanism by which all nodes in the network agree on the same version of the truth. A consensus algorithm allows nodes on the system to trust that a given piece of data is valid, and that it has been synchronized with all other nodes.

Cryptocurrency: A digital currency built upon cryptographic protocols.

Decentralized Application (DAPP): A decentralized application built on top of a blockchain based system.

Secure Cryptographic Hash Functions: A secure cryptographic hash function is a hash function that preserves one-wayness—easy to compute, but virtually impossible to reverse engineer.

Cryptographic Keys: The use of symmetric (same) keys and asymmetric (public-private) key pairs for the use of signing and verifying transactions.

Merkle Tree Root: The root of a Merkle tree (binary hash tree). The root is the result of all leafs hashed together to a single hash.

10.3.1 Anatomy of the Blockchain

Components of the block's header:

1. **Version**: The version of block validation rules it follows.
2. **Previous Block Hash**: The hash of the previous block in the blockchain.
3. **Merkle Root Hash**: The root of all transactions hashes in a block.
4. **Timestamp**: The Unix epoch time the block was mined.
5. **Bits**: Encoded version of the target threshold.
6. **Nonce**: Arbitrary number that can only be used once.
7. **Transaction Count**: Total count of transactions contained within this block.

In Fig. 10.2, we show the basic architecture of the blockchain. A blockchain is very similar to a linked list—each block contains a pointer to the previous block. A key difference in blockchain is that each block contains a *hash pointer* to the previous block. A hash pointer contains two things: A pointer, or reference to the location of the previous block, and the cryptographic hash of that block. Storing the cryptographic hash of the previous block allows us to verify that the block we are pointing to has not been tampered with. To verify a block, we simply compare our stored hash pointer with the previous block's hash and make sure they are equal.

10.3.2 Understanding a Block's Hash

Cryptographic hash functions are an important aspect of blockchain's security. For this reason, let us take a look at how block hashes are calculated and how they are used in preventing an attack. To calculate the hash, three inputs are used: Previous block hash, the Merkle root hash, and the nonce. These values are processed by the

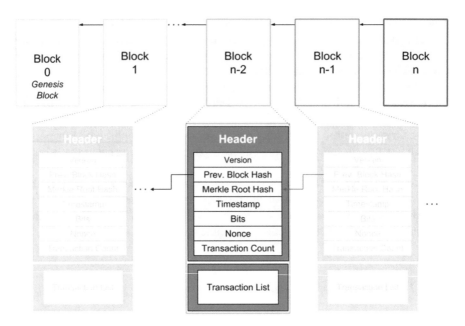

Fig. 10.2 Anatomy of the Blockchain

SHA-256 cryptographic hashing algorithm. The output is the block hash—a fixed size output that uniquely represents all of the block's contents.

In Bitcoin, hashing is performed by miners, and the hash produced must be lower than the target hash set by the network. To find a hash meeting this criteria, miners try different nonce values and check if the output hash is lower than the target, while the previous block hash and Merkle root hash remain the same. Miners do this iteratively until a valid hash is found. Because of this, the mining process consumes a lot of power and compute resources. This procedure is how the miners create *proof-of-work*. In Fig. 10.3 we illustrate how the block hashes are calculated.

To understand how it works, consider a scenario where an attacker attempts to pay themselves some Bitcoins by modifying one of the blocks in the chain. Imagine they attempt to add a fake transaction to block 1, claiming that someone has sent them some coins. Upon changing the transaction list, the hacker will be forced to update the Merkle root hash. Because the block's hash is dependent on the Merkle root hash, if the Merkle root hash is altered, then we must recalculate the block's hash. But that is not so easy. In Bitcoin, it takes considerable compute power to mine one block. So the attacker would then have to invest power and time recalculating the block they maliciously altered. Once the attacker has calculated the new hash, then they have to figure out a way to make the block a legitimate part of the blockchain.

This is where hash pointers play a key role. For the attacker to alter any block in the chain, they also have to change every other block that follows! Why? Because every subsequent block points to the previous block—block 2 contains a hash

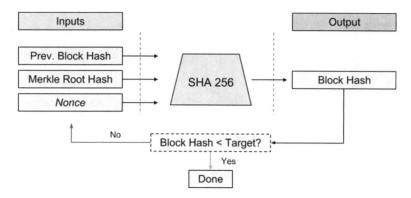

Fig. 10.3   Block SHA256 calculation

pointer to block 1. But our attacker was forced to recalculate the hash for block 1, so a comparison with the hash pointer in block 2 will fail. To avoid this, the attacker must change the hash pointer in block 2 to match the block hash of the new (malicious) block 1. But changing the hash pointer in block 2, changes block 2's hash. Thus, the attacker would have to recalculate the hash for block 2 as well. Once they change block 2, same would have to be done for block 3, and block 4 …etc. During this time, the network has still been progressing while the attacker is spending time altering past blocks. Time and cost used in such an attack are expensive and pointless as long as the attacker holds less than 51% of the network's compute power. It is this combination of proof-of-work and hash pointers that trumps 51% attacks and is considered to be the fundamental security feature of Bitcoin's blockchain.

10.3.3   Lifecycle of a Transaction

To understand how transactions are executed in a blockchain, let us consider an example scenario where Alice sends Bob .5 Bitcoin (BTC). In order for the transaction to take effect and be accepted into the blockchain, the following main steps need to be completed (Fig. 10.4):

Let us assume Alice's current balance is: 10BTC and Bob's is 2BTC.

1. **Alice agrees to *send* Bob .5BTC**

    (a) Alice initiates a transaction using Bob's Bitcoin address. While Bob's identity is not linked to his Bitcoin address, Bob may create a new address for every new transaction to minimize tracking of his activity.

    (b) Bitcoin is pseudo-anonymous, meaning Bob's transactions are not fully obfuscated, and if his address is exposed in connection to his identity, then there are tools that can potentially track all of his past activity on Bitcoin.

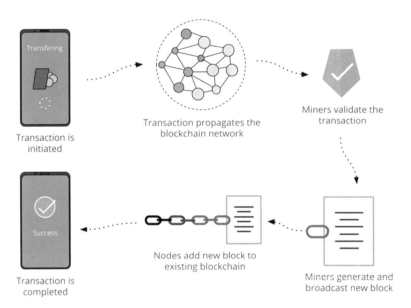

Fig. 10.4 How a Bitcoin transaction is executed

2. **Alice** *generates a transaction*

   (a) When Alice broadcasts a transaction to the blockchain network, the message notes that Alice should now have .5 less BTC and Bob should gain .5BTC. In reality, no coin or asset is actually transferred (there is no digital coin that actually exists in the form of bits), instead, only records of transactions are recorded in the blockchain's ledger. In order for the transaction to be broadcast securely, Alice signs that the transaction is legitimate. This verifies that no one is trying to withdraw coins out of her wallet without permission. Alice signs the message using her private key to Bob's public key, thus only Bob can spend these coins.

3. **Alice's wallet or interface into the Bitcoin network will now propagate the transaction to known peers**

   (a) Once the transaction has been generated and is valid, Alice's wallet or interface to the network will propagate the transaction to her known peers. These nodes will in turn propagate it to their peers upon validating the transaction. This mechanism is called *flooding*.

4. **Miners receive the transaction, and** *validate* **it, ensuring that it has not been corrupted or tampered with.**

   (a) Miners will use the consensus rules to validate the transactions making sure there is no double spending and that each address associated with the transaction exists.

5. **Miners include the transaction into a block, and apply the consensus algorithm (proof-of-work in case of Bitcoin) to *mine* a new block.**

   (a) Transactions are then added to the new block in order of precedence. Transactions are added in descending order based off of their fees. Each transaction usually contains a fee that is paid to the miner. Once the miner receives the previous block in the network, they will start mining the newest block.

6. **Once a new block is mined, miners then broadcast the new block to be *added to the blockchain* by all other nodes in the network.**

   (a) The miner will propagate its new block to the network and begin the process all over again with new transactions.

10.4 Features of Blockchain

A blockchain provides key benefits that have never been possible before. These benefits stem from the clever combination of novel and existing technologies that allow the community to build innovative blockchain based solutions. In this section we cover some of the important features that a blockchain provides and discuss why they are important in IoT.

10.4.1 Consensus Algorithms in IoT

Blockchains can be considered "trustless" because they provide a mechanism to validate that data being added to the blockchain is legitimate. To achieve this, all nodes need a way of agreeing on the correct version of the truth. The algorithms used to reach an agreement are referred to as "consensus algorithms." For example, Bitcoin uses the *"Proof of Work"* (PoW) algorithm, but as we will see in this section, PoW is not the only algorithm that exists; there are many, and all of them offer different advantages and disadvantages. In IoT, it is essential that the consensus algorithms used can meet certain security, energy consumption, and computational requirements. In this section, we introduce a short list of the most prominent consensus algorithms, and examine their viability in IoT solutions.

Byzantine Generals Problem Before diving into different consensus algorithms, let us further define the goal of a consensus algorithm. In July 5th, 1982, Leslie Lamport, Robert Shostak, and Marshall Pease published a paper named "The Byzantine Generals Problem." From the original paper:

> *...imagine that several divisions of the Byzantine army are camped outside an enemy city, each division commanded by its own general. The generals can communicate with one another only by messenger. After observing the enemy, they must decide upon a common*

plan of action. However, some of the generals may be traitors, trying to prevent the loyal generals from reaching agreement. The generals must decide on when to attack the city, but they need a strong majority of their army to attack at the same time. The generals must have an algorithm to guarantee that (A) all loyal generals decide upon the same plan of action ... (B) A small number of traitors cannot cause the loyal generals to adopt a bad plan...The loyal generals will all do what the algorithm says they should, but the traitors may do anything they wish. The algorithm must guarantee condition A regardless of what the traitors do. The loyal generals should not only reach agreement, but should agree upon a reasonable plan.

In the case of blockchain, the generals are the nodes in the distributed network, and the messages are the communications, or transactions, across blockchain network. In short, how do all truthful network nodes reach a consensus on the validity of a new transaction even if there exists a certain percentage of malicious or faulty nodes? A **Byzantine Fault Tolerant** system is one that can tolerate the Byzantine Generals Problem.

Proof-of-Work (PoW)   Proof-of-Work algorithms require "miners" to solve a very complex cryptographic puzzle to try to prove that the current transactions on the blockchain are valid. This is the consensus algorithm used in Bitcoin. All miners receive transactions and begin a race to "mine" a new block. The first "miner" to solve this puzzle correctly, wins and receives an incentive in return. In Bitcoin, "miners" receive Bitcoins as a reward. The reward is halved every 210,000 blocks. In PoW, nodes trust the longest chain—the one with the most blocks added to it by other miners. Thus PoW is safe as long as 51% of the compute power is owned by honest miners.

In PoW, solving the puzzle consumes a lot of computational power and takes considerable amount of time to complete. Thus, adding new blocks translates to high energy costs and low amount of transactions per second. In IoT, both present a big challenge. First, the actual sensors/devices on the network will not be interfacing with computation and consensus. The main gateway and fog domain will most likely be in-charge of computation and consensus as they can manage memory and power in a more sustainable fashion. Sensors will primarily rely on sending information to the fog and dealing with identity management between peers. PoW could potentially work with IoT devices, but there would have to be a strict separation of compute nodes and light clients (sensors) throughout the network. We argue that while a feasible algorithm for IoT, it is not a good choice for IoT solutions due to the large computational and energy consumption requirements that will have to be introduced into the networks.

Proof of Stake (PoS)   Proof of Stake does not require expensive compute resources to mine blocks. Instead, PoS uses a validation process based on the amount of coins that you already own. If you own 1% of the stake in the blockchain, then you will have a 1% chance of getting chosen to create, or "mint," a block. Thus, simply by having a stake in the system, you can be chosen to "mint" a block. The idea is that the more value you have at stake in the system, the less likely you will be willing to create a malicious block. If a block is invalidated by the rest of the network, then you lose your stake. This action will fall into an invalidation period, where the consensus for that transaction may be taken over by fellow peers, but your validity will drop among the nodes.

We argue that PoS would be a good fit for IoT because does not suffer from PoW energy drawbacks and does not require high computational capabilities. With PoS, a possible drawback is that a node with more stake has more control of the network; and this control can continue growing because the node with the most stake is more likely to be chosen to mint a block. In permissioned blockchains this should not be a problem, but more research is needed to understand the effects of PoS in permissioned and permissionless IoT blockchains.

Proof of Activity (PoA) Similar to PoW, Proof of Activity requires miners to mine a new block, the only difference being that the transactions on the network are not required to be part of the new block, the mining is done for the sole purpose of solving a cryptographic puzzle. Once a new block is found, a similar validation to PoS is performed. The block is broadcasted to a group of chosen validators for them to sign the new block. The likelihood a new validator is chosen is similar to that of PoS, the more stake they own in the network, the more likely they will be chosen to sign the new block. Proof of Activity suffers from the same drawbacks as PoW. Because of this, it is probably not a good choice for IoT applications.

Proof of Elapsed Time (PoET): Proof of Elapsed Time is a bit different than the other consensus algorithms mentioned so far. PoET was developed by Intel and is a proposed contribution to the open-source Hyperledger blockchain project. At a high level, PoET essentially works by assigning each node a random wait time, the validator with the shortest wait time "wins" and gets to mine the next block. The algorithm is considered to be "lottery algorithm"—the probability of being selected is proportional to the amount of resources contributed. This consensus algorithm has advantages in that it is much more energy efficient than PoW and does not require expensive hardware. On the other hand, it requires Intel processors to run it (requires trusted execution environment on the CPU), in which case it requires trust in Intel's hardware, which many say goes against the decentralization of trust concept. As far as IoT devices are concerned, we believe that PoET would be a good option for private IoT blockchains. This is because there is no need to have high compute power, or expensive hardware, and is also power efficient (Table 10.2).

While not an exhaustive list of consensus algorithms (and there are many), it is easy to see that at the heart of a blockchain is the consensus algorithm that glues the whole system together. Each consensus algorithm will have its own advantages and disadvantages depending on the use case; different industries and applications will apply different consensus depending on requirements such as scalability, transactions per second, and if the system will be permissioned or permissionless.

Table 10.2 Consensus algorithms in IoT

Consensus Algorithm	Description	IoT compatibility
Proof-of-Work	Computation is needed to solve cryptographic puzzle to ensure consensus.	No
Proof of Stake	Ability to mint a new block is proportional to the stake in the blockchain network.	Yes
Proof of Activity	Computation is needed to solve cryptographic puzzle to only known validators who are active.	No
Proof of Elapsed Time	Use of random time intervals that determine which node is the current miner.	Yes

10.4.2   Cryptography

What makes blockchains trustworthy and secure is its underlying mechanisms based on cryptography, signed keys, and digital signatures. While Bitcoin has been exposed to various attacks in the past, it is worth noting that the ledger itself, or the blockchain, has never itself been knowingly hacked. In the past, Bitcoin hacks targeted Bitcoin wallets or Bitcoin exchange websites instead. Let us consider Bitcoin's cryptographic elements as an example, and see how they are used to maintain the block chain's integrity. Bitcoin's cryptographic components are mainly composed of:

- **Secure Hash Algorithm (SHA-256)**: Cryptographic hash functions are a set of mathematical functions that output unique outputs for unique inputs. The input can be of any size, and the output is always a fixed size—256 bits (32 bytes) in the case of SHA-256. If any one bit of the input is changed, the cryptographic hash function outputs a completely different and unpredictable output. Secure cryptographic hashing preserves one-wayness, that is, you can easily produce a hash from a given input, but it is extremely difficult to generate the input to the hash by only knowing the hashed output value. How difficult? SHA-256 is used for most functions including integrity, block-chaining, and hashcash cost function calculations.
- **Elliptic Curve Digital Signature Algorithm (ECDSA)**: ECDSA is used to create cryptographic keys that can derive addresses for use within the blockchain. Each ECDSA algorithm calls a specific curve to be used for key generation, which enables efficient computation.

Cryptography is at the heart of why the blockchain is so revolutionary. Everything from consensus algorithms, to encryption, to the immutability aspects of the blockchain are due to the underlying cryptography. This is a fundamental key in unlocking the potential to IoT, as different devices need to engage in transactions with trustless entities and devices on a constant basis.

10.4.3   Decentralized

Having a decentralized architecture can propel IoT applications to be realized at a wide scale. Currently, IoT systems mostly depend on client/server or publish subscribe architectures [7]. Centralized architectures require expensive infrastructure with high compute and storage capabilities. In addition, they present a form of centralized control that can be act as a single point of failure or the target of a security attack. Publish subscribe architectures can also have a few drawbacks with scalability and security. If devices could perform secure transactions using a peer-to-peer paradigm, it would greatly reduce the cost, transaction time, and probability of service interruption.

The blockchain is composed of a decentralized, distributed network of nodes that participate in transactions and maintenance of the network. This is the core concept behind blockchain. All transactions are peer-to-peer and are tracked by all of the participating nodes in a network. Blockchain networks have a reliability factor of $(n - 1)$—if any node fails, or drops from the network, there is no interruption to service. The network always maintains availability and fault tolerance. Decentralization in IoT is a very attractive alternative to previous architectures, but there are still many challenges, and no clear consensus on how to best take advantage of blockchains decentralized nature in IoT (Fig. 10.5).

10.4.4   Transparency and Trust

The use of a public ledger allows all nodes on the network to see the entire history of the given blockchain. This opens access to the history of data on the chain, giving transparency to all transactions. The trust that is built up within the network is maintained through the use of the public ledger and gossip protocol. Each node always

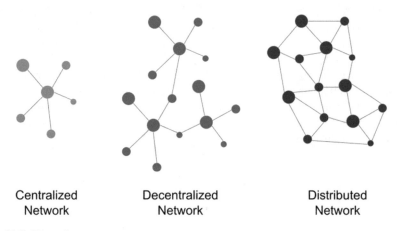

| Centralized Network | Decentralized Network | Distributed Network |

Fig. 10.5 Network types

knows of its nearest neighbors and new nodes. Through each node gossiping to one another, they learn of new transactions. Utilizing the public ledger and protocol instills trust within each node as each node is responsible for one another. This decentralization mechanism holds the nodes responsible for the integrity of the network.

10.4.5   Permissioned, Permissionless, and Consortium

Permissionless blockchains, such as Bitcoin, are designed so that anyone can join and participate in the network without having to establish their identity. There is no need to verify a given user through some sort of identity management system. The only identity needed is the user's public key. In contrast, permissioned ledgers are primarily used in private applications where strong indicators of identity are required to join the network. Permissioned ledgers are preferred among B2B and B2C enterprises. There are usually multiple layers of validation before enrollment to the network is verified. The use of regulators, as seen in IBM and Linux Foundation's Hyperledger, is used to ensure all users meet various requirements on the network. Other blockchains such as Ethereum give one the option to setup the network as permissioned, permissionless, or consortium. Consortium blockchains are very similar to permissioned blockchains. The key difference being that in a consortium, new participants are authenticated by a predetermined group of private entities.

10.4.6   Smart Contracts

Originally introduced by Nick Szabo in 1994, smart contracts consist of small computer programs that contain—embedded in their code—an agreement between two entities. This contract is then distributed across the blockchain and is responsible for facilitating the execution, verification, and enforcement of an agreement between seller and buyer. Essentially, a smart contract is just a digital, auto-enforceable version of a traditional paper-based contract. Ethereum is the most popular blockchain system with embodied smart contracts. It has a current market cap of more than $35 billion as of November 2017. As we will see in later sections, smart contracts allow devices in IoT to negotiate and execute previously agreed actions automatically, enabling a new set of functions and use cases for IoT solutions.

10.4.7 *Advantages and Disadvantages*

There has been much debate over the use of blockchain technology and its possible applications. In most use cases, traditional back end infrastructures offer a good solution to existing problems. Yet, the industry is beginning to move to a more decentralized infrastructure to improve security and trust between users and the rest of the network. While blockchain presents a lot of promise, it is not a silver bullet—blockchain does not solve all security and privacy concerns, it is only part of the solution. With every new technology there are advantages and disadvantages, and blockchain is just one part of a complex technology stack.

Blockchain technology has multiple disadvantages that have decreased its adoption rate. An often-overlooked challenge is that the technology is initially difficult to understand and adopt. Trying to get people to use blockchain applications is a difficult task, which brings disadvantages as people believe it is an unnecessary precaution for a network.

Scalability is another widely debated challenge. As an example, there has been much debate over scaling in regard to Bitcoin, which brought about a fork in the chain to allow larger than 1 MB block sizes. People felt that this size limitation does not scale with the adoption of Bitcoin and transactions will take longer and longer to be validated and added to the main chain. There has also been other discussions of the scalability of Ethereum with the nature of storing everything within various Merkle roots, where over time downloading the full chain will be much larger than Bitcoin's full chain (as of April 2018 its around 180 GB). To avoid similar storage issues, people—especially users on mobile devices—use Simplified Payment Verification (SPV) nodes which allow them to not run a full node and use filters to only grab the information that they need. This will rise over the next years as well as the use of Lightning network and other off-chain protocols.

Other disadvantages include the size of the network and limiting the control of nodes. Whether one is building a permissionless, permissioned, or consortium blockchain, limits will have to be set on the admin privileges of nodes, so that the network does not gravitate towards a "centralized" paradigm. With this, there is always the risk of a Sybil (51%) attack on the network. As these are definitely important disadvantages, there are also a great deal of positives from the technology.

Blockchains bring about a new way to enable privacy and security between parties through cryptographic principles. The cryptographic principles employed ensure the handling of assets to be controlled only by the one who hold the private key. The decentralized nature enables all users to share the responsibility for the integrity of the network. Blockchains use an immutable ledger, once something is added into the ledger it cannot be changed or altered. This allows for a fully trustworthy system as we can trust that it will not be manipulated. There is no more "middle-man" or centralized authority that holds all of the information. Every node on the network holds a copy of the ledger which allows for confirmations, validity, and for a truly trustless system to survive. Blockchains are fairly simple to bootstrap once they are implemented. Furthermore, it is a new way to envision technology and the next frontier of Internet.

10.5   Blockchain Applications in IoT

From financial services, to government services, to peer-to-peer transactions, companies around the world are working to integrate blockchain into our everyday lives. Currently, there is no consensus on exactly how blockchain might transform different industries. Thus, in this section we introduce different IoT applications, and examine how integrating a blockchain might transform these use cases.

10.5.1   M2M Transactions

According to Cisco, it is estimated that there will be 26 billion connected devices on the Internet by 2020. M2M interactions are essential for the true potential of IoT to be realized. Multiple challenges still need to be addressed for M2M interactions to truly flourish in IoT, including connectivity standards, lightweight security protocols, and ensuring data privacy; aspects which are covered in Chap. 4 of this book. While there are technical challenges in implementing M2M interactions, "smart contracts" introduce a solution to a fundamental M2M challenge: what protocol do the devices in the IoT utilize to negotiate and execute M2M transactions?

Smart contracts will be heavily used in IoT. You can imagine a vending machine that can automatically order certain items, and pay for the transaction through the agreement of a smart contract. All of this can be accomplished without the need for a central server, or other central entity. The contract would be automatically negotiated, executed, and enforced by the blockchain network.

10.5.2   Energy Management

Blockchain and smart contracts show potential promise in the energy sector. As mentioned in the IoT Verticals chapter of this book, IoT energy use cases include energy monitoring through smart meters and IoT energy management in the connected home. Through these mechanisms, power providers can collect more data on energy patterns, and adjust power plant performance and predictability.

As the grid gets smarter and more capable, homes will be able to not only consume energy, but also provide energy that they generate through solar, wind, or any other means. Potentially, homes could use smart contracts on a blockchain to negotiate energy exchanges, and execute energy transfer from one home to another automatically. Payments for the renewable energy transfer would be bought and sold via a blockchain network.

10.5.3   Supply Chain Management

The supply chain is often a complex set of interactions among a long chain of different vendors. Tracking a set of shipped goods is often a convoluted task that requires information from many parties. IoT has already begun to provide more insights that enables companies to collect data as their goods travel the globe. Sensors provide temperature data, location data and more; giving companies new found control and quality assurance that did not exist before.

Blockchain can be used to simplify the expensive logistics involved when shipping products around the world. By using smart contracts, shipments can be tracked at each stage. Every time a product arrives to a location, that product can be scanned, and a contract would be executed between the two vendors exchanging goods. This would enable an open and verifiable history of where the product was handed off, its condition, and if the contract terms were met (time, date, temperature…etc.). This eliminates the need for each stakeholder to independently track an asset in their own database, a database that provides no transparency, collaboration, or verification with all other stakeholders in the supply chain.

In addition, using a blockchain network to track products can provide more transparency and accountability in trading. Consumers will be able to track and understand where their products came from, and how the product arrived to their doorstep. For example, according to the Mintel Press Office, only 26% of consumers trust organic food labels, and only 13% believe that organic foods are highly regulated. Having better insight into where food was grown, how it was processed, and how it arrived to the store is important to consumers. There are already exist IoT solutions in agriculture that aim to improve quality of food by means of yield monitoring, optimal seeding, optimal water usage, and more. There also exist IoT solutions in supply chain management to monitor conditions and track of goods as they are transported from the source to the store. Adding all of the information collected via IoT to a consumer accessible blockchain would provide the consumer a secure, trackable, and tamper proof way of understanding where their goods were sourced from, and how they got to their store; increasing the trust between consumer and producer (Fig. 10.6).

10.5.4   Healthcare

Healthcare is considered one of the most important verticals for IoT. Intelligent wearable devices present new ways to monitor non-critical patients remotely while clearing up room in hospitals for more critical patients. The healthcare industry is already adopting real-time tracking of medical devices, personnel, and patients. That said, there are still critical challenges in the collection, management, and distribution of patient data that blockchain has potential to provide solutions for.

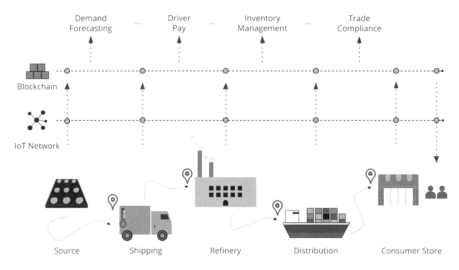

Fig. 10.6 Blockchain in supply chain

The main limitation that blockchains can help improve is around the collection and storage of patient data. According to the centers for medicare and medicaid services, these records hold information such as demographics, progress notes, problems, medications, vital signs, past medical history, immunizations, laboratory data, and radiology reports. Currently, these electronic health records (EHRs) operate in largely in silos; each medical facility collects, maintains, and stores its own medical records for each patient. This creates a high potential for duplication of data while also preventing the cross validation, verification, and data accuracy. Blockchain may allow for all medical records to be stored and shared in a decentralized manner, ensuring one verifiable, non-immutable source of information on any patient, to any authorized provider.

Having EHRs on a blockchain would provide a mechanism that would enable:

1. **IoT Data Exchange**: The ability for M2M medical data management would open the doors for a secure and viable way for patient's data to be monitored remotely by medical staff. The blockchain allows for these M2M interactions to happen automatically, and would ensure a secure data transfer while preventing duplication of data. In addition, when IoT sensors can exchange data through the blockchain, data is protected from tampering and single sources of failure can be eliminated.

2. **Data Interoperability**: The potential to create a single EHR system is an opportunity that the entire industry is excited about. It is so important, that according to the Premier Healthcare Alliance, sharing data across organizations could save hospitals about 93 billion dollars over 5 years alone. A system like that on blockchain would contain a single version of patients records and would be shareable, traceable, anonymized and would put the patient in control of what records could be accessed and by whom.

3. **Drug and Treatment Management**: In 2015, a study conducted by the American Journal of Managed care found that 76.9% of patients that participated in the study had at least one medication discrepancy in their medication lists. In addition to errors, there are also issues with ensuring that control substances such as opioids are not abused or that a patient is not a victim of fraud. A shared EHR system would allow pharmacies and medical staff to ensure that a patient is not prescribed more than once, and would provide a clean record of substances taken in the past. The power of shared data, along with new smart labels that leverage the power of IoT to remind patients of when to take their prescriptions along with a track of when the medicines are taken, would provide very useful data not just for doctors, but also for the machine learning algorithms that are trying to provide more specialized care.

There are of course a lot of challenges that still need to be addressed such as privacy and access management to name a few. Additionally, blockchain's adoption in the healthcare sector will largely depend on the cooperation of healthcare providers, who currently depend on a large array of proprietary software solutions and established IT infrastructures.

10.5.5   Retail

IoT is already used in the retail—enabling the tracking of products in stores, automating and tracking product delivery, and allowing for more beneficial loyalty programs. The blockchain can further these advances, which can result in a better customer experience by increasing consumer trust and improving consumer reward programs.

Product Authenticity: The authenticity of a product is difficult to identify, and can result in damage to brands and declining sales. An IP Commission Report on U.S Intellectual Property mentions that the cost of counterfeit goods to the U.S economy could be anywhere between $225 and $600 billion annual U.S dollars. Blockchain along with IoT solutions would provide the consumer with a clear and direct insight into the entire history of the product—from where their product was manufactured to how it arrived to the store. Such transparency drastically lowers the possibility of a merchant or consumer unknowingly buying a product that is not genuine.

Loyalty Programs: Currently, many loyalty programs work in silos, and do not work together to benefit the consumer. IoT solutions are allowing retailers to enhance the customer experience by collecting data on customer patterns and behaviors. Blockchain could allow for one universal loyalty program that a consumer can use at any store or to buy any service. This way, all the companies get access to consumer behavior while providing more value and savings for the customer.

Inventory Tracking: IoT solutions have allowed retailers full visibility into their products and merchandise—along with the ability to track product performance and stocking levels through digitized inventory and supply chain. Like in other use cases, blockchain introduces a way to track a product at every point in the supply chain and in-store, providing an accurate, and up-to date, trail of where the product is.

10.5.6   *Automotive and Transportation*

The automotive industry is going through transformations unlike the ones seen in the last few decades. From electric vehicles to autonomous vehicles, the industry is going through significant technological changes. These changes represent vast opportunities for drivers, manufacturers, and other stakeholders such as insurance providers and dealerships. Companies have already discovered how IoT can improve services, efficiency and can even provide real-time visibility into vehicle functions. As in other use cases, blockchain can augment IoT to create an array of potential benefits.

M2M Microtransactions: One of the most important use cases that blockchain can securely enable is M2M microtransactions. Vehicles would be able to automatically negotiate and pay for a wide array of services. Services like finding or reserving a parking spot automatically, or negotiating a faster lane if the person in the car is in a hurry, or automatic payments at a gas station or charging station; just to name a few. All of these M2M interactions could be negotiated and automatically executed through smart contracts. As the industry moves towards autonomous vehicles, these M2M transactions will become ever more crucial.

Vehicle Dynamic Ecosystem: IoT, analytics, artificial intelligence, and blockchain are redefining how vehicles will be owned and cared for. IoT is enabling manufacturers to collect and track more data about their vehicles, improving in-vehicle experience, maintenance downtime, and quality. Logging sensor data in vehicles to a blockchain based system enables the automobile ecosystem to view all of the same data about a particular vehicle, a set of vehicles (specific model), or even a brand of vehicles. This means that regulators, manufacturers, insurance providers, etc. all see the same exact data on a vehicle. Data that cannot be modified and is reliable opening the door for opportunities for new business models. Insurance providers could automatically provide dynamic pricing based on driving behaviors, and even automate the insurance claim process as soon as an accident is detected. Manufacturers could automatically use that same data to run analytics on their vehicles to extract patterns and possible issues early (allowing for a more proactive recall and maintenance schedule). Even auto financing and title transfers could be done in a much faster, transparent and verifiable way through the blockchain.

10.5.7 Smart City

According to a past World Urban Prospects report, 54% of the world's population lived in cities as of 2014, and that number is expected to grow to about 66% by 2050. With such population growth, cities have already begun developing smart cities to cope with growing challenges and provide more benefits for their citizens. Around the world, there are hundreds of smart city pilots taking place. IoT solutions are being used to digitize the world around us and improve things like transportation, air/water quality, energy management, and public safety.

Blockchain and IoT: To support the evolution of smart cities, the blockchain can be combined with current IoT solutions. Blockchain can accelerate the adoption of energy microgrids by providing a billing system for automatic negotiation and execution of energy distribution. It can be used to automate water supply management by implementing smart contracts that continuously track and manage water distribution so that it happens in the most efficient manner. Air and water quality can also be improved by implementing blockchain systems to record and share data from sensors installed all around the city.

Governance and digital services: Another way that blockchains can influence smart cities is through the digitization of citizen records and government services—with the potential to practically eliminate paperwork across government agencies and services. Here are some examples of potential services or solutions:

- **Civil Registration**: Can be used for record keeping of each citizen. A blockchain would make these records secure, tamper proof (reducing fraud) and shareable among a variety of stakeholders with needed access to the data.
- **Citizen Identity**: Holding the digital identities for each citizen on a blockchain. Digital management of one's identity through blockchain could eliminate a lot of paperwork and make government services much faster and more efficient.
- **Governance**: Digitizing all records and transactions would transform the efficiency of government agencies. Currently, all records are maintained in silos, making sharing of data across agencies hard and inefficient. Not only would a blockchain improve efficiency, but also would increase transparency and visibility into processes.

10.5.8 Identity, Authentication, and Access Management

Most application stacks require a form of authentication. This topic has been researched and implemented through handshake protocols, key escrow, and various cryptographic modules. For IoT, with respect to blockchain technology, there will have to be a primary use of identity management. When a new device is added to the network, the use of key escrow will have to establish their identity on the network.

This exchange and generation of keys can be determined within the secure enclave of the device's hardware—removing the risk of attacks via ports, wireless, and Bluetooth capabilities.

Once the identity of the device has been setup, it would broadcast its public key/address to the network for others to know of its presence. The use of public key/asymmetric cryptography adds a benefit to the network, where all we need to know is your public key. Similar to Bitcoin, there could be time sensitive intervals set in place, where the device generates new addresses—thus never using the same address twice. This can promote anonymity to malicious observers and sway predictive analysis by attackers.

Another alternative to identity management is to use an escrow or auditing node. This node can be in-charge of asset management and communicating to others when a new node has joined or established itself to the network. In a sense, they will work as a Directory Server similar to BitTorrent type peer-to-peer networks. This allows for easily addresses the key-value store of asset management which could be mapped to public addresses.

As blockchain evolves and starts being used in IoT frameworks, the identity of each device and model on the network will become significantly more important. Each individual device will be granted access via identity and key management. The key management will need to be controlled via the hardware on each device where the actual access is done through software. As IoT networks are quite large, the entire infrastructure will need to uphold strong cryptographic modules to maintain identity management. While actual key storage is done through HSMs on each device.

10.5.9   Other Blockchain IoT Applications

While it is still too early to tell which IoT solutions the blockchain will revolutionize, the ones mentioned in earlier sections constitute the use cases with the most notable traction. Other notable use cases include:

Decentralized DNS: Provides a more secure Internet that is decentralized and not easily hijackable, potentially preventing past attacks on IoT devices. Examples include Namecoin and EmerDNS, already available through browser extensions.

Legal Contracts: Provides a system where things such as ownership registries, notary services, taxes, and even voting could be performed on a blockchain.

Insurance: Insurance on a blockchain could affect multiple industries. In automotive, insurance can turn into an on-demand and dynamic policy system based on information that is retrieved in real time from sensors in your car. The same could be done for other property like your home.

Sharing Economy: Companies like Slock.it have created a platform that allows anyone to share anything with others. Using blockchain, they can lock and unlock physical assets based on predetermined smart contracts, giving anyone temporary access to any physical asset.

10.6 Blockchain Security in IoT

When it comes to blockchain technology, there are normal security risks that modern day technological infrastructures face every day. Yet, blockchain technology also holds an important security risk which involves key management. As mentioned in earlier sections, one's private keys are the ultimate key/password to obtain your information and assets. Whoever holds those private keys holds your identity. Throughout this section we will discuss the advantages and disadvantages of security within blockchains and how that relates IoT.

10.6.1 Trust Between Nodes

Decentralization allows for a trustless mechanism to perform consensus among nodes while adhering to one's privacy and truthfulness. Do you have what you say you have? Based off of your connection, known past activity, can we correctly identify you? All of these questions and more need to be asked when building a blockchain based system. The elimination of a single point of failure is a huge win for all stakeholders. However, this now brings attack vectors to all nodes. Especially in the realm of IoT, we need to carefully consider the protocols between applications and nodes. All messaging between nodes should be secure and private. There should be no possibility of 51% attacks or node compromises. Sybil attacks, also known as a 51% take over of a network, are one of the attacks that are looked at when it comes to consensus and node propagation. The use of keys, messaging systems, and gossip protocols can help protect against this as there are multiple layers of verification before any information that a node posts to the network is accepted and added to the chain. To ensure all nodes are safe, we need to maintain trust between them.

A node first joins the network by bootstrapping off of some discovery peers. These peers are hard coded into the blockchain code-base. These nodes may be the major nodes that help uphold the network, or just the main nodes that we started with. Once they connect they start gossiping between one another to propagate information throughout the network and add blocks to the chain. If an attacker could control a node within this network, they could potentially take over the entire network or propagate faulty information to sway or alter the chain in some malicious manner. By posting invalid transactions and possibly using another malicious node to accept it could be catastrophic. Luckily, most security measures will be built into the blockchain network when developed. The trustless nature of a decentralized network allows for a consensus to take place among nodes before things are committed to the chain or propagated. If a faulty transaction is propagated, another node will realize that this does not match the chain and has not been seen by other nodes. Nodes could add in delays between propagation to make sure that n of m other nodes have verified this transaction or information that was propagated to validate the peer.

10.6.2   Malicious Activity and Cryptographic Principles

If malicious activity does happen to take place within a blockchain, the hope is that nodes and users will easily be able to verify given information based off of the secure cryptographic nature of information through the use of secure cryptographic hash functions or elliptic curve cryptography. If faulty information is introduced into the network, then the actual hash of that information will be different than what is recorded in the main chain. From basic verification of hashes, we can easily distinguish the integrity of the data (as seen in Sect. 10.3). Also, the specific curves that are used in the elliptic cryptography modules are specific and used for a reason. It is highly advised to use NIST approved and known cryptographic modules. Never attempt to write proprietary cryptographic libraries. Most of the time, these will not be tested as in-depth as NIST approved libraries and have the potential for collision. Collisions within cryptographic modules can lead to stolen keys, and overall compromise of the blockchain.

Attacks and hacks have been taking place within the blockchain industry over the past few months of 2017. Most have been from ICO's or "Initial Coin Offerings" for Ethereum's ERC20 tokens. Others have been from exchanges and hardware wallets. These attacks mainly occur from exploiting bugs in smart contract code or finding flaws in the safeguarding of private keys. As we have mentioned throughout this chapter, private keys are the holy grail of wallets and blockchain identity. To maintain security and privacy when it comes to keys is to ensure a proper key management and escrow process. Key management can be taken care of in software or hardware. The use of HSMs (Hardware Security Modules) can move the overhead of key escrow and processing to a hardware device to ensure privacy, security, and proper authentication mechanism for nodes. When maintaining your key management and escrow in software, there are more attack vectors exposed. Some core wallets within the blockchain space keep keys in "keyfiles" or a file that is held in local storage. This can be attacked from any sort of malware from phishing attacks to visiting a malicious web site that installs loggers onto your system. A way to protect keys when they are stored through software is by using multi-signature wallets. Multi-signature wallets need more than one user to have access to the wallet. By using a n of m or a majority of the users to allows access to a wallet means that if 1 key is compromised the entire wallet is not lost.

10.6.3   IoT Security and Blockchain Advantages

For most IoT devices, they rely on a central entity to send them information or alert them of security risks. By moving these responsibilities to individual nodes that are decentralized, it theoretically makes these devices "smart" devices. By using a blockchain for IoT, the security level and fundamentals will greatly increase, and it will put the messaging and alerting functions within each devices protocol layer.

When dealing with the multiple layers within blockchain technology, we commonly focus upon consensus. Consensus algorithms are what allow the decentralized network to obtain a "trustless" model. As mentioned in earlier sections there are various types of consensus algorithms to base your blockchain on when building its network and infrastructure. The types of attacks vary in regard to consensus algorithm. For example, PoW deals with miners providing enough computation to gain rewards, enabling the blockchain to grow. In this type of blockchain, the attacker would have to perform a Sybil attack to compromise the network. In a PoS blockchain, the attacker would have to take control of the actual digital asset or sway the market, as nodes use the asset as a proving point within the network. Attacks within PoS blockchains deal much more with attack vectors of change in currency, whereas PoW deals with Dos/DDoS and Sybil attacks. PoS takes a different approach to the normal attack vectors that the security industry has seen over the years. In regard to IoT, the use of PoS would be a great benefit as each device could "mint" its own token in order to pay into the blockchain or protocol of their nature. This will protect them from Sybil attacks which could happen on a specific protocol layer and maintain consensus among $n + m$ IoT devices.

Blockchain offers many security advantages for any desired application or system. Yet blockchain technology is not the end-all-be-all answers for all applications. Blockchain technology should only be used for use cases that require high security, privacy, and a peer-to-peer nature in regard to networking. IoT can greatly benefit from blockchain technology as it will be able to secure the protocol layer and information that is broadcasted between devices and networks. As blockchain technology grows, so will the attack vectors. There will always be phishing attempts, punycode domains, and smart contract hacks. As time progresses the security space will evolve to build out standards and proper testing methodologies for blockchain technology. The importance of key management, node propagation, messaging, and consensus is what upholds the privacy and security within blockchain technology. Attackers will always try to outsmart your system, so be aware of your technology when building and implementing it in both a secure manner in regard to IoT and blockchain technology.

10.7 Summary

Blockchain is expanding to new industries every day, and has the possibility to propel IoT forward. This potential is greatly due to the technology's foundation in cryptography and the mechanisms by which it addresses the Byzantine Generals Problem. Blockchain presents key features such as decentralization, security, and trust—all important aspects in IoT solutions. A handful of use cases in M2M, energy management, supply chain management, healthcare, retail and transportation display a picture of a fast-emerging technology within various industries. Lastly it is important to consider the challenges being faced by blockchain, such as scalability, privacy, and anonymity. While blockchain is not the answer to all the challenges in

IoT, it should be clear to appreciate why the hype exists—the technology presents many new possibilities that are only beginning to gain traction.

Problems and Exercises

1. What is the double spending problem in digital currencies?
2. Describe what a "Merkle Tree" is? How is it used in Bitcoin?
3. In Sect. 10.3.2 we mention hash pointers, and how they are key to immutability of the blockchain. Keeping that in mind, what are other features of blockchains that work with hash pointers to maintain immutability?
4. What are they key characteristics provided by the blockchain? Explain what they are, and why they are important for adoption in IoT solutions.
5. What is a hash function and how does it work? What is the difference between a hash and a cryptographic hash function? Provide an example of how cryptographic hashes are used in a blockchain (any blockchain will suffice as an example).
6. What is a hash collision? Does Bitcoin suffer from the probability of hash collisions?
7. Consider a scenario where there a potential double spend attempt by a malicious actor in Bitcoin. Explain how the blockchain works to reject such attempt and what the malicious actor would have to do in order to fool all other honest nodes.
8. In table format, describe centralized, decentralized, and distributed network architectures.
9. Perform a search and mention five companies that are currently working on blockchain + IoT solutions. Describe their solutions and how IoT and Blockchain are being combined. Make sure to include at least one start up and at least one established company.
10. What type of records can be kept in a blockchain?
11. In Sect. 10.4 we describe some consensus algorithms. Research consensus algorithms for blockchain and name an algorithm that we did not mention in this section. Is it good for IoT? Explain why or why not.
12. What is Elliptic Curve Cryptography and how does it benefit the use of keys within blockchain technology?
13. Describe a Sybil attack and other types of attack vectors that could take place on a blockchain.
14. Blockchains all start from a genesis block and then maintain a block height as the chain grows. Describe the importance of block heights as timestamps and lookups within Merkle Trees.
15. Describe the difference between permissioned, permissionless, and consortium blockchains. What type do you think best fits a blockchain involving IoT devices.
16. What is the difference between a smart contract and multi-sig address?

References

1. Swan, Melanie. Blockchain: Blueprint for a New Economy. O'Reilly Media, Inc., 2015.
2. Nakamoto, Satoshi. "Bitcoin: A peer-to-peer electronic cash system." (2008): 28.
3. K. Christidis and M. Devetsikiotis, "Blockchains and Smart Contracts for the Internet of Things", IEEE Access, vol. 4, pp. 2292-2303, 2016.
4. Chaum, David. "Blind signatures for untraceable payments." Advances in cryptology. Springer US, 1983.
5. Gupta, Vinay. "A Brief History of Blockchain." Harvard Business Review, 5 Apr. 2017, hbr.org/2017/02/a-brief-history-of-blockchain.
6. Szabo, Nick. "Formalizing and Securing Relationships on Public Networks." First Monday, vol. 2, no. 9, Jan. 1997, https://doi.org/10.5210/fm.v2i9.548.
7. Croman, Kyle, et al. "On Scaling Decentralized Blockchains." Financial Cryptography and Data Security Lecture Notes in Computer Science, 2016, pp. 106–125., https://doi.org/10.1007/978-3-662-53357-4_8.
8. Dickson, Ben. "Decentralizing IoT Networks through Blockchain." TechCrunch, TechCrunch, 28 June 2016, techcrunch.com/2016/06/28/decentralizing-iot-networks-through-blockchain/.
9. I. Crigg and K. Griffith, "A Quick History of Cryptocurrencies BBTC — Before Bitcoin", Bitcoin Magazine, 2014. [Online]. Available: https://bitcoinmagazine.com/articles/quick-history-cryptocurrencies-bbtc-bitcoin-1397682630/.
10. Antonopoulos, Andreas M. "Mastering Bitcoin: Programming the Open Blockchain". O'Reilly Media Inc. 2017
11. N. Kshetri, "Can Blockchain Strengthen the Internet of Things?", IT Professional, vol. 19, no. 4, pp. 68-72, 2017.
12. V. Nordahl and M. Rao, "Blockchain Cryptography", My Blockchain Blog, 2017. [Online]. Available: https://www.myblockchainblog.com/blog/blockchain-cryptography.
13. K. Lewis, "Blockchain: Four blockchain use cases transforming business", Internet of Things blog, 2017. [Online]. Available: https://www.ibm.com/blogs/internet-of-things/iot-blockchain-use-cases/.
14. N. Murty, S. Ananthasayanam, A. Singh, R. Malhotra, V. Vaid and A. Madan, Blockchain: The next innovation to make our cities smarter. PWC, 2018, pp. 22-30.
15. A. Castor, "A (Short) Guide to Blockchain Consensus Protocols – CoinDesk", CoinDesk, 2017. [Online]. Available: https://www.coindesk.com/short-guide-blockchain-consensus-protocols/.
16. Z. Witherspoon, "A Hitchhiker's Guide to Consensus Algorithms – Hacker Noon", Hacker Noon, 2018. [Online]. Available: https://hackernoon.com/a-hitchhikers-guide-to-consensus-algorithms-d81aae3eb0e3.
17. C. Hammerschmidt, "Consensus in Blockchain Systems. In Short. – Chris Hammerschmidt – Medium", Medium, 2017. [Online]. Available: https://medium.com/@chrshmmmr/consensus-in-blockchain-systems-in-short-691fc7d1fefe.
18. D. Mingxiao, M. Xiaofeng, Z. Zhe, W. Xiangwei and C. Qijun, "A review on consensus algorithm of blockchain", 2017 IEEE International Conference on Systems, Man, and Cybernetics (SMC), 2017.
19. L. Lamport, R. Shostak and M. Pease, "The Byzantine Generals Problem", ACM Transactions on Programming Languages and Systems, vol. 4, no. 3, pp. 382-401, 1982.
20. F. Tschorsch and B. Scheuermann, "Bitcoin and Beyond: A Technical Survey on Decentralized Digital Currencies", IEEE Communications Surveys & Tutorials, vol. 18, no. 3, pp. 2084-2123, 2016.
21. A. Bahga and V. Madisetti, "Blockchain Platform for Industrial Internet of Things", Journal of Software Engineering and Applications, vol. 09, no. 10, pp. 533-546, 2016.

Chapter 11
Industry Organizations and Standards Landscape

11.1 Overview

The IoT industry landscape is crowded with different standards bodies and organizations chipping away at various aspects of the technology. As is typically the case early on in the technology cycle, some of the organizations are tackling the same problem and hence a subset of the standards that they are proposing are overlapping and competing for mainstream adoption. This creates confusion in a vast and multifaceted industry and inevitably slows down product development, as vendors do not want to take bets on standards that may never take off in the market (think Betamax vs. VHS in the early video format war days).

Some of the industry organizations focus their efforts on a specific IoT vertical, whereas others are involved in defining crosscutting technologies that apply across various IoT applications and verticals. Furthermore, not all organizations are actively defining their own standards; rather some are promoting harmony and alignment among others, which define and ratify standards.

What is common across all these standards is that they are all being based on (or migrating to) a common normalization layer, the IP network layer, which guarantees system interoperability while accommodating a multitude of link layer technologies, in addition to a plethora of application protocols. IP constitutes the thin waist of the proverbial hourglass that is the IoT's protocol stack (refer to Fig. 11.1). The diversity in Physical and Link layer standards is a manifestation of the IoT challenges and requirements that impact that layer of the protocol stack, as was discussed in Chap. 5 (Sect. 5.1.1). By the same token, the large number of Application layer standards is a reflection of the many industry verticals and applications (as discussed in Chap. 9) that IoT enables.

In this chapter, we will provide an overview of the key IoT standards defining organizations and the various protocols that they have been defining or promoting. Our focus will be on standards operating at the Physical, Data Link, Network, and

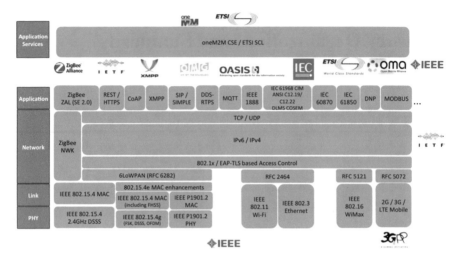

Fig. 11.1 IoT standards landscape

Transport layers of the OSI model presented in Chap. 2. We will also touch upon a select subset of standards efforts operating at the Application layer of the model. As can be seen in Fig. 11.1, such efforts are numerous, industry vertical specific and require expert domain knowledge in the associated industry or application (e.g., IEC 61968, ANSI C12.19/C12.22, DLMS/COSEM are Smart Grid standards).

11.2   IEEE (Institute of Electrical and Electronics Engineers)

IEEE is a well-established technology standards body, which, among other things, had defined the standards for Ethernet and wireless Local Area Networks (LANs). Given its legacy and expertise in physical and link layer network technologies, the IEEE embarked on defining a number of physical and link layer standards for IoT. These include the 802.15.4 family of low-power wireless protocols, which were discussed in Sect. 5.1.2.1, the 802.11ah long-range Wi-Fi standard discussed in Sect. 5.1.2.3, as well as the 1901 power line communications standards. The latter define technologies for carrying network data, in addition to Alternating current (AC), over conventional electric wiring.

Beyond the efforts on standardizing physical and link layer technologies, IEEE kicked off the IoT Initiative as a platform for the technical community to collaborate on technologies that advance the IoT. Adjunct to this initiative, many IoT related standards activities had been completed or are underway. We will go through an overview of these activities next.

11.2.1 IEEE 1451 Series

The IEEE 1451 series addresses smart transducers, which are defined as devices that convert a physical measurement into an electrical signal, or vice versa. Transducers include sensors or actuators that we discussed in Chap. 3. The standards define communication interfaces for interconnecting smart transducers to networks or external systems via either wired or wireless mechanisms. Among the main elements of these standards is the definition of the Transducer Electronic Data Sheets (TEDS). The TEDS is associated with every smart transducer. It provides relevant technical data pertaining to the transducer in a standard format. Such data includes the device identity, type, accuracy, calibration, or other manufacturer-related information, etc. The standards define common mechanisms by which a transducer can communicate its associated TEDS to the connected network or system. TEDS may be implemented in one of two ways. They can be embedded onboard within the transducer itself, typically on some memory component such as EEPROM. Alternatively, a virtual TEDS can be implemented as an off-board data file that is stored in some component separated from the transducer albeit accessible to the instrument or system connected to the transducer. Virtual TEDS allows the extension of the TEDS standard to legacy sensors and devices where onboard or embedded memory may not exist.

11.2.2 IEEE 1547 Series

The IEEE 1547 series addresses Smart Grid, and in particular handling distributed resources in electric power systems. The standard defines technical requirements for interconnecting distributed generators and energy storage systems to electric power systems. Examples of such generators include fuel cells, photovoltaic, micro-turbine, reciprocating engines, wind generators, large turbines, and other local generators. The technology helps utilities tap into surplus electricity from alternative and renewable energy sources. Furthermore, the IEEE 1547 series deals with various facets of renewable energy, including micro-grids (IEEE 1547.4) and secondary networks for distributed resources (IEEE 1547.6).

11.2.3 IEEE 1609 Series

The IEEE 1609 series addresses intelligent transportation systems (ITS) and focuses on Wireless Access in Vehicular Environments (WAVE). The series defines the architecture, services, and interfaces to enable secure vehicle-to-vehicle and vehicle to roadside infrastructure wireless communication. The standard enables applications that include vehicle safety, enhanced navigation, traffic management, automated tolling and more. The IEEE 1609 series specifies standards for communication security (IEEE 1609.2), WAVE connection management (IEEE 1609.3), and Layer 3 through Layer 7 operation across multiple channels on top of IEEE 802.11p.

11.2.4   IEEE 1888 Series

The IEEE 1888 series focuses on ubiquitous green community control networks. It describes remote control architecture for buildings, digital communities, and metropolitan networks. The standard defines the data formats between systems as well as the data exchange protocol that interconnects various components, including gateways, storage systems, and application units over an IP network. This network provides open interfaces for public administration/service, property management, and individual service. The interfaces enable central management, remote surveillance, and collaboration.

11.2.5   IEEE 1900 Series

IEEE 1900 series focuses on dynamic spectrum access radio systems and networks. One of the main goals of this series is to improve spectrum utilization. To that effect, the standard explores architectures and interfaces for dynamic spectrum access in the TV whitespace frequency bands, as well as management systems for optimization of radio resource usage, spectrum access control, and compliance with regional regulations aimed at protecting broadcast systems. The standard also defines policy language and architectures for managing dynamic spectrum access among distributed heterogeneous devices.

11.2.6   IEEE 2030 Series

IEEE 2030 series focuses on the smart grid, including electric vehicle infrastructure. It defines a reference model for smart grid interoperability including the three pillars of energy, information, and communications technologies. The standard addresses applications for electric vehicles and associated support infrastructure used for personal and mass transit. Furthermore, the standard covers energy storage systems that are integrated with the electric power infrastructure and relevant test procedures for these systems.

11.2.7   IEEE 2040 Series

The IEEE 2040 series focuses on connected, automated, and intelligent vehicles. The series defines an overview and architectural framework (IEEE 2040), taxonomy and definitions (IEEE 2040.1), as well as testing and verification (IEEE 2040.2) standards. The series leverages existing standards where applicable.

11.2.8   IEEE 11073 Series

The IEEE 11073 series of standards focuses on point-of-care medical device communication and personal health device communication. The standard enables interoperability between medical devices and external computer systems. It defines information models to guarantee semantic interoperability between communicating medical devices. It also specifies a tree hierarchy for modeling the device and its relevant information: measurements, physiological and technical alerts, as well as contextual data.

11.2.9   IEEE 2413 Series

The IEEE 2413 series defines an architectural framework for the IoT, including descriptions of various IoT verticals, definitions of their associated abstractions and identification of commonalities across those verticals. The standard establishes a reference model for IoT domain verticals and an architecture that defines the building blocks and common elements.

11.3   IETF

The IETF has been instrumental in defining and standardizing Internet technologies, including IPv4 and IPv6 as well as numerous routing protocols (e.g., OSPF, RIP, PIM, BGP), application protocols (e.g., HTTP, LDAP, SMTP), and security protocols (e.g., TLS, IPSec, IKE). In 2006, work started in the IETF on a number of IoT standards. The initial scope centered on enabling IP on top of IEEE 802.15.4 wireless networks, but has expanded beyond that over time. Currently, there are five IETF working groups focusing on IoT related technologies. We will discuss their work next.

11.3.1   ROLL

The Routing over Low Power and Lossy networks (ROLL) working group focuses on routing issues for Low Power and Lossy Networks (LLNs). LLNs typically comprised of embedded devices with limited power, memory, and processing resources that are interconnected by a variety of link technologies. LLNs cover a multitude of applications such as building automation, smart homes, smart health care, industrial monitoring, environmental monitoring, asset tracking, smart grid, etc. The ROLL working group is concerned with defining routing requirements for a subset of the aforementioned applications: industrial (RFC 5673), connected home (RFC 5826), building automation (RFC 5867), and urban sensor networks (RFC 5548). The working group is approaching these requirements by defining an IPv6 architecture that enables scalable networks of constraint devices to communicate with high reliability. Routing security and manageability (e.g., autonomic configuration) are among the key issues that ROLL is looking into.

ROLL analyzed the particular routing protocol requirements of LLNs, starting with the constraints that these protocols must adhere to. The following constraints were identified, which stem from the constrained nature of the nodes in LLNs:

- Protocols need to operate with minimal amount of state.
- Protocols must be optimized for efficiency, i.e., saving energy, memory, and processing power.
- Protocols must support unicast and multicast application traffic patterns.
- Protocols must be very efficient in encoding information to operate with very small link layer maximum transfer unit (MTU) size.

The ROLL working group evaluated existing routing protocols to examine whether they could operate within the confines of the above constraints. The following protocols were analyzed: OSPF (RFC2328), IS-IS (RFC1142), RIP (RFC2453), OLSR (RFC3626), TBRPF (RFC3684), AODV (RFC3561), DSR (RFC4728), DYMO and OLSv2 (RFC7181). Based on this analysis, the working group determined that none of the existing protocols meets the requirements of LLNs. As a result, the working group defined a new protocol, RPL, which was discussed in Sect. 5.2.2.2.

11.3.2   Core

The Constrained RESTful Environments (CORE) working group focuses on defining a framework for RESTful applications running over constrained IP networks. These applications include applications to monitor simple sensors (e.g., temperature sensors or power meters), to control actuators (e.g., valves or light switches) and to remotely manage devices. Such applications are typical of several IoT verticals such as home and building automation and Smart Grid. The applications are forced to operate under the same set of constraints that define LLNs, namely: limitations on memory, processing power, and energy as well as high loss rates and small packet sizes. In addition, the applications must deal with the fact that nodes are typically powered off and wake up for a short period of time.

The framework defined by the working group assumes a general operating paradigm for applications where network nodes run embedded web services and are responsible for resources (e.g., sensors or actuators) that can be queried or manipulated by remote nodes. Furthermore, nodes may publish local resource changes to remote nodes that have subscribed to receive notifications. CORE has defined the CoAP protocol, which was discussed in Sect. 5.3.5.1, to support this application framework.

One of the key challenges to applications running in these constrained environments is security. The working group's scope includes selecting viable approaches for security bootstrapping to handle secure service discovery, distribution of security credentials, and application-specific node configuration.

11.3.3   6LowPAN

The IPv6 over Low-Power Wireless Personal Area Networks (6LowPAN) working group focused on enabling IPv6 over IEEE 802.15.4 networks. The group started its work in 2005 and concluded in 2014 after working through the following goals:

First, defining a fragmentation and reassembly layer to allow adaptation of IPv6 to IEEE 802.15.4 links. This is because the link protocol data units may be as small as 81 bytes, which is much smaller than that the minimum IPv6 packet size of 1280 bytes.

Second, introduce an IPv6 header compression mechanism to avoid excessive fragmentation and reassembly, since the IPv6 header alone is 40 bytes long, without optional headers.

Third, specify methods for IPv6 address stateless auto configuration to reduce the provisioning overhead on the end nodes.

Fourth, examine mesh routing protocol suitability to 802.15.4 networks, especially in light of the packet size constraints.

Finally, investigate the suitability of existing network management protocols and mechanisms in terms of meeting the requirements for minimal configuration and self-healing as well as meeting the constraints in processing power, memory, and packet size.

The working group produced six standards: 6LowPAN problem statement document (RFC4919), IPv6 adaptation layer and header format specification (RFC4944), IPv6 header compression specification (RFC6282), 6LowPAN use cases and applications document (RFC6568), IPv6 routing requirements document (RFC6606), and IPv6 neighbor discovery optimization specification (RFC6775).

11.3.4   6TisCH

This working group is chartered with enabling IPv6 over the Time Slotted Channel Hopping (TSCH) mode of IEEE 802.15.4e. The target network comprised of Low Power and Lossy Networks (LLNs) connected through a common backbone via LLN Border Routers (LBRs). The focus of the working group is on defining an architecture that describes the design of 6TiSCH networks in terms of the component building blocks and protocol signaling flows. The working group will also produce an information model that describes the management requirements of 6TiSCH network nodes, together with a data model mapping for an existing protocol, such as Concise Binary Object Representation (CBOR) over the Constrained Application Protocol (CoAP). In addition, the working group will define a minimal and a best practice 6TiSCH configuration that provides guidance on how to construct a 6TiSCH network using the Routing Protocol for LLNs (RPL) and static TSCH schedule. Finally, the working group may produce implementation and co-existence guides to help accelerate the industry.

11.3.5   ACE

The Authentication and Authorization for Constrained Environments (ACE) working group is tasked with producing use cases and requirements for authentication and authorization in IoT, as well as defining protocol mechanisms that can address these requirements, and are capable of running on constrained IoT devices. The scope of the work is limited to RESTful architectures running the Constrained Application Protocol (CoAP) over Datagram Transport Layer Security (DTLS). Hence, the working group is looking to provide a standardized solution for authentication and authorization to enable a client's authorized access to REST resources hosted on a server. Both client and server are assumed to be constrained devices. The access will be facilitated by a non-constrained authorization server. The working group will evaluate existing protocol mechanisms for suitability and applicability to constrained environments, and will advise on any required restrictions, changes, or gaps.

11.4   ITU

The International Telecommunication Union (ITU) is a United Nations (UN) specialized agency with over 190 member states and over 700 industry members in addition to universities as well as research and development institutes. It has been heavily involved in the definition and development of telecommunication standards.

ITU published one of the first reports on "The Internet of Things" in 2005 and has been involved in IoT since then, producing multiple standards documents in this space, as discussed next.

Recommendation ITU-T Y.2060, *Overview of the Internet of Things,* provides a definition of IoT, terming it: "*A global infrastructure for the Information Society, enabling advanced services by interconnecting (physical and virtual) things based on, existing and evolving, interoperable information and communication technologies.*" It describes the concept and scope of IoT, discussing its fundamental characteristics and high-level requirements, and providing a detailed overview of the IoT reference model. Additionally, the standard discusses the IoT ecosystem and accompanying business models.

Recommendation ITU-T Y.2061, *Requirements for support of machine-oriented communication applications in the NGN environment*, offers a description of machine-oriented communication applications in next-generation network (NGN) environments; covering the NGN extensions, additions, and device capabilities required to support MOC applications.

Recommendation ITU-T Y.2062, *Framework of object-to-object communication for ubiquitous networking in an NGN environment*, discusses the concept and high-level architectural model of such communication, and provides a mechanism to identify objects and enable communications between them.

Recommendation ITU-T Y.2063, *Framework of Web of Things*, specifies the functional architecture including conceptual and deployment models for the Web of Things. The standard also provides an overview of service information flows and use cases in home control.

Recommendation ITU-T Y.2069, *Terms and definitions for Internet of Things*, specifies the terms and definitions relevant to the Internet of things (IoT) from an ITU-T perspective, in order to clarify the Internet of Things and IoT related activities.

ITU has multiple study groups looking into various aspects of IoT: Study Group 11 started activity in July 2014 and is looking into application programmatic interfaces and protocols for IoT as well as IoT testing. Study Group 13 focuses on the networking aspects of IoT. Study Group 15 looks at Smart Grid and home networks. Study Group 16 focuses on IoT applications including eHealth. Study Group 17 is looking at the security and privacy protection aspects of IoT. In addition, there are multiple focus groups looking at topics including smart cities, water management, and connected cars.

11.5 IPSO Alliance

The "Internet Protocol for Smart Objects" (IPSO) Alliance is an open non-profit special interest group that promotes the use of the IP protocol to connect smart objects (i.e., *Things*) to the network. It was formed in 2008 and includes members from technology and communication companies in addition to industry verticals companies (e.g., energy). The alliance complements the work of other standards defining bodies, such as the IETF, IEEE, and ETSI, by promoting IoT technologies through publishing whitepapers and hosting webinars, interoperability events, and challenges.

The interoperability events have helped in advancing IP technologies for IoT by providing a vendor-neutral forum to test evolving IoT technologies and providing feedback to the standards bodies defining them in order to fix potential issues that affect interoperability. For instance, in one of the interoperability events held in conjunction with the IETF, a number of issues related to early versions of RPL were communicated back to the Routing over Low Power and Lossy Networks (ROLL) working group in order to improve the developing drafts.

IPSO has published the IPSO Application Framework, which defines a representational state transfer RESTful design for use in IP smart objects for Machine-to-Machine applications. It specifies a set of REST interfaces that may be used by a Thing to represent its available resources and to interact with other Things and remote applications. The framework was extended to cover a wide range of use cases and to more precisely describe the parameters of smart objects during an interoperability event held during IETF 84 in Vancouver, Canada.

11.6   OCF

The Open Connectivity Foundation (OCF) is an industry group that focuses on developing standards and certification for IoT devices based on the IETF CoAP protocol. It was formed in July 2014 by Intel, Broadcom, and Samsung Electronics under the name of the Open Interconnect Consortium. The consortium changed its name to OCF in February 2016. It currently has more than 80 member companies including General Electric, Cisco Systems, Microsoft, and Qualcomm. The OCF is defining a framework for easy device discovery and trusted connectivity between things. In September 2015, it released the first version of the specification of this framework. OCF is also working on open source reference implementation of the specification, which is called "IoTivity."

11.7   IIC

The Industrial Internet Consortium is a non-profit organization that aims to accelerate the development and adoption of interconnected machines and devices, intelligent analytics, and people at work. It was founded by AT&T, Cisco, General Electric, IBM, and Intel in March 2014. IIC does not develop standards for IoT; rather, it provides requirements to other standards defining organizations. IIC focuses on creating use cases, reference architectures, frameworks, and test-beds for real IoT applications across varying industrial environments. IIC also states among its goals to facilitate open forums for sharing and exchanging real-world ideas, practices, and insights, in addition to building confidence around new and innovative approaches to security. The work of the IIC does not include consumer IoT, rather it is targeted at business verticals such as energy, healthcare, transportation, and manufacturing.

11.8   ETSI

The European Telecommunication Standards Institute (ETSI) is an independent non-profit standards defining organization. ETSI was among the very first organizations to develop a set of standards that define a complete horizontal service layer for M2M communications.

The ETSI M2M standards specify architectural components for IoT including: devices (things), gateways with associated interfaces, applications, access technologies as well as the M2M Service Capabilities Layer (middleware). They also include security, traffic scheduling, device discovery, and lifecycle management features. These standards, which were released in 2012, include:

- Requirements in ETSI TS 102 689
- Functional architecture in ETSI TS 102 690
- Interface definitions in ETSI TS 102 921

ETSI is also looking into various applications of M2M technologies, including: smart appliances, smart metering, smart cities, smart grid, eHealth, intelligent transportation systems, and wireless industrial automation.

11.9   oneM2M

In July 2012, seven standards development organizations (TIA and ATSI from USA, ARIB and TTC from Japan, CCSA from China, ETSI from Europe and TTA from Korea) launched a global organization to jointly define and standardize the common horizontal functions of the IoT Application Services layer under the umbrella of the oneM2M Partnership Project (http://www.onem2m.org). The founders agreed to transfer and stop their own overlapping IoT Application Service layer work. The partnership has grown to include, in addition to the seven standards bodies, five global information and communications technology forums and more than 200 companies. oneM2M states among its objectives the development of the following:

- Use cases and requirements for a common set of Application Services capabilities.
- Service architecture and Protocols/APIs/standard objects based on this architecture (open interfaces & protocols).

- Security and privacy aspects (authentication, encryption, integrity verification).
- Reachability and discovery of applications.
- Interoperability, including test and conformance specifications.
- Collection of data for accounting (to be used for billing and statistical purposes).
- Identification and naming of devices and applications.
- Information models and data management (including store and publish/subscribe functionality).
- Management aspects (including remote management of entities).

Among the work items being undertaken by oneM2M, the effort on Abstractions and Semantics Enablement will be key to achieving application level interoperability for IoT, as was discussed in Chap. 4. This area of Semantics remains a major gap in the overall IoT standardization journey.

11.10   AllSeen Alliance

The AllSeen Alliance was formed in December 2013 as a Linux Foundation Collaboration Project.

It is an open non-profit consortium that aims to promote the IoT based on the AllJoyn open source project. AllJoyn is an open, secure, and programmable software framework for connectivity and services. It enables devices to discover, connect, and interact directly with other AllJoyn-enabled products. The project was originally created by Qualcomm and released into the open source domain.

It consists of an open source software development kit (SDK) and code base of service frameworks that enable basic IoT functions such as discovery, onboarding, connection management, message routing, and security, thereby ensuring interoperability among systems.

11.11   Thread Group

The Thread working group was formed in July 2014 and included Google's Nest subsidiary, Samsung, ARM Holdings, Freescale, Silicon Labs, Big Ass Fans, and the lock company Yale. The purpose of the group is to promote Thread as the protocol for the connected home and certify products that support this protocol. The Thread protocol is a closed-documentation royalty-free protocol that runs on top of IEEE 802.15.4 and 6LowPAN. It adds functions such as security, routing, setup, and device wakeup to maximize battery life. Thread competes with other protocols already in this space such as Bluetooth Smart, Z-Wave, and ZigBee.

11.12 ZigBee Alliance

The ZigBee Alliance was formed in 2002 by Motorola, Philips, Invensys, Honeywell, and Mitsubishi to develop, maintain, and publish the ZigBee standard. Since then, the alliance has grown to include over 170 participant members and over 230 adopter companies, including ABB, Fujitsu, British Telecom, Huawei, Cisco, etc. The alliance publishes "application profiles" that enable vendors to create interoperable products. The initial ZigBee specification focused on home automation, but the scope has since expanded to include large building automation, retail applications, and health monitoring.

Most of the protocol specifications are based on the IEEE 802.15.4 radio, even though the more recent Smart Energy specifications are no longer tied to 802.15.4.

The initial protocols standardized by the alliance were based on the standard IEEE 802.15.4 MAC/PHY, but defined a ZigBee specific stack that includes the networking and services layer, through the full application layer. Since those beginnings, the ZigBee Alliance has undertaken a constant effort to increase the interoperability with the Internet Protocol suite, which renders ZigBee as one of the protocols that are capable of adapting to different market segments. In 2013, the ZigBee Alliance released ZigBee IP, an IoT solution based on IPv6, RPL, and 6LowPAN.

11.13 TIA

The Telecommunications Industry Association (TIA) develops industry standards for information and communication technologies, and represents over 400 companies in this domain. The TIA TR-50 engineering committee was launched in 2009 to develop application programmatic interface (API) standards for the monitoring and bi-directional communication between smart devices and other devices, applications, or networks. The committee includes many industry players, including Alcatel Lucent, AT&T, CenturyLink, Cisco, Ericsson, ILS Technology, Intel, LG, Nokia Siemens Networks, Numerex, Qualcomm, Sprint, Verizon, and Wyless. Even pre-dating TR-50, TIA was involved in M2M standards, with several of its engineering committees having worked on smart device communications, including TR-45 (Mobile and Personal Communications Systems Standards), TR-48 (Vehicular Telematics), TR-49 (Healthcare ICT) and through its work on the Third Generation Partnership Project 2 (3GPP2).

11.14   Z-Wave Alliance

The Z-Wave Alliance is an industry consortium of over 300 companies creating IoT products and service over the Z-Wave protocol. Z-Wave is a short-range wireless protocol, initially developed by a small Danish company called Zensys. Z-Wave is a vertically integrated protocol, which runs over its own radio. Z-Wave's physical and media access layers were ratified by the International Telecommunication Union (ITU) as the international standard G.9959. Z-Wave is often considered to be the main competitor to ZigBee, but unlike ZigBee, it only focuses on home environment applications.

11.15   OASIS

OASIS is a non-profit consortium that drives the development, convergence, and adoption of open standards for the global information society. OASIS produces standards for security, Internet of Things, cloud computing, energy, content technologies, emergency management, and other areas.

There are three technical committees in OASIS involved in defining IoT technologies:

The Advanced Message Queuing Protocol (AMQP) technical committee is standardizing the AMQP protocol, a secure, reliable, and open Internet protocol for handling business messaging.
The Message Queuing Telemetry Transport (MQTT) technical committee is standardizing the MQTT protocol, a lightweight publish/subscribe reliable messaging transport protocol suitable for communication in M2M/IoT contexts where a small code footprint is required and/or network bandwidth is at a premium.
The Open Building Information Exchange (oBIX) technical committee is defining technologies to enable mechanical and electrical control systems in buildings to communicate with enterprise applications.

11.16   LoRa Alliance

The LoRa Alliance is an open, non-profit association to standardize Low Power Wide Area Networks (LPWAN) using the LoRa protocol (LoRaWAN). The alliance was announced in January 2015, and initial members include IoT solution providers Actility, Cisco, Eolane, IBM, Kerlink, IMST, MultiTech, Sagemcom, Semtech, and Microchip Technology, as well as telecom operators: Bouygues Telecom, KPN, SingTel, Proximus, Swisscom, and FastNet (part of Telkom South Africa). The LoRA protocol provides long-range wireless connectivity for devices at low bit rates (from 0.3 to 50 kbps) with low-power consumption for battery-powered devices. LoRaWAN transceivers can communicate over distances of more than 100 km (62 miles) in favorable environments, 15 km (9 miles) in typical semi-rural environments and more than 2 km (1.2 miles) in dense urban environments.

The LoRa alliance claims that the scope of applications where LPWAN's are applicable is endless, but indicates that the main applications driving current network deployments are intelligent building, supply chain, Smart City, and agriculture.

11.17   Gaps and Standards Progress Scorecard

The road to a standards-based IoT is well underway. The industry has made significant strides towards converging on the IP network protocol as the common basis for IoT communication protocols. Multiple Physical and Link layer standards have been defined to address the requirements of constrained devices, which are limited in both compute capacity and available power. Some work remains at these layers, particularly with regard to adding support for determinism and time-sensitive applications. At the Network layer, the gaps are relatively limited and manifest in the need to add support for routing over Time Slotted Channel Hopping (TCSH) link technologies. The lion's share of the gaps exists at the Application Protocols and Application Services layers. The former is currently characterized by a multitude of competing and largely functionally overlapping standards. No clear winner has emerged; especially as the industry adoption remains highly fragmented. The latter is currently in a state where the industry has more or less rallied around a common forum, namely oneM2M, and an initial standard has been released, which defines the Common Services Entities and Common Services Functions. However, at the time of this writing, the market acceptance and adoption of the standard remain unknown. In addition, the released standard is only a first step towards standardization as the area of Semantics remains largely unchartered territory. Figure 11.2 summarizes the progress scorecard for IoT industry standards.

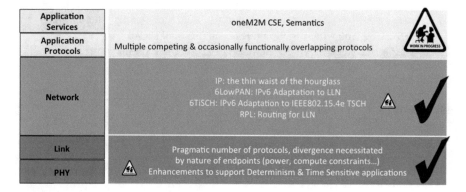

Fig. 11.2 IoT standards progress scorecard

11.18   Summary

In this chapter, we started with an overview of the IoT standardization landscape and then provided an overview of the main standards defining organizations involved in IoT and a snapshot of the projects that they are undertaking. We covered the following industry organizations: IEEE, IETF, ITU, IPSO Alliance, OCF, IIC, ETSI, oneM2M, AllSeen Alliance, Thread Group, ZigBee Alliance, TIA, Z-Wave Alliance, OASIS, and LoRa Alliance. Finally, we presented a summary of the standards gaps and provided a scorecard of the progress to the time of this writing.

Problems and Exercises

1. Name three established networking standards bodies involved in defining technology standards for IoT?
2. Which devices does IEEE 1451 series address? What does it specifically define? What does TEDS provide for IEEE 1451 devices? Provide specific examples.
3. What are the two mays to implement TEDS?
4. What does the IEEE 1888 standard define?
5. What constraints should routing protocols adhere to in order to meet the requirements of LLNs, as analyzed by the IETF ROLL workgroup?
6. Which RESTful protocol, defined by the IETF CORE workgroup, extends RESTful architectures to constrained devices? Why is REST applicable here?
7. What is the role of the IPSO Alliance among IoT standards organizations?
8. What two standards bodies are developing competing wireless technologies for home automation?
9. What is the scope of the standards being developed by oneM2M?
10. What IoT verticals does the work of the IIC encompass?
11. The LoRA Alliance standardizes the LoRA protocol. Describe the data rate and range characteristics of the technology?
12. Is the IoT standards landscape well defined? What is the net result of this on the industry?
13. Where does the industry stand on the road to a standards-based IoT? State the gaps per protocol layer.
14. Name two IoT Application Protocols that are being standardized by OASIS. Describe what function does each protocol serve.
15. Is the ZigBee stack based on the Internet Protocol? Explain.

References

1. Wobschall, D., IEEE 1451 – A Universal Transducer Protocol Standard, https://eesensors.com/media/wysiwyg-docs-pdfs/ESP16_Atest.pdf
2. Li, B., et al., Wireless Access for Vehicular Environments, http://www.mehrpouyan.info/Projects/Group%205.pdf

3. Khattab, A., et al. An Overview of IEEE Standardization Efforts for Cognitive Radio Networks, IEEE International Symposium on Circuits and Systems (ISCAS), May 2015.

4. Kasparick, M., et al. New IEEE 11073 Standards for Interoperable, Networked Point-of-Care Medical Devices, 37th Annual International Conference of the IEEE Engineering in Medicine and Biology Society (EMBC), August 2015.

5. Logvinov, O., Standard for an Architectural Framework for the Internet of Things (IoT), http://grouper.ieee.org/groups/2413/Intro-to-IEEE-P2413.pdf.

6. IETF, ROLL Working Group charter, https://datatracker.ietf.org/wg/roll/charter/

7. Vasseur, JP., Terms Used in Routing for Low-Power and Lossy Networks, RFC 7102, January 2014.

8. Levis, P., et al., Overview of Existing Routing Protocols for Low-Power and Lossy Networks, draft-ietf-roll-protocols-survey-07, work in progress, April 2009.

9. IETF, CORE Working Group charter, https://datatracker.ietf.org/wg/core/charter/

10. Kushalnagar, N., et al., IPv6 over Low-Power Wireless Personal Area Networks (6LoWPANs): Overview, Assumptions, Problem Statement, and Goals, RFC4919, August 2007.

11. Montenegro, G., et al., Transmission of IPv6 Packets over IEEE 802.15.4 Networks, RFC4944, September 2007.

12. Hui J., et al., Compression Format for IPv6 Datagrams over IEEE 802.15.4-Based Networks, RFC6282, September 2011.

13. Kim E., et al., Design and Application Spaces for IPv6 over Low-Power Wireless Personal Area Networks (6LoWPANs), RFC6568, April 2012.

14. Kim E., et al., Problem Statement and Requirements for IPv6 over Low-Power Wireless Personal Area Network (6LoWPAN) Routing, RFC6606, May 2012.

15. Shelby, Z., et al., Neighbor Discovery Optimization for IPv6 over Low-Power Wireless Personal Area Networks (6LoWPANs), RFC6775, November 2012.

16. IETF, 6TiSCH Working Group Charter, https://datatracker.ietf.org/wg/6tisch/charter/

17. IETF, ACE Working Group Charter, https://datatracker.ietf.org/wg/ace/charter/

18. Zavazava, C., ITU Work on the Internet of Things, Presentation at ICTP Workshop, March 2015.

19. IPSO Alliance, http://www.ipso-alliance.org/

20. IETF, IPSO Alliance Successfully Demonstrates Internet of Things Interoperability, IETF Journal, October 2012.

21. Open Connectivity Foundation, http://openconnectivity.org/

22. ETSI IoT, http://www.etsi.org/technologies-clusters/technologies/internet-of-things

23. oneM2M, http://www.onem2m.org/about-onem2m/why-onem2m

24. Thread Group, http://www.threadgroup.org/

25. Mazhelis O., et al., "Internet of Things Market, Value Networks, and Business Models: State of the Art Report", 2013.

26. Z-Wave Alliance, http://z-wavealliance.org/

27. TIA TR-50, http://www.tiaonline.org/all-standards/committees/tr-50

28. OASIS, https://www.oasis-open.org/committees/tc_cat.php?cat=iot

29. IBM LoRaWAN Press Release, https://www03.ibm.com/press/us/en/pressrelease/46287.wss

30. LoRa Alliance, https://www.lora-alliance.org/

Chapter 12
The Role of Open Source in IoT

12.1 The Open Source Movement

Open source in the computer industry is the publishing of source code or hardware design, with associated licensing that permits the reuse, modification, improvement, and potential commercialization under favorable terms. Example of favorable distribution terms includes the following criteria:

- Free Distribution: Any party may sell or give away the open-source component as part of a larger system without being obligated to pay a royalty or other fee for such sale.
- Source Code/Design: The source code or design must be distributed and made publicly available.
- Derived Works: Derivation and modification of the original open-source component are allowed under the original licensing terms.
- No Discrimination: The license must not discriminate against any person, group, or a field of business, academics, or research.
- No Packaging Restrictions: The open-source component is not limited to be used as part of a specific distribution or product and is not precluded from being used with other open-source or closed-source components.
- Technology Neutral: There are no assumptions or conditions favoring a specific technology or interface.

While any system can potentially be released under an open source license by its owner, successful open source projects have associated communities of interest that are integral to their success. Such communities are typically geographically distributed and rely on electronic platforms for collaboration. These platforms ensure process compliance, source code management, issue tracking, and continuous integration and test.

© The Author(s), under exclusive license to Springer Nature Switzerland AG 2022
A. Rayes, S. Salam, *Internet of Things from Hype to Reality*,
https://doi.org/10.1007/978-3-030-90158-5_12

The development lifecycle of an open-source activity is quite different from the proprietary development cycle. Building a critical mass with an engaged open source community is a critical factor in successful adoption of a project. The ability of a community to garner interest and passion is an indicator of their engagement and potential for providing the advocacy necessary for successful market adoption.

That takes time. On the other hand, if a company decides to create a product, they will staff the project accordingly, and progress in the early phases of the project will be achieved much faster but the rate of progress will remain relatively constant over time.

However, with open source, once the community is fully engaged, the rate of progress can rapidly accelerate and the project can potentially progress at a rate that can far outpace closed source development. This is referred to as the "crowdsourcing" effect. According to Howe [5], crowdsourcing is *the act of a company or institution taking a function once performed by employees and outsourcing it to an undefined (and generally large) network of people in the form of an open call."* Without a doubt, open source is one of the most successful forms of crowdsourcing in the software development industry. Figure 12.1 shows how the crowdsourcing effect impacts the speed of development.

Like other initiatives, the open source movement has certain disadvantages. For example, the leadership of the project does not have control over the contributors. If a key developer decides to move on to another project, there is very little that the coordinators of the open source organization can do. They cannot nominate or recruit another leader unless one comes forward. Another issue is focusing the energy of the contributors in the right direction. If a group of people were to make a contribution that is not in line with the original goal or intent of the project, there are only two options: either the leadership rejects the contribution or they allow it. If they reject the contribution, they will lose the potential contributors. If they allow the contribution, they risk diluting the original impact of their open source project.

There are many open-source success stories: Linux, Apache Hadoop and HTTP server, MySQL, Google Chrome, OpenOffice, Android, and Java to name a few. The days of viewing open source as a fad are long gone. Open source is how modern organizations and increasingly more traditional organizations build software. Large

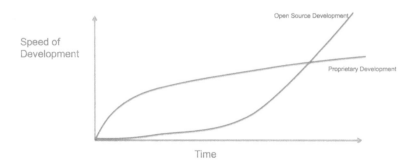

Fig. 12.1 The crowdsourcing effect on the speed of development

corporations are embracing open source and intend to use it in production. Recently, John Donovan, CTO at AT&T mentioned that, today, open source products represent about 5% of their infrastructure. They plan for that number to reach 50% by 2020. The open source high-speed train is in motion and there is no turning back.

12.2 Why Open Source?

There are numerous reasons driving individuals, corporations, small businesses, non-profits, government agencies, and other organizations to consume, publish, collaborate on, or support open source. We will discuss the main drivers here.

12.2.1 Drivers for Open Source Consumers

The reasons driving individuals and organizations to leverage and use open source projects are many, and can be attributed to the following:

Business Efficiency: Many technical problems already have open source solutions available. Hence, instead of wasting time and resources reinventing the wheel, open source consumers can use the best-of-breed solution and focus their efforts on working to address yet-unsolved challenges. These are the types of challenges that add value to their business or mission. This enables a shift from low-value work to high-value work.

Best-of-Breed Solution: Evidence shows that open source software has better quality compared to closed source [2]. With a closed source system, bugs can potentially be detected and resolved by only the employees of the company developing that system. Whereas open source provides clear advantages here: First, it presents the opportunity to tap into a larger pool of contributors and leverage the knowledge of the world's best engineers, not just those on a company's payroll. Second, open source systems are hardened through exposure to a wide array of use cases, not just the one that the original developer intended. This helps in surfacing issues and corner cases much more rapidly compared to traditional test and quality assurance processes baked into typical engineering/development pipelines.

Lower Total Cost of Ownership (TCO): Whether employing open source or closed source systems, certain costs, such as training, maintenance, and support, are sunk costs that have to be paid. In the case of closed source commercial systems, these costs are baked into the equipment price or licensing fees. What sets open source systems apart is the generally lower up-front cost (you do not pay for the right to use the underlying intellectual property). The cost center is shifted from licensing to customization and integration. This generally yields a lower total cost of ownership compared to proprietary and closed systems.

Cost	Open source	Proprietary
Licensing	No	Yes
Training	Yes	Yes
Maintenance	Yes	Yes
Support	Yes	Yes

Modern, Nimble Development Processes: Open source projects go hand-in-hand with online collaboration tools and platforms that enable distributed, asynchronous, and lock-free electronic workflows. These workflows enable rapid development and allow for more frequent releases. This provides the adopters of open source systems with the required system capabilities without the typical long lead times associated with more traditional corporate processes. This applies not only to new feature functionality, but also to bugs and security vulnerabilities. With access to the source code, the adopters of open source systems can often apply patches, or fixes, at their own convenience, without being gated by the release cycles of a specific vendor.

12.2.2   Drivers for Open Source Contributors

Open source contributors include both individuals and large corporations. There are many moral and participatory motivations that drive individuals to contribute to open source projects. While acknowledging the importance of those motives and contributions, in this section, we will only focus on the drivers that encourage large corporations to engage in open source projects.

Workforce Multiplier: Open source provides a platform for scaling a development organization's workforce. This happens in two ways: First, when a community comes together to solve a shared challenge, the human capital that becomes dedicated to work on the problem can quickly eclipse what could have been possible in a close corporate setting. Also, the diversity of that capital has been proven to correlate to the degree of innovation and quality of ideas generated. Second, the incubators of the open source system receive peer review and feedback from the community of adopters, who effectively act as "for free" testers of the open source system. This helps improve the original product and bring it to a level of quality and maturity that a small group of developers would have trouble achieving on their own.

Better Product Architecture: Open source generally leads to well-architected systems that are designed with modularity, maintainability, and flexibility in mind. This is because open source systems, by their nature, are built for a wide array of use cases, environments, and users. Hence, technical shortcuts that typically lure developers who are working on proprietary systems, e.g., due to scheduling constraints or laser-focus on a specific use case, generally do not manifest in open source projects. Over the long run, this results in greater flexibility and lower customization costs when comparing open source with closed source systems. This is the reason why some software engineering pundits advocate for architecting all software, even proprietary or internal code, as if it were open source.

Great Advertising: Contributors and shepherds of successful open source projects are perceived as industry thought leaders. This bestows upon them the ability to shape the conversation around a particular software problem and allows them to associate their brand with the preferred solution. In a way, this solution becomes the de facto standard for the associated technology. For example, 37Signals is known for creating Ruby on Rails. GitHub is known for creating Hubot.

Customer Feedback and Trust: Open source offers companies a direct line of interaction with their most passionate customers. It empowers those customers to have a collective powerful voice in the technology development process. The feedback that a company receives can better guide its product development priorities and roadmap decisions, in addition to improving the overall product quality. Furthermore, open source increases transparency which helps promote the customer's trust in a corporation's software.

Attracting and Vetting Talent: Open source allows a corporation to showcase to the developer community the interesting challenges that it is trying to solve, and how it is looking at solving them. Open source developers can casually contribute to projects, to learn how the organization works, and what it is like to develop solutions for a particular set of challenges. If they are engaged and enthused, the likelihood of them applying for a job at the corporation will be much higher than if the organization were a black box. Similarly, the corporation can see firsthand the quality of the contributed code of prospective employees, which provides better confidence in their capabilities than a typical interview process.

12.3   Open Source vs. Standards

Promoting interoperability through standards is achieved in a very different way compared to open source. Standards organizations come in a continuum of sizes, from the large and well-established international bodies such as IEEE or ITU to the more nimble and usually scope-focused organizations. Smaller organizations tend to have less procedures and target specific problem domains. Regardless of the size of the organization, companies approach them in the same way: they bring their technology and try to turn it into a standard. This usually results in long debates, power struggles, and eventually negotiations, which lead to the creation of a document. This process may take years to conclude. If the company fails to include its technology into a specification, it may try somewhere else, in a different organization.

In the case of IoT, the situation is more complex. The behavior described above is possible but, since IoT is a green field, some companies may claim that the existing standard bodies do not have the specific skill set or expertise required to realize a new IoT standard. This may result in the creation of a new organization, specifically designed to address one of the IoT verticals such as industrial automation.

However, even if the scene has changed, the format remains more or less the same. A credible standards organization needs to have rules and processes in place to ensure quality and openness. This also applies to IoT standards organizations (Chap. 10). Therefore, the development cycle of IoT standards is on track to match the pace of other technologies in "legacy" standards bodies, and this is to be expected. There needs to be a requirements definition phase, a scoping phase, a debate phase, a drafting phase, a review phase and finally a voting or some sort of consensus to sanction the work. Eventually, when the standard draft is stable enough, companies can develop to it, which may add several months of delay before a final stable implementation sees the light of day.

In the open-source world however, things can proceed at a much faster rate. A group of developers write source code, they submit it to an existing project if there is one. The code is peer reviewed. If it does not cause any regressions in the system operation and follows the best practice coding guidelines, the code is integrated. No one can block a contribution on the grounds that their company is doing things differently, or because there is a better way to implement. If there is, then code must be submitted by those making such claims. Eventually, the end-users will vote by evaluating the code and its functionality. Some user may feel compelled to fix bugs so their company can use the product, and other users benefit instantly.

Of course, the leap of faith a company may take by giving away the implementation of their technology is a substantial barrier to overcome. But the key to success in open source is to add a "secret sauce" that complements the public domain functions. The open source project then becomes a vehicle to get immediate feedback on a way to do things, ignite the spark of curiosity, and attract potential developers and partners. With a common basis built, new proprietary improvements can be added on top of the public domain code. This brings all the players to a higher common ground, which is beneficial for everybody, the producer and the consumer.

12.4 Open Source Partnering with Standards

As we saw earlier, the way companies approach open source and standards is very different. However, since open source is beneficial for companies, standard bodies quickly realized that they could use open source efforts for their benefit. After all, what the consumer needs is not a 300 page document describing in mundane details how a system should be implemented. Consumers want to have real products in their hand, with real functionalities to use and evaluate in their own business or home environments. This is not something that they get out of the usually dry reading of a standards document. "Code is King" and having some code, which implements a standard is a very powerful combination. The standard represents an agreement between several parties and the code is the proof that the system on paper does indeed work.

Table 12.1 Examples of open source initiatives for IoT

Standards organization or project	Open source implementation
Open Interconnect Consortium	IoTVity (Linux Foundation)
oneM2M	IoTDM (Linux Foundation), OCEAN, OM2M (Eclipse)
Allseen Alliance	AllJoyn
ZigBee® Alliance (IEEE)	Zboss, Open-ZB, NS2, OpNet
CoAP (IETF)	Californium (Eclipse)
MQTT (OASIS)	Mosquitto.org, Paho (Eclipse)
ZWave (Z-Wave Alliance)	openZwave
DASH7 (Alliance)	OSS-7, OpenTag
Modbus (Schneider)	libmodbus.org
BACnet (ASHRAE)	Wacnet
KNX (ISO)	Linknx and Webknx2

Therefore, it is now becoming a must-have for a project under development in a standards body to be associated with some form of open source effort. Following are some examples related to IoT (Table 12.1).

12.5   A Tour of Open-Source Activities in IoT

As mentioned previously, the IoT open source community is quite active. There are several open projects, some are backed by consortiums of large industry players, others are backed by just a single startup. Large or small, they all aim at facilitating the deployment of IoT solutions. But, unfortunately, they are not compatible with each other. Some of the larger efforts are attempting to bridge the gap and connect with other overlapping communities or projects.

The list below is far from being exhaustive. It is merely meant to provide an overview of active projects, which have the potential to make a difference in the IoT space. The list is organized per the IoT reference model presented in Fig. 1.5.

12.5.1   IoT Devices

12.5.1.1   Hardware

Arduino

Arduino is both a hardware specification for interactive electronics and a set of software that includes an Integrated Development Environment (IDE) and the Arduino programming language. Arduino is "a tool for making computers that can sense and control more of the physical world than your desktop computer." The organization behind it offers a variety of electronic boards, starter kits, robots and related products for sale, and many other groups have used Arduino to build IoT-related hardware and software products of their own.

GizmoSphere

GizmoSphere is an open source development platform for the embedded design community; the site includes code downloads and hardware schematics along with free user guides, specification sheets, and other documentation.

Tinkerforge

Tinkerforge is a system of open source stackable microcontroller building blocks. It allows the control of motors and reading out sensors with the following programming languages: C, C++, C#, Object Pascal, Java, PHP, Python, and Ruby over a USB or Wi-Fi connection on Windows, Linux, and Mac OS X. All of the hardware is licensed under CERN OHL (CERN Open Hardware License).

BeagleBoard

BeagleBoard offers credit-card sized computers that can run Android and Linux. Because they have very low-power requirements, they are a good option for IoT devices. Both the hardware designs and the software they run are open source, and BeagleBoard hardware (often sold under the name BeagleBone) is available through a wide variety of distributors.

12.5.1.2   Operating Systems

Contiki

Contiki is an open source operating system for networked, memory-constrained systems with a particular focus on low-power wireless Internet of Things devices. Examples of where Contiki is used include street lighting systems, sound monitoring for smart cities, radiation monitoring systems, and alarm systems. Other key features include highly efficient memory allocation, full IP networking, very low-power consumption, dynamic module loading, and more. Supported hardware platforms include Redwire Econotags, Zolertia z1 motes, ST Microelectronics development kits and Texas Instruments chips and boards. Paid commercial support is available.

Raspbian

While the Raspberry Pi is not an open source project, many components of its OS are. Raspbian is a free operating system based on Debian optimized for the Raspberry Pi hardware.

RIOT

This 1.5 kB embedded OS bills itself as "the friendly operating system for the Internet of Things." It fits in the category of Contiki and TinyOS. Forked from the FeuerWhere project, RIOT debuted in 2013. It aims to be both developer- and resource-friendly. It supports multiple architectures, including MSP430, ARM7, Cortex-M0, Cortex-M3, Cortex-M4, and standard x86 PCs.

12.5.2   IoT Services Platform

12.5.2.1   Eclipse IoT Project

Eclipse is sponsoring several different projects surrounding IoT. They include application frameworks and services; open source implementations of IoT protocols, including MQTT CoAP, OMA-DM, and OMA LWM2M; and tools for working with Lua, which Eclipse is promoting as an ideal IoT programming language. Eclipse-related projects include:

- Paho provides client implementations of the MQTT protocol.
- Mihini is an embedded Lua runtime providing hardware abstraction and other services.
- Koneki provides tools for embedded Lua developers.
- Eclipse SCADA is a complete Java/OSGi-based SCADA system which provides communication, monitoring, GUI and other capabilities.
- Kura is a Java/OSGi-based M2M container for gateways. It has support for Modbus, CANbus, MQTT, and other protocols.
- Mosquitto is a lightweight server implementation of the MQTT and MQTT-SN protocols written in C.
- Ponte bridges IoT protocols (MQTT and CoAP) to the Web.
- Smarthome provides a complete set of services for home automation gateways.
- OM2M implements the ETSI M2M standard.
- Californium is a Java implementation of the CoAP protocol, which includes DTLS for security.
- Wakaama is an implementation of LWM2M written in C.
- Krikkit is a rules system for programming edge devices.
- Concierge is a lightweight implementation of OSGi Core R5.

12.5.2.2   Kinoma

The Kinoma group's hardware and software prototyping solutions help developers, programmers, and designers rapidly create connected products. Owned by Marvell, the Kinoma software platform encompasses three different open source projects. Kimona Create is a DIY construction kit for prototyping electronic devices. Kimona Studio is the development environment that works with Create and the Kinoma Platform Runtime. Kimona Connect is a free iOS and Android app that links smartphones and tables with IoT devices.

12.5.2.3   OneM2M the Linux Foundation and Eclipse

The purpose and goal of oneM2M is to develop technical specifications, which address the need for a common M2M Service Layer that can be readily embedded within various hardware and software. oneM2M positions itself as a cross vertical platform. This means that it will be well suited for various sectors such as industrial, energy, home, etc. These specifications are being implemented as open source projects at the Linux Foundation (IoTDM), Eclipse (oM2M), and OCEAN.

12.5.2.4   Open Interconnect Consortium (OIC)

The goal of OIC is to enable application developers and device manufacturers to deliver interoperable products across Android, iOS, Windows, Linux, Tizen, and more. The Linux Foundation hosts a project called IoTvity, which provides open source code for OIC. At the time of this writing, OIC and oneM2M are specifying gateway functions to bridge the two domains.

12.5.2.5   IT6.eu, OpenIoT, and IoTSyS

The European Union is actively financing the development of IoT research. OpenIoT and IoTSyS are examples. The OpenIoT website explains that the project is "an open source middleware for getting information from sensor clouds, without worrying what exact sensors are used." It aims to enable cloud-based "sensing as a service."

IoTSyS is an IoT middleware providing a communication stack for smart devices. It supports multiple standards and protocols, including IPv6, oBIX, 6LoWPAN, Constrained Application Protocol, and Efficient XML Interchange.

12.5.2.6   DeviceHive

This project offers a data collection facility for connecting IoT devices. It includes easy-to-use web-based management software for creating devices, applying security rules, and monitoring devices. The website offers sample projects built with DeviceHub, and it also has a "playground" section that allows users to use DeviceHub online to see how it works.

12.5.2.7   IoT Toolkit

The group behind this project is working on a variety of tools for integrating multiple IoT-related sensor networks and protocols. IoT Toolkit implements HTTP/ REST, CoAP, and MQTT protocols and acts as a stateful bridge between these different protocols.

The primary project is a Smart Object API, but the group is also working on an HTTP-to-CoAP Semantic mapping, an application framework with embedded software agents and more.

Note there is a difference between open source efforts implementing a standard (such as oneM2M and OIC) versus open source efforts trying to realize a middleware implementation with their own data models and protocols. We expect the industry to be more likely to embrace the former.

12.6   Conclusions

There are many aspects to IoT (device, transport, data aggregation and collection, big data, etc.), this translates to a large number of standards and slow progress. Most of the standards are backed by an open source activity. It is now becoming clear that the industry wants to see working code in addition to seeing concise documents describing a technology. The open-source community has preceded the standards in most cases, proposing working solutions to real problems.

Therefore there are two classes of open-source activities in IoT: one backed by a standard, and those evolving by themselves. The latter group is of course more agile and can offer solutions without the overhead of standard development procedures. However, in many cases, there is no domination of one group over the others. This leads to the conclusion that, eventually, a combination of standard plus associated open source will be the long-term solutions the industry will adopt.

Problems and Exercises

1. What is open source? What are the key benefits to the producer and users?
2. Why open source platform is appealing to platform for developers? Why appealing application consumers (companies and individuals)?
3. List three downsides for open source projects.

4. List two main disadvantages of open-source projects.
5. Linux is a well know open source project. List three other examples of successful open source projects.
6. Name three examples of IoT open source activities.
7. Are there major differences between standards and open source developments? If so, what is the key difference (i.e., what's the deliverables/outcomes of standard bodies and they are for Open Source)? What is the relationship between the two?
8. Name three standards which are implemented in open source.
9. When was the open source label developed? Who developed it?
10. A license defines the rights and obligations that a licensor grants to a licensee. Does need open source provide licenses to its users? What do such license impose?
11. Certification often helps to build higher user confidence. Are there certifications issued for open source? If so, name two examples.
12. Global Desktop Project is an example of Open Source initiative developed by the United Nation University. What does it do?
13. What are the main phases of IoT standard development cycles? What are the main phases of IoT open source developments? What are they key differences?
14. It is said that a key to success in open source is to add a "secret sauce" that complements the public domain functions? Why is it the case? Can you provide an example?
15. What is meant by "Code is King" in open source?

References

1. Open Source Definition, The Open Source Initiative, https://opensource.org/osd
2. Wheeler, D., Why Open Source Software, http://www.dwheeler.com/oss_fs_why.html, July 2015.
3. http://ben.balter.com/2012/06/26/why-you-should-always-write-software-as-open-source/
4. Preston-Werner, T., Open Source (Almost) Everything, http://tom.preston-werner.com/2011/11/22/open-source-everything.html, November 2011.
5. Jaksic, M., et al., Proceedings of the XIII International Symposium SymOrg 2012, June 2012.
6. Wright, S., et al., Open source and standards: The role of open source in the dialogue between research and standardization, 2014 IEEE Globecom Workshops (GC Wkshps), December 2014.
7. Github Hubot: http://hubot.github.com
8. M. St. Laurent, Understanding Open Source and Free Software Licensing. O'Reilly Media. p. 4. May 2008, ISBN 9780596553951.
9. W. T. Verts, "Open source software". World Book Online Reference Center. Archived from the original on January 1, 2011.
10. R. Rothwell, "Creating wealth with free software". Free Software Magazine, September 2008.
11. E. S. Raymond, "Goodbye free software, hello open source", August 2—8, Online: http://www.catb.org/esr/open-source.html
12. Arduino: Online: http://www.arduino.cc/

13. GizmoSphere: Online: https://en.wikipedia.org/w/index.php?title=GizmoSphere&action=edit &redlink=1
14. Tinkerforge: https://en.wikipedia.org/wiki/Tinkerforge
15. BeagleBoard: http://beagleboard.org/
16. AllJoyn: https://allseenalliance.org/developer-resources/alljoyn-open-source-project
17. Contiki: http://www.contiki-os.org/
18. Raspbian: http://raspbian.org/
19. RIOT: http://riot-os.org/
20. Eclipse IoT Project: http://iot.eclipse.org/
21. Kinoma: http://www.marvell.com/kinoma/
22. oneM2M: http://www.onem2m.org
23. IoTDM: Open-source Project implementing oneM2M specification, https://wiki.opendaylight. org/view/IoTDM:Main
24. OIC (Open Interconnect Consortium): http://openinterconnect.org
25. OpenIoT: http://openiot.eu/
26. IoT Toolkit: https://github.com/connectIOT/iottoolkit/wiki/IoT-Toolkit-Overview-and-links
27. T. Pham, Verint Systems Inc. and Matthew B. Weinstein and Jamie L. Ryerson. "Easy as ABC: Categorizing Open Source Licenses"; www.IPO.org. June 2010.

Appendix A

Glossary

6LowPAN
IPv6 over Low-power Wireless Personal Area Networks (IETF).

6TiSCH
IPv6 over Time Slotted Channel Hopping mode of IEEE 802.15.4 (IETF).

AAA
Authentication, authorization, and accounting. See also TACACS+ and RADIUS (Various).

AAAA
Authentication, Authorization, Accounting, and Auditing (Various).

Access Modes
The security appliance CLI uses several command modes. The commands available in each mode vary. See also user EXEC mode, privileged EXEC mode, global configuration mode, command-specific configuration mode (Cisco).

ACE
Access Control Entry. Information entered into the configuration that lets you specify what type of traffic to permit or deny on an interface. By default, traffic that is not explicitly permitted is denied (Cisco).

ACL
Access Control List. A collection of ACEs. An ACL lets you specify what type of traffic to allow on an interface. By default, traffic that is not explicitly permitted is denied. ACLs are usually applied to the interface which is the source of inbound traffic. See also rule, outbound ACL (Cisco).

A. Rayes, S. Salam, *Internet of Things from Hype to Reality*,
https://doi.org/10.1007/978-3-030-90158-5

Actuators

"An actuator is a type of motor that is responsible for moving or controlling a mechanism or system. It is operated by a source of energy, typically electric current, hydraulic fluid pressure, or pneumatic pressure, and converts that energy into motion. An actuator is the mechanism by which a control system acts upon an environment. The control system can be simple (a fixed mechanical or electronic system), software-based (e.g., a printer driver, robot control system), a human, or any other input" (Wikipedia).

Address

An address is used for locating and accessing—"talking to"—a *Device*, a *Resource*, or a *Service*. In some cases, the ID and the Address can be the same, but conceptually they are different (IoT-A).

Address Resolution Protocol (ARP)

Address Resolution Protocol. A low-level TCP/IP protocol that maps a hardware address, or MAC address, to an IP address. An example hardware address is 00:00:a6:00:01:ba. The first three groups of characters (00:00:a6) identify the manufacturer; the rest of the characters (00:01:ba) identify the system card. ARP is defined in RFC 826 (Cisco).

Address Translation

The translation of a network address and/or port to another network address/or port. See also IP address, interface PAT, NAT, PAT, Static PAT, xlate (Cisco).

ADN

Application Dedicated Node. oneM2M compliant device (i.e., Thing) with restricted functionality (oneM2M).

AES

Advanced Encryption Standard. A symmetric block cipher that can encrypt and decrypt information. The AES algorithm is capable of using cryptographic keys of 128, 192, and 256 bits to encrypt and decrypt data in blocks of 128 bits. See also DES (Cisco).

AH

Authentication Header. An IP protocol (type 51) that can ensure data integrity, authentication, and replay detection. AH is embedded in the data to be protected (a full IP datagram, for example). AH can be used either by itself or with ESP. This is an older IPSec protocol that is less important in most networks than ESP. AH provides authentication services but does not provide encryption services. It is provided to ensure compatibility with IPSec peers that do not support ESP, which provides both authentication and encryption. See also encryption and VPN. Refer to the RFC 2402 (Cisco).

AMQP

Advanced Message Queuing Protocol (Various).

Application Software
"Software that provides an application *service* to the *user*. It is specific to an application in the multimedia and/or hypermedia domain and is composed of programs and data" ([ETSI-ETR173]).

Architectural Reference Model
The IoT-A architectural reference model follows the definition of the IoT reference model and combines it with the related IoT reference architecture. Furthermore, it describes the methodology with which the reference model and the reference architecture are derived, including the use of internal and external stakeholder requirements (IoT-A).

Architecture
"The fundamental organization of a *system* embodied in its components, their relationships to each other, and to the environment, and the principles guiding its design and evolution" ([IEEE-1471-2000]).

Architecture Vision
"A high-level, aspirational *view* of the target *architecture*" ([TOGAF9]).

ARP
Address Resolution Protocol. A low-level TCP/IP protocol that maps a hardware address, or MAC address, to an IP address. An example hardware address is 00:00:a6:00:01:ba. The first three groups of characters (00:00:a6) identify the manufacturer; the rest of the characters (00:01:ba) identify the system card. ARP is defined in RFC 826 (Cisco).

ASA
Adaptive Security Algorithm. Used by the security appliance to perform inspections. ASA allows one-way (inside to outside) connections without an explicit configuration for each internal system and application (Cisco).

ASDM
Adaptive Security Device Manager. An application for managing and configuring a single security appliance (Cisco).

ASN
Application Service Node. Fully featured oneM2M compliant device (oneM2M).

Association
An association establishes the relation between a *service* and *resource* on the one hand and a *Physical Entity* on the other hand (IoT-A).

Asymmetric Encryption
Also called public key systems, asymmetric encryption allows anyone to obtain access to the public key of anyone else. Once the public key is accessed, one can send an encrypted message to that person using the public key. See also encryption, public key (Cisco).

Augmented Entity
The composition of a *Physical Entity* together with its *Virtual Entity* (IoT-A).

Authentication
Cryptographic protocols and services that verify the identity of users and the integrity of data. One of the functions of the IPSec framework. Authentication establishes the integrity of datastream and ensures that it is not tampered with in transit. It also provides confirmation about the origin of the datastream. See also AAA, encryption, and VPN (Cisco).

Authentication
Authentication ensures that the entities involved in any operation are who they claim to be. A masquerade attack or an impersonation attack usually targets this requirement where an entity claims to be another identity (Various).

Authorization
Authorization: ensures that entities have the required control permissions to perform the operation they request to perform (Various).

Auto Applet Download
Automatically downloads the WebVPN port-forwarding applet when the user first logs in to WebVPN (Cisco).

Auto-signon
This command provides a single sign-on method for WebVPN users. It passes the WebVPN login credentials (username and password) to internal servers for authentication using NTLM authentication, basic authentication, or both (Cisco).

Availability
Availability refers to characteristic of a system or subsystem that is continuously operational for a desirably long period of time. It is typically measured relative to "100% operational" or "never failing." A widely-held but difficult-to-achieve standard of availability for a system or product is known as "five 9s" (available 99.999% of the time in a given year) availability (Various).

AVB
Audio Video Bridging. IEEE standards for supporting time sensitive audio/video streams over wireless Ethernet networks. Also known as Time Sensitive Networking (Various).

Backup Server
IPSec backup servers let a VPN client connect to the central site when the primary security appliance is unavailable (Cisco).

Backward Secrecy
Backward Secrecy: ensures that any new object that joins the network will not be able to understand the communications that were exchanged prior to joining the network (Various).

BGP
Border Gateway Protocol. BGP performs interdomain routing in TCP/IP networks. BGP is an Exterior Gateway Protocol, which means that it performs routing between multiple autonomous systems or domains and exchanges routing and access information with other BGP systems. The security appliance does not support BGP. See also EGP (Cisco).

BI
Business Intelligence (Authors).

Blockchain
A decentralized public ledger that records all transactions within a given network (Authors).

BLT stream
Bandwidth Limited Traffic stream. Stream or flow of packets whose bandwidth is constrained (Cisco).

Bluetooth
Short-range wireless protocol usually used to connect input/output electronic accessories and peripherals.

BOOTP
Bootstrap Protocol. Let us diskless workstations boot over the network as is described in RFC 951 and RFC 1542 (Cisco).

BPDU
Bridge Protocol Data Unit. Spanning-Tree Protocol hello packet that is sent out at configurable intervals to exchange information among bridges in the network. Protocol data unit is the OSI term for packet (Cisco).

Business Logic
Goal or behavior of a system involving Things serving a particular business purpose. Business Logic can define the behavior of a single Thing, a group of Things, or a complete business process (IoT-A).

CA
Certificate Authority, Certification Authority. A third-party entity that is responsible for issuing and revoking certificates. Each device with the public key of the CA can authenticate a device that has a certificate issued by the CA. The term CA also refers to software that provides CA services. See also certificate, CRL, public key, RA (Cisco).

Cache
A temporary repository of information accumulated from previous task executions that can be reused, decreasing the time required to perform the tasks. Caching stores frequently reused objects in the system cache, which reduces the need to perform repeated rewriting and compressing of content (Cisco).

Carrousel Attack
This attack targets the Network layer in the OSI stack and can be launched if the routing protocol supports source routing, where the object generating the packets can specify the whole routing path of the packets it wishes to send to the fog device (Various).

CBC
Cipher Block Chaining. A cryptographic technique that increases the encryption strength of an algorithm. CBC requires an initialization vector (IV) to start encryption. The IV is explicitly given in the IPSec packet (Cisco).

certificate
A signed cryptographic object that contains the identity of a user or device and the public key of the CA that issued the certificate. Certificates have an expiration date and may also be placed on a CRL if known to be compromised. Certificates also establish non-repudiation for IKE negotiation, which means that you can prove to a third party that IKE negotiation was completed with a specific peer (Cisco).

CHAP
Challenge Handshake Authentication Protocol (Cisco).

CIFS
Common Internet File System. It is a platform-independent file sharing system that provides users with network access to files, printers, and other machine resources. Microsoft implemented CIFS for networks of Windows computers, however, open source implementations of CIFS provide file access to servers running other operating systems, such as Linux, UNIX, and Mac OS X (Cisco).

CLI
Command line interface. The primary interface for entering configuration and monitoring commands to the security appliance (Cisco).

Client update
Let us update revisions of clients to which the update applies; provide a URL or IP address from which to get the update; and, in the case of Windows clients, optionally notify users that they should update their VPN client version (Cisco).

client/server computing
Distributed computing (processing) network systems in which transaction responsibilities are divided into two parts: client (front end) and server (back end). Also called distributed computing. See also RPC (Cisco).

CoAP
Constrained Application Protocol (Various).

command-specific configuration mode
From global configuration mode, some commands enter a command-specific configuration mode. All user EXEC, privileged EXEC, global configuration, and command-specific configuration commands are available in this mode. See also global configuration mode, privileged EXEC mode, user EXEC mode (Cisco).

Communication Model
The communication model aims at defining the main communication paradigms for connecting elements, as, in the IoT-A case, defined in the domain model. This model provides a set of communication rules to build interoperable stacks, together with insights about the main interactions among the elements of the domain model (IoT-A).

Compression
The process of encoding information using fewer bits or other information-bearing units than an unencoded representation would use. Compression can reduce the size of transferring packets and increase communication performance (Cisco).

Confidentiality
Confidentiality ensures that the exchanged messages can be understood only by the intended entities (Various).

Consensus Algorithm
A consensus algorithm allows nodes on the network to trust that a given piece of data is valid, and that it has been synchronized with all other nodes (Authors).

Consortium Blockchain
A blockchain where the network is controlled by a certain set of nodes (Authors).

Constrained Network
A constrained network is a network of devices with restricted capabilities regarding storage, computing power, and/or transfer rate (IoT-A).

Container
Light-weight virtualization construct where the underlying operating system kernel is common among members (Authors).

Content Rewriting/Transformation
Interprets and modifies applications so that they render correctly over a WebVPN connection (Cisco).

Controller
Anything that has the capability to affect a *Physical Entity*, like changing its state or moving it (IoT-A).

cookie
A cookie is an object stored by a browser. Cookies contain information, such as user preferences, to persistent storage (Cisco).

CORE
IETF Constrained RESTful Environments workgroup (IETF).

CPU
Central Processing Unit. Main processor (Cisco).

CRC
Cyclical Redundancy Check. Error-checking technique in which the frame recipient calculates a remainder by dividing frame contents by a prime binary divisor and compares the calculated remainder to a value stored in the frame by the sending node (Cisco).

Credentials
A credential is a record that contains the authentication information (credentials) required to connect to a resource. Most credentials contain a user name and password (IoT-A).

CRM
Customer Relation Management (Authors).

Crypto map
A data structure with a unique name and sequence number that is used for configuring VPNs on the security appliance. A crypto map selects data flows that need security processing and defines the policy for these flows and the crypto peer that traffic needs to go to. A crypto map is applied to an interface. Crypto maps contain the ACLs, encryption standards, peers, and other parameters necessary to specify security policies for VPNs using IKE and IPSec. See also VPN (Cisco).

Cryptocurrency
A digital currency built upon cryptographic protocols (Authors).

Cryptography
Encryption, authentication, integrity, keys, and other services used for secure communication over networks. See also VPN and IPSec (Cisco).

CSE
Common Services Entity. In oneM2M architecture, the middleware layer that sits in between applications (Application Entity) and the underlying network services (Network Services Entity) (Various).

Data confidentiality
Describes any method that manipulates data so that no attacker can read it. This is commonly achieved by data encryption and keys that are only available to the parties involved in the communication (Cisco).

Data integrity
Describes mechanisms that, through the use of encryption based on secret key or public key algorithms, allow the recipient of a piece of protected data to verify that the data has not been modified in transit (Cisco).

Data origin authentication
A security service where the receiver can verify that protected data could have originated only from the sender. This service requires a data integrity service plus a key distribution mechanism, where a secret key is shared only between the sender and receiver (Cisco).

DDS RTPS
Distribute Data Service Real Time Publish and Subscribe Protocol (Various).

Decryption
Application of a specific algorithm or cipher to encrypted data so as to render the data comprehensible to those who are authorized to see the information. See also encryption (Cisco).

Denial of Sleep Attack
Denial of Sleep: Different data link layer protocols were proposed to reduce the power consumption of smart objects by switching them into sleep whenever they are not needed. Examples of these protocols include S-MAC [3] and T-MAC [4] protocols (Various).

DES
Data encryption standard. DES was published in 1977 by the National Bureau of Standards and is a secret key encryption scheme based on the Lucifer algorithm from IBM. Cisco uses DES in classic crypto (40-bit and 56-bit key lengths), IPSec crypto (56-bit key), and 3DES (triple DES), which performs encryption three times using a 56-bit key. 3DES is more secure than DES but requires more processing for encryption and decryption. See also AES, ESP (Cisco).

Device
Technical physical component (hardware) with communication capabilities to other IT systems. A *device* can be either attached to or embedded inside a *Physical Entity*, or monitor a *Physical Entity* in its vicinity (IoT-A).

DHCP
Dynamic Host Configuration Protocol. Provides a mechanism for allocating IP addresses to hosts dynamically, so that addresses can be reused when hosts no longer need them and so that mobile computers, such as laptops, receive an IP address applicable to the LAN to which it is connected (Cisco).

Digital certificate
See certificate (Cisco).

Digital Certificate Or Pubic Key Certificate
In cryptography, a public key certificate (also known as a digital certificate or identity certificate) is an electronic document used to prove ownership of a public key. The certificate includes information about the key, information about its owner's identity, and the digital signature of an entity that has verified the certificate's contents are correct. If the signature is valid, and the person examining the certificate trusts the signer, then they know they can use that key to communicate with its owner.

In a typical public-key infrastructure (PKI) scheme, the signer is a certificate authority (CA), usually a company that charges customers to issue certificates for them. In a web of trust scheme, the signer is either the key's owner (a self-signed certificate) or other users ("endorsements") whom the person examining the certificate might know and trust.

Certificates are an important component of Transport Layer Security (TLS, sometimes called by its older name SSL, Secure Sockets Layer), where they prevent an attacker from impersonating a secure website or other server. They are also used in other important applications, such as email encryption and code signing (see also, PKI and X509) (Wikipedia).

Digital Entity
Any computational or data element of an IT-based system (IoT-A).

Discovery
Discovery is a *service* to find unknown *resources/entities/services* based on a rough specification of the desired result. It may be utilized by a human or another *service*. Credentials for authorization are considered when executing the discovery (IoT-A).

DN
Distinguished Name. Global, authoritative name of an entry in the OSI Directory (X.500) (Cisco).

DNS
Domain Name System (or Service). An Internet service that translates domain names into IP addresses (Cisco).

Docker
Open source project that provides a packaging framework to simplify the portability and automate the deployment of applications in Containers (Authors).

DODAG
Destination Oriented Directed Acyclic Graph (IETF).

Domain Model
"A domain model describes objects belonging to a particular area of interest. The domain model also defines attributes of those objects, such as name and identifier. The domain model defines relationships between objects such as 'instruments produce data sets'. Besides describing a domain, domain models also help to facilitate correlative use and exchange of data between domains" ([CCSDS 312.0-G-0]).

DoS
Denial of Service. A type of network attack in which the goal is to render a network service unavailable (Cisco).

DSL
Digital subscriber line. Public network technology that delivers high bandwidth over conventional copper wiring at limited distances. DSL is provisioned via modem pairs, with one modem located at a central office and the other at the customer site. Because most DSL technologies do not use the whole bandwidth of the twisted pair, there is room remaining for a voice channel (Cisco).

DSP
digital signal processor. A DSP segments a voice signal into frames and stores them in voice packets (Cisco).

DSS
Digital Signature Standard. A digital signature algorithm designed by The US National Institute of Standards and Technology and based on public-key cryptography. DSS does not do user datagram encryption. DSS is a component in classic crypto, as well as the Redcreek IPSec card, but not in IPSec implemented in Cisco IOS software (Cisco).

Dynamic NAT
See NAT and address translation (Cisco).

EGP
Exterior Gateway Protocol. Replaced by BGP. The security appliance does not support EGP. See also BGP (Cisco).

EIGRP
Enhanced Interior Gateway Routing Protocol. The security appliance does not support EIGRP (Cisco).

Encryption
Application of a specific algorithm or cipher to data so as to render the data incomprehensible to those unauthorized to see the information. See also decryption (Cisco).

ESMTP
Extended SMTP. Extended version of SMTP that includes additional functionality, such as delivery notification and session delivery. ESMTP is described in RFC 1869, SMTP Service Extensions (Cisco).

ESP
Encapsulating Security Payload. An IPSec protocol, ESP provides authentication and encryption services for establishing a secure tunnel over an insecure network. For more information, refer to RFCs 2406 and 1827 (Cisco).

EVPN
Ethernet Virtual Private Networks, an IETF solution standardized in RFC 7432 (IETF).

Failover, Failover mode
Failover lets you configure two security appliances so that one will take over operation if the other one fails. The security appliance supports two failover configurations, Active/Active failover and Active/Standby failover. Each failover configuration has its own method for determining and performing failover. With Active/Active failover, both units can pass network traffic. This lets you configure load balancing on your network. Active/Active failover is only available on units running in multiple context mode. With Active/Standby failover, only one unit passes traffic while the other unit waits in a standby state. Active/Standby failover is available on units running in either single or multiple context mode (Cisco).

FCAPS
(see NMS) Fault, Configuration, Accounting, Performance, and Security management (Authors).

FFD
IEEE 802.15.4 Full-Function Device. Implements all of the functions of the IEEE 802.15.4 communication stack (IEEE).

Flash, Flash memory
A nonvolatile storage device used to store the configuration file when the security appliance is powered down (Cisco).

Flooding Attack
The adversary can flood the neighboring nodes with dummy packets and request them to deliver those packets to the fog device, where devices waste energy receiving and transmitting those dummy packets (Various).

Forward Secrecy and Backward Secrecy
Forward Secrecy: ensures that when an object leaves the network, it will not understand the communications that are exchanged after its departure. Backward Secrecy:
ensures that any new object that joins the network will not be able to understand the
communications that were exchanged prior to joining the network (Various).

Freshness
Freshness: ensures that the data is fresh. Replay attacks target this requirement
where an old message is replayed in order to return an entity into an old state
(Various).

FTP
File Transfer Protocol. Part of the TCP/IP protocol stack, used for transferring files
between hosts (Cisco).

Gateway
A Gateway is a forwarding element, enabling various local networks to be connected (IoT-A).

Global Configuration Mode
Global configuration mode lets you to change the security appliance configuration.
All user EXEC, privileged EXEC, and global configuration commands are available
in this mode. See also user EXEC mode, privileged EXEC mode, command-specific
configuration mode (Cisco).

Global Storage
Storage that contains global information about many *entities of interest*. Access to
the *global storage* is available over the *Internet* (IoT-A).

GMT
Greenwich Mean Time. Replaced by UTC (Coordinated Universal Time) in 1967 as
the world time standard (Cisco).

GPRS
General packet radio service. A service defined and standardized by the European
Telecommunication Standards Institute. GPRS is an IP-packet-based extension of
GSM networks and provides mobile, wireless, data communications (Cisco).

GRE
Generic Routing Encapsulation described in RFCs 1701 and 1702. GRE is a tunneling protocol that can encapsulate a wide variety of protocol packet types inside IP
tunnels, creating a virtual point-to-point link to routers at remote points over an IP
network. By connecting multiprotocol subnetworks in a single-protocol backbone
environment, IP tunneling using GRE allows network expansion across a single
protocol backbone environment (Cisco).

GSM
Global System for Mobile Communication. A digital, mobile, radio standard developed for mobile, wireless, voice communications (Cisco).

GTP
GPRS tunneling protocol. GTP handles the flow of user packet data and signaling information between the SGSN and GGSN in a GPRS network. GTP is defined on both the Gn and Gp interfaces of a GPRS network (Cisco).

Host
The name for any device on a TCP/IP network that has an IP address. See also network and node (Cisco).

Host/network
An IP address and netmask used with other information to identify a single host or network subnet for security appliance configuration, such as an address translation (xlate) or ACE (Cisco).

HTTP
Hypertext Transfer Protocol. A protocol used by browsers and web servers to transfer files. When a user views a web page, the browser can use HTTP to request and receive the files used by the web page. HTTP transmissions are not encrypted (Cisco).

HTTPS
Hypertext Transfer Protocol Secure. An SSL-encrypted version of HTTP (Cisco).

IANA
Internet Assigned Number Authority. Assigns all port and protocol numbers for use on the Internet (Cisco).

ICMP
Internet Control Message Protocol. Network-layer Internet protocol that reports errors and provides other information relevant to IP packet processing (Cisco).

Identifier (ID)
Artificially generated or natural feature used to disambiguate things from each other. There can be several Ids for the same *Physical Entity*. The set of Ids is an attribute of a *Physical Entity* (IoT-A).

Identity
Properties of an entity that makes it definable and recognizable (IoT-A).

IDS
Intrusion Detection System. A method of detecting malicious network activity by signatures and then implementing a policy for that signature (Cisco).

IETF
The Internet Engineering Task Force. A technical standards organization that develops RFC documents defining protocols for the Internet (Cisco).

IGMP
Internet Group Management Protocol. IGMP is a protocol used by IPv4 systems to report IP multicast memberships to neighboring multicast routers (Cisco).

IKE

Internet Key Exchange. IKE establishes a shared security policy and authenticates keys for services (such as IPSec) that require keys. Before any IPSec traffic can be passed, each security appliance must verify the identity of its peer. This can be done by manually entering preshared keys into both hosts or by a CA service. IKE is a hybrid protocol that uses part Oakley and part of another protocol suite called SKEME inside ISAKMP framework. This is the protocol formerly known as ISAKMP/Oakley, and is defined in RFC 2409 (Cisco).

ILS

Internet Locator Service. ILS is based on LDAP and is ILSv2 compliant. ILS was developed by Microsoft for use with its NetMeeting, SiteServer, and Active Directory products (Cisco).

IMAP

Internet Message Access Protocol. Method of accessing e-mail or bulletin board messages kept on a mail server that can be shared. IMAP permits client e-mail applications to access remote message stores as if they were local without actually transferring the message (Cisco).

implicit rule

An access rule automatically created by the security appliance based on default rules or as a result of user-defined rules (Cisco).

IMSI

International Mobile Subscriber Identity. One of two components of a GTP tunnel ID, the other being the NSAPI. See also NSAPI (Cisco).

Information Model

"An information model is a representation of concepts, relationships, constraints, rules, and operations to specify data semantics for a chosen domain of discourse. The advantage of using an information model is that it can provide sharable, stable, and organized structure of information requirements for the domain context.

The information model is an abstract representation of entities which can be real objects such as devices in a network or logical such as the entities used in a billing system. Typically, the information model provides formalism to the description of a specific domain without constraining how that description is mapped to an actual implementation. Thus, different mappings can be derived from the same information model. Such mappings are called data models" ([AutoI]).

Infrastructure Services

Specific services that are essential for any IoT implementation to work properly. Such services provide support for essential features of the IoT (IoT-A).

Integrity

Integrity ensures that the exchanged messages were not altered/tampered by a third party (Various).

Interface
"Named set of operations that characterize the behaviour of an entity" ([OGS]).

interface
The physical connection between a particular network and a security appliance (Cisco).

interface ip_address
The IP address of a security appliance network interface. Each interface IP address must be unique. Two or more interfaces must not be given the same IP address or IP addresses that are on the same IP network (Cisco).

Internet
"The *Internet* is a global *system* of interconnected computer networks that use the standard *Internet* protocol suite (TCP/IP) to serve billions of *users* worldwide. It is a network of networks that consists of millions of private, public, academic, business, and government networks of local to global scope that are linked by a broad array of electronic and optical networking technologies. The *Internet* carries a vast array of information *resources* and *services*, most notably the inter-linked hypertext documents of the World Wide Web (WWW) and the infrastructure to support electronic mail.

Most traditional communications media, such as telephone and television *services*, are reshaped or redefined using the technologies of the *Internet*, giving rise to *services* such as Voice over *Internet* Protocol (VoIP) and IPTV. Newspaper publishing has been reshaped into Web sites, blogging, and web feeds. The *Internet* has enabled or accelerated the creation of new forms of *human* interactions through instant messaging, *Internet* forums, and social networking sites.

The *Internet* has no centralized governance in either technological implementation or policies for access and usage; each constituent network sets its own standards. Only the overreaching definitions of the two principal name spaces in the *Internet*, the *Internet*-protocol address space and the domain-name *system*, are directed by a maintainer organization, the *Internet* Corporation for Assigned Names and Numbers (ICANN). The technical underpinning and standardization of the core protocols (IPv4 and IPv6) is an activity of the *Internet* Engineering Task Force (IETF), a non-profit organization of loosely affiliated international participants that anyone may associate with by contributing technical expertise" (Wikipedia).

Internet of Things (IoT)
IoT is the network of things, with device identification, embedded software intelligence, sensors, and connectivity connecting people and things over the Internet at anytime, anyplace, with anything and anyone (Authors).

Interoperability
"The ability to share information and services. The ability of two or more systems or components to exchange and use information. The ability of systems to provide and receive services from other systems and to use the services so interchanged to enable them to operate effectively together" ([TOGAF 9]).

intf *n*
Any interface, usually beginning with port 2, that connects to a subset network of your design that you can custom name and configure (Cisco).

intranet
Intranetwork. A LAN that uses IP. See also network and Internet (Cisco).

IoT Service
Software component enabling interaction with resources through a well-defined interface. Can be orchestrated together with non-IoT services (e.g., enterprise services). Interaction with the service is done via the network (IoT-A).

IP
Internet Protocol. IP protocols are the most popular nonproprietary protocols because they can be used to communicate across any set of interconnected networks and are equally well suited for LAN and WAN communications (Cisco).

IP address
An IP protocol address. A security appliance interface ip_address. IP version 4 addresses are 32 bits in length. This address space is used to designate the network number, optional subnetwork number, and a host number. The 32 bits are grouped into four octets (8 binary bits), represented by 4 decimal numbers separated by periods, or dots. The meaning of each of the four octets is determined by their use in a particular network (Cisco).

IP pool
A range of local IP addresses specified by a name, and a range with a starting IP address and an ending address. IP Pools are used by DHCP and VPNs to assign local IP addresses to clients on the inside interface (Cisco).

IPS
Intrusion Prevention Service. An in-line, deep-packet inspection-based solution that helps mitigate a wide range of network attacks (Cisco).

IPSec
IP Security. A framework of open standards that provides data confidentiality, data integrity, and data authentication between participating peers. IPSec provides these security services at the IP layer. IPSec uses IKE to handle the negotiation of protocols and algorithms based on local policy and to generate the encryption and authentication keys to be used by IPSec. IPSec can protect one or more data flows between a pair of hosts, between a pair of security gateways, or between a security gateway and a host (Cisco).

ISAKMP
Internet Security Association and Key Management Protocol. A protocol framework that defines payload formats, the mechanics of implementing a key exchange protocol, and the negotiation of a security association. See IKE (Cisco).

IS-IS

Intermediate System to Intermediate System. A routing protocol used as the control plane for IP and next generation Ethernet networks (Authors).

ISP

Internet Service Provider. An organization that provides connection to the Internet via their services, such as modem dial in over telephone voice lines or DSL (Cisco).

ISV

Independent Software Vendors (ISV) (Authors).

key

A data object used for encryption, decryption, or authentication (Cisco).

LAN

Local area network. A network residing in one location, such as a single building or campus. See also Internet, intranet, and network (Cisco).

layer, layers

Networking models implement layers with which different protocols are associated. The most common networking model is the OSI model, which consists of the following seven layers, in order: physical, data link, network, transport, session, presentation, and application (Cisco).

LDAP

Lightweight Directory Access Protocol. LDAP provides management and browser applications with access to X.500 directories (Cisco).

Ledger

A shared and distributed history of all transactions within the blockchain (Authors).

LISP

Locator/Identifier Separation Protocol, an IETF solution standardized in RFC 6830 (IETF).

LLN

Low Power and Lossy Networks (IETF).

Local Storage

Special type of *resource* that contains information about one or only a few *entities* in the vicinity of a *device* (IoT-A).

Location Technologies

All technologies whose primary purpose is to establish and communicate the location of a *device*, e.g., GPS, RTLS, etc. (IoT-A).

Look-up

In contrast to *discovery*, *look-up* is a *service* that *addresses* exiting known *resources* using a key or *identifier* (IoT-A).

LoRaWAN Network Architecture
LoRaWAN network architecture is typically laid out in a star-of-stars topology in which gateways are a transparent bridge relaying messages between end-devices and a central network server in the backend. Gateways are connected to the network server via standard IP connections while end-devices use single-hop wireless communication to one or many gateways. All endpoint communication is generally bi-directional, but also supports operation such as multicast enabling software upgrade over the air or other mass distribution messages to reduce the on air communication time (LoRa Alliance).

LoRaWAN™
LoEa WAN is a Low Power Wide Area Network (LPWAN) specification intended for wireless battery operated Things in regional, national, or global network. LoRaWAN target key requirements of Internet of Things such as secure bi-directional communication, mobility, and localization services. This standard will provide seamless interoperability among smart Things without the need of complex local installations and gives back the freedom to the user, developer, businesses enabling the roll out of IoT (LoRa Alliance).

LPN
Low-Power Network (LPN) or Low-Power Wide-Area Network (LPWAN) is a type of wireless telecommunication network designed to allow long-range communications at a low bit rate among things (connected objects), such as sensors operated on a battery (Wikipedia).

LPWAN
Low-Power Wide-Area Network (LPWAN) or Low-Power Network (LPN) is a type of wireless telecommunication network designed to allow long-range communications at a low bit rate among things (connected objects), such as sensors operated on a battery (Wikipedia).

M2M (also referred to as machine to machine)
"The automatic communications between *devices* without *human* intervention. It often refers to a *system* of remote *sensors* that is continuously transmitting data to a central *system*. Agricultural weather sensing *systems*, automatic meter reading and *RFID* tags are examples" ([COMPDICT-M2M]).

MAN
Metropolitan Area Network. A network for a city or metro area (Various).

Mask
A 32-bit mask that shows how an Internet address is divided into network, subnet, and host parts. The mask has ones in the bit positions to be used for the network and subnet parts, and zeros for the host part. The mask should contain at least the standard network portion, and the subnet field should be contiguous with the network portion (Cisco).

Merkle Tree
A data structure where each leaf of the tree is a hash of data and the root is the hash of all its children hashes (Authors).

Microcontroller

"A *microcontroller* is a small computer on a single integrated circuit containing a processor core, memory, and programmable input/output peripherals. Program memory in the form of NOR flash or OTP ROM is also often included on chip, as well as a typically small amount of RAM. *Microcontrollers* are designed for embedded applications, in contrast to the microprocessors used in personal computers or other general purpose applications.

Microcontrollers are used in automatically controlled products and *devices*, such as automobile engine control *systems*, implantable medical *devices*, remote controls, office machines, appliances, power tools, and toys. By reducing the size and cost compared to a design that uses a separate microprocessor, memory, and input/output *devices*, *microcontrollers* make it economical to digitally control even more *devices* and processes. Mixed signal *microcontrollers* are common, integrating analog components needed to control non-digital electronic *systems*" (Wikipedia).

Miner

A node that generates new blocks for the blockchain through the work of computation and using the given consensus algorithm (Authors).

Mode

See Access Modes (Cisco).

MQTT

Message Queuing Telemetry Transport, an application layer protocol (OASIS).

MS

Mobile Station. Refers generically to any mobile device, such as a mobile handset or computer that is used to access network services. GPRS networks support three classes of MS, which describe the type of operation supported within the GPRS and the GSM mobile wireless networks. For example, a Class A MS supports simultaneous operation of GPRS and GSM services (Cisco).

MTU

Maximum transmission unit, the maximum number of bytes in a packet that can flow efficiently across the network with best response time. For Ethernet, the default MTU is 1500 bytes, but each network can have different values, with serial connections having the smallest values. The MTU is described in RFC 1191 (Cisco).

Multicast

Multicast refers to a network addressing method in which the source transmits a packet to multiple destinations, a multicast group, simultaneously. See also PIM, SMR (Cisco).

NAT

Network Address Translation. Mechanism for reducing the need for globally unique IP addresses. NAT allows an organization with addresses that are not globally unique to connect to the Internet by translating those addresses into a globally routable address space (Cisco).

Network
In the context of security appliance configuration, a network is a group of computing devices that share part of an IP address space and not a single host. A network consists of multiple nodes or hosts. See also host, Internet, intranet, IP, LAN, and node (Cisco).

Network resource
Resource hosted somewhere in the network, e.g., in the cloud (IoT-A).

Next-Generation Networks (NGN)
"Packet-based network able to provide telecommunication *service*s and able to make use of multiple broadband, QoS-enabled transport technologies and in which *service*-related functions are independent from underlying transport-related technologies" ([ETSI TR 102 477]).

NFC
Near Field Communication (Various).

NMS
Network management system: a software system responsible for managing a network. It includes: Fault, Configuration, Accounting, Performance and Security management (known as FCAPS). NMS communicates with Element Management Systems (EMS), agents and/or the network devices themselves to collect data, push updates or help keep track of network statistics and resources (Authors).

Node
Devices such as routers and printers that would not normally be called hosts. See also host, network (Cisco).

Non Repudiation
Non Repudiation: ensures that an entity cannot deny an action that it has performed (Various).

nonvolatile storage, memory
Storage or memory that, unlike RAM, retains its contents without power. Data in a nonvolatile storage device survives a power-off, power-on cycle or reboot (Cisco).

NTP
Network time protocol (Cisco).

OASIS
Organization for the Advancement of Structured Information Standards (Various).

object grouping
Simplifies access control by letting you apply access control statements to groups of network objects, such as protocol, services, hosts, and networks (Cisco).

Observer
Anything that has the capability to monitor a *Physical Entity*, like its state or location (IoT-A).

OEM
Original Equipment Manufacturers (Authors).

On-device Resource
Resource hosted inside a *Device* and enabling access to the *Device* and thus to the related *Physical Entity* (IoT-A).

Ontology
Ontology is the philosophical study of the nature of being, or reality, as well as the basic categories of being and their relations. Ontology engineering offers a direction towards solving the inter-operability problems brought about by semantic obstacles, i.e., the obstacles related to the definitions of business terms and software classes. Ontology engineering is a set of tasks related to the development of ontologies for a particular domain (Wikipedia).

Open source
Open source in the computer industry is the sharing of source code or hardware design, with the permission to reuse, modify, and improve at no cost (Various).

Operator
The operator owns administration rights on the services it provides and/or on the entities it owns, is able to negotiate partnerships with equivalent counterparts and define policies specifying how a service can be accessed by users (IoT-A).

OSI
Open Systems Interconnection (Authors).

OSPF
Open Shortest Path First. OSPF is a routing protocol for IP networks. OSPF is a routing protocol widely deployed in large networks because of its efficient use of network bandwidth and its rapid convergence after changes in topology. The security appliance supports OSPF (Cisco).

outbound
Refers to traffic whose destination is on an interface with lower security than the source interface (Cisco).

PAN
Personal Area Network. A network comprising electronic accessories/peripherals or wearable devices (Various).

Passive Digital Entities
A digital representation of something stored in an IT-based system (IoT-A).

PCE
Path Computational Element. A server dedicated to running network path computation calculations. Typically used in network traffic engineering applications (IETF).

Permissioned Blockchain
A private blockchain with strong understanding of identity management and nodes within the network (Authors).

Permissionless Blockchain
A public blockchain that allows anyone to join the network and participate (Authors).

Perspective (also referred to as architectural perspective)
"Architectural perspective is a collection of activities, checklists, tactics and guidelines to guide the process of ensuring that a *system* exhibits a particular set of closely related quality properties that require consideration across a number of the *system*'s architectural *views*" ([ROZANSKI2005]).

Physical Entity
Any physical object that is relevant from a user or application perspective (IoT-A).

PIM
Protocol Independent Multicast. PIM provides a scalable method for determining the best paths for distributing a specific multicast transmission to a group of hosts. Each host has registered using IGMP to receive the transmission. See also PIM-SM (Cisco).

Ping
An ICMP request sent by a host to determine if a second host is accessible (Cisco).

PIX
Private Internet eXchange. The Cisco PIX 500-series security appliances range from compact, plug-and-play desktop models for small/home offices to carrier-class gigabit models for the most demanding enterprise and service provider environments. Cisco PIX security appliances provide robust, enterprise-class integrated network security services to create a strong multilayered defense for fast changing network environments (Cisco).

PKI
A public key infrastructure is a set of roles, policies, and procedures needed to create, manage, distribute, use, store, and revoke digital certificates and manage public-key encryption. The purpose of a PKI is to facilitate the secure electronic transfer of information for a range of network activities such as e-commerce, Internet banking, and confidential email. It is required for activities where simple passwords are an inadequate authentication method and more rigorous proof is required to confirm the identity of the parties involved in the communication and to validate the information being transferred.

In cryptography, a PKI is an arrangement that binds public keys with respective identities of entities (like persons and organizations). The binding is established through a process of registration and issuance of certificates at and by a certificate authority (CA). Depending on the assurance level of the binding, this may be carried out by an automated process or under human supervision (Wikipedia).

Port
A field in the packet headers of TCP and UDP protocols that identifies the higher level service which is the source or destination of the packet (Cisco).

PPP
Point-to-Point Protocol. Developed for dial-up ISP access using analog phone lines and modems (Cisco).

PPTP
Point-to-Point Tunneling Protocol. PPTP was introduced by Microsoft to provide secure remote access to Windows networks; however, because it is vulnerable to attack, PPTP is commonly used only when stronger security methods are not available or are not required. PPTP Ports are pptp, 1723/tcp, 1723/udp, and pptp. For more information about PPTP, see RFC 2637. See also PAC, PPTP GRE, PPTP GRE tunnel, PNS, PPTP session, and PPTP TCP (Cisco).

Privacy
Information Privacy is the interest an individual has in controlling, or at least significantly influencing, the handling of data about themselves (IoT-A).

Proxy-ARP
Enables the security appliance to reply to an ARP request for IP addresses in the global pool. See also ARP (Cisco).

public key
A public key is one of a pair of keys that are generated by devices involved in public key infrastructure. Data encrypted with a public key can only be decrypted using the associated private key. When a private key is used to produce a digital signature, the receiver can use the public key of the sender to verify that the message was signed by the sender. These characteristics of key pairs provide a scalable and secure method of authentication over an insecure media, such as the Internet (Cisco).

Public Key Certificate
See Digital Certificate

QoS
Quality of service. Measure of performance for a transmission system that reflects its transmission quality and service availability (Cisco).

RADIUS
Remote Authentication Dial-In User Service. RADIUS is a distributed client/server system that secures networks against unauthorized access. RFC 2058 and RFC 2059 define the RADIUS protocol standard. See also AAA and TACACS+ (Cisco).

Reference Architecture
A reference architecture is an architectural design pattern that indicates how an abstract set of mechanisms and relationships realizes a predetermined set of requirements. It captures the essence of the architecture of a collection of systems. The main purpose of a reference architecture is to provide guidance for the development of architectures. One or more reference architectures may be derived from a common reference model, to address different purposes/usages to which the Reference Model may be targeted (IoT-A).

Reference Model
"A reference model is an abstract framework for understanding significant relationships among the entities of some environment. It enables the development of specific reference or concrete architectures using consistent standards or specifications supporting that environment. A reference model consists of a minimal set of unifying concepts, axioms and relationships within a particular problem domain, and is independent of specific standards, technologies, implementations, or other concrete details. A reference model may be used as a basis for education and explaining standards to non-specialists" ([OASIS-RM]).

Refresh
Retrieve the running configuration from the security appliance and update the screen. The icon and the button perform the same function (Cisco).

Requirement
"A quantitative statement of business need that must be met by a particular *architecture* or work package" ([TOGAF9]).

Resolution
Service by which a given *ID* is associated with a set of *Addresses* of information and interaction *Services*. Information services allow querying, changing, and adding information about the thing in question, while interaction services enable direct interaction with the thing by accessing the *Resources* of the associated *Devices*. Based on a priori knowledge (IoT-A).

Resource
Computational element that gives access to information about or actuation capabilities on a *Physical Entity* (IoT-A).

REST or RESTful
Representational State Transfer. The architectural paradigm for the World Wide Web employing the HTTP protocol (Various).

RFC
Request for Comments. RFC documents define protocols and standards for communications over the Internet. RFCs are developed and published by IETF (Cisco).

RFD
IEEE 802.15.4 Reduced Function Device. Implements minimal subset of the protocol stack, and is typically battery powered (IEEE).

RFID
"The use of electromagnetic or inductive coupling in the radio frequency portion of the spectrum to communicate to or from a tag through a variety of modulation and encoding schemes to uniquely read the *identity* of an RF Tag" ([ISO/IEC 19762]).

RIP
Routing Information Protocol. Interior gateway protocol (IGP) supplied with UNIX BSD systems. The most common IGP in the Internet. RIP uses hop count as a routing metric (Cisco).

ROLL
IETF Routing over Low Power and Lossy Networks workgroup (IETF).

RPL
Routing Protocol for Low Power and Lossy Networks, a distance vector routing protocol for IoT standardized in RFC6550 (IETF).

RSA
A public key cryptographic algorithm (named after its inventors, Rivest, Shamir, and Adelman) with a variable key length. The main weakness of RSA is that it is significantly slow to compute compared to popular secret-key algorithms, such as DES. The Cisco implementation of IKE uses a Diffie–Hellman exchange to get the secret keys. This exchange can be authenticated with RSA (or preshared keys). With the Diffie–Hellman exchange, the DES key never crosses the network (not even in encrypted form), which is not the case with the RSA encrypt and sign technique. RSA is not public domain, and must be licensed from RSA Data Security (Cisco).

RSH
Remote Shell. A protocol that allows a user to execute commands on a remote system without having to log in to the system. For example, RSH can be used to remotely examine the status of a number of access servers without connecting to each communication server, executing the command, and then disconnecting from the communication server (Cisco).

RSU
Road Side Unit (Various).

RTP
Real-Time Transport Protocol. Commonly used with IP networks. RTP is designed to provide end-to-end network transport functions for applications transmitting real-time data, such as audio, video, or simulation data, over multicast or unicast network services. RTP provides such services as payload type identification, sequence numbering, timestamping, and delivery monitoring to real-time applications (Cisco).

RTSP
Real Time Streaming Protocol. Enables the controlled delivery of real-time data, such as audio and video. RTSP is designed to work with established protocols, such as RTP and HTTP (Cisco).

rule
Conditional statements added to the security appliance configuration to define security policy for a particular situation. See also ACE, ACL, NAT (Cisco).

running configuration
The configuration currently running in RAM on the security appliance. The configuration that determines the operational characteristics of the security appliance (Cisco).

SA

Security association. An instance of security policy and keying material applied to a data flow. SAs are established in pairs by IPSec peers during both phases of IPSec. SAs specify the encryption algorithms and other security parameters used to create a secure tunnel. Phase 1 SAs (IKE SAs) establish a secure tunnel for negotiating Phase 2 SAs. Phase 2 SAs (IPSec SAs) establish the secure tunnel used for sending user data. Both IKE and IPSec use SAs, although SAs are independent of one another. IPSec SAs are unidirectional and they are unique in each security protocol. A set of SAs are needed for a protected data pipe, one per direction per protocol. For example, if you have a pipe that supports ESP between peers, one ESP SA is required for each direction. SAs are uniquely identified by destination (IPSec end-point) address, security protocol (AH or ESP), and Security Parameter Index. IKE negotiates and establishes SAs on behalf of IPSec. A user can also establish IPSec SAs manually. An IKE SA is used by IKE only, and unlike the IPSec SA, it is bidirectional (Cisco).

Satoshi Nakamoto

Pseudonym for the person or group of people who created Bitcoin (Authors).

SCL

Services Capability Layer. A set of common application services standardized by ETSI TS 102690 (Authors).

secret key

A secret key is a key shared only between the sender and receiver. See key, public key (Cisco).

Security

The correct term is 'information security' and typically information security comprises three component parts:

Confidentiality. Assurance that information is shared only among authorized persons or organizations. Breaches of confidentiality can occur when data is not handled in a manner appropriate to safeguard the confidentiality of the information concerned. Such disclosure can take place by word of mouth, by printing, copying, e-mailing or creating documents and other data, etc.;

Integrity. Assurance that the information is authentic and complete. Ensuring that information can be relied upon to be sufficiently accurate for its purpose. The term 'integrity' is used frequently when considering information security as it represents one of the primary indicators of information security (or lack of it). The integrity of data is not only whether the data is 'correct', but whether it can be trusted and relied upon;

Availability. Assurance that the systems responsible for delivering, storing, and processing information are accessible when needed, by those who need them ([ISO27001]).

security context
You can partition a single security appliance into multiple virtual firewalls, known as security contexts. Each context is an independent firewall, with its own security policy, interfaces, and administrators. Multiple contexts are similar to having multiple stand-alone firewalls (Cisco).

security services
See cryptography (Cisco).

Selective-Forwarding Attack
This attack takes place in the case when the object cannot send its generated packets directly to the fog device but must rely on other objects that lie along the path towards the fog device to deliver those packets (Various).

Semantics
The study of meaning. It focuses on the relation between signifiers, like words, phrases, signs, and symbols, and what they stand for their denotation (Wikipedia).

Sensor
A sensor is a special *Device* that perceives certain characteristics of the real world and transfers them into a digital representation (IoT-A).

serial transmission
A method of data transmission in which the bits of a data character are transmitted sequentially over a single channel (Cisco).

Service
"Services are the mechanism by which needs and capabilities are brought together" ([OASIS-RM]).

SI
Systems Integrators (Authors).

Sinkhole Attack
In this attack, a malicious object claims that it has the shortest-path to the fog device which attracts all neighboring objects that do not have the transmission capability to reach the fog device to forward their packets to that malicious object and count on that object to deliver their packets (Various).

SIP
Session Initiation Protocol. Enables call handling sessions, particularly two-party audio conferences, or "calls." SIP works with SDP for call signaling. SDP specifies the ports for the media stream. Using SIP, the security appliance can support any SIP VoIP gateways and VoIP proxy servers (Cisco).

site-to-site VPN
A site-to-site VPN is established between two IPSec peers that connect remote networks into a single VPN. In this type of VPN, neither IPSec peer is the destination or source of user traffic. Instead, each IPSec peer provides encryption and authentication services for hosts on the LANs connected to each IPSec peer. The hosts on each LAN send and receive data through the secure tunnel established by the pair of IPSec peers (Cisco).

SMO
Systems Management Overview (Authors).

SMTP
Simple Mail Transfer Protocol. SMTP is an Internet protocol that supports email services (Cisco).

SNMP
Simple Network Management Protocol. A standard method for managing network devices using data structures called Management Information Bases (Cisco).

spoofing
A type of attack designed to foil network security mechanisms such as filters and access lists. A spoofing attack sends a packet that claims to be from an address from which it was not actually sent (Cisco).

SQL*Net
Structured Query Language Protocol. An Oracle protocol used to communicate between client and server processes (Cisco).

SSH
Secure Shell. An application running on top of a reliable transport layer, such as TCP/IP, that provides strong authentication and encryption capabilities (Cisco).

SSL
Secure Sockets Layer. A protocol that resides between the application layer and TCP/IP to provide transparent encryption of data traffic (Cisco).

SSN
Semantic Sensor Network (Various).

Stakeholder (also referred to as system stakeholder)
"An individual, team, or organization (or classes thereof) with interests in, or concerns relative to, a *system*" ([IEEE-1471-2000]).

stateful inspection
Network protocols maintain certain data, called state information, at each end of a network connection between two hosts. State information is necessary to implement the features of a protocol, such as guaranteed packet delivery, data sequencing, flow control, and transaction or session IDs. Some of the protocol state information is sent in each packet while each protocol is being used. For example, a browser connected to a web server uses HTTP and supporting TCP/IP protocols. Each protocol layer maintains state information in the packets it sends and receives. The security appliance and some other firewalls inspect the state information in each packet to verify that it is current and valid for every protocol it contains. This is called stateful inspection and is designed to create a powerful barrier to certain types of computer security threats (Cisco).

Storage
Special type of *Resource* that stores information coming from *resources* and provides information about *Entities*. They may also include *services* to process the
information stored by the *resource*. As *Storages* are *Resources*, they can be deployed
either on-device or in the network (IoT-A).

STP
Spanning Tree Protocol. A protocol to create a loop-free Ethernet topology
(Authors).

Stretch Attack
This attack targets the Network layer in the OSI stack. If the routing protocol supports source routing, then a malicious object can send the packets that it is supposed
to report to the fog device through very long paths rather than the direct and short
ones as illustrated in Fi (Various).

subnetmask
See mask (Cisco).

System
"A collection of components organized to accomplish a specific function or set of
functions" ([IEEE-1471-2000]).

TACACS+
Terminal Access Controller Access Control System Plus. A client-server protocol
that supports AAA services, including command authorization. See also AAA,
RADIUS (Cisco).

Tag
Label or other physical object used to identify the *Physical Entity* to which it is
attached (IoT-A).

TAPI
Telephony Application Programming Interface. A programming interface in Microsoft Windows that supports telephony functions (Cisco).

TCP
Transmission Control Protocol. Connection-oriented transport layer protocol that
provides reliable full-duplex data transmission (Cisco).

TCP Intercept
With the TCP intercept feature, once the optional embryonic connection limit is
reached, and until the embryonic connection count falls below this threshold, every
SYN bound for the effected server is intercepted. For each SYN, the security appliance responds on behalf of the server with an empty SYN/ACK segment. The security appliance retains pertinent state information, drops the packet, and waits for the
client acknowledgment. If the ACK is received, then a copy of the client SYN segment is sent to the server and the TCP three-way handshake is performed between
the security appliance and the server. If this three-way handshake completes, may
the connection resume as normal. If the client does not respond during any part of
the connection phase, then the security appliance retransmits the necessary segment
using exponential back-offs (Cisco).

TDP

Tag Distribution Protocol. TDP is used by tag switching devices to distribute, request, and release tag binding information for multiple network layer protocols in a tag switching network. TDP does not replace routing protocols. Instead, it uses information learned from routing protocols to create tag bindings. TDP is also used to open, monitor, and close TDP sessions and to indicate errors that occur during those sessions. TDP operates over a connection-oriented transport layer protocol with guaranteed sequential delivery (such as TCP). The use of TDP does not preclude the use of other mechanisms to distribute tag binding information, such as piggybacking information on other protocols (Cisco).

Telnet

A terminal emulation protocol for TCP/IP networks such as the Internet. Telnet is a common way to control web servers remotely; however, its security vulnerabilities have led to its replacement by SSH (Cisco).

TFTP

Trivial File Transfer Protocol. TFTP is a simple protocol used to transfer files. It runs on UDP and is explained in depth in RFC 1350 (Cisco).

Thing

Generally speaking, any *physical object*. In the term *"Internet of Things"* however, it denotes the same concept as a *Physical Entity* (IoT-A).

TID

Tunnel Identifier (Cisco).

TLS

Transport Layer Security. A future IETF protocol to replace SSL (Cisco).

TMN

Telecommunications Management Network (TMN) of IUT-T (Authors).

TMN

Telecommunications Management Network of IUT-T (Various).

Traffic policing

The traffic policing feature ensures that no traffic exceeds the maximum rate (bits per second) that you configure, thus ensuring that no one traffic flow can take over the entire resource (Cisco).

TSCH

Time Slotted Channel Hopping. A mode of IEEE 802.15.4 networks (IEEE).

TSN

Time Sensitive Networking. See also AVB (IEEE).

TSP

TAPI Service Provider. See also TAPI (Cisco).

UDP

User Datagram Protocol. A connectionless transport layer protocol in the IP protocol stack. UDP is a simple protocol that exchanges datagrams without acknowledgments or guaranteed delivery, which requires other protocols to handle error processing and retransmission. UDP is defined in RFC 768 (Cisco).

UMTS

Universal Mobile Telecommunication System. An extension of GPRS networks that moves toward an all-IP network by delivering broadband information, including commerce and entertainment services, to mobile users via fixed, wireless, and satellite networks (Cisco).

Unconstrained Network

An unconstrained network is a network of devices with no restriction on capabilities such as storage, computing power, and/or transfer rate (IoT-A).

Unicast RPF

Unicast Reverse Path Forwarding. Unicast RPF guards against spoofing by ensuring that packets have a source IP address that matches the correct source interface according to the routing table (Cisco).

URL

Uniform Resource Locator. A standardized addressing scheme for accessing hypertext documents and other services using a browser. For example, http://www.cisco.com (Cisco).

User

A *Human* or any *Active Digital Entity* that is interested in interacting with a particular physical object. (IoT-A).

User EXEC mode

User EXEC mode lets you to see the security appliance settings. The user EXEC mode prompt appears as follows when you first access the security appliance. See also command-specific configuration mode, global configuration mode, and privileged EXEC mode (Cisco).

Vampire Attack

This attack exploits the fact that the majority of IoT objects have a limited battery lifetime where a malicious user misbehaves in a way that makes devices consume extra amounts of power so that they run out of battery earlier causing a service disruption (Various).

VANET

Vehicular ad-Hoc Network (Various).

VAR

Value-added Resellers (Authors).

View
"The representation of a related set of concerns. A *view* is what is seen from a *viewpoint*. An architecture *view* may be represented by a model to demonstrate to stakeholders their areas of interest in the architecture. A*view* does not have to be visual or graphical in nature" ([TOGAF 9]).

Viewpoint
"A definition of the perspective from which a *view* is taken. It is a specification of the conventions for constructing and using a *view* (often by means of an appropriate schema or template). A *view* is what you see; a*viewpoint* is where you are looking from—the vantage point or perspective that determines what you see" ([TOGAF 9]).

Virtual Entity
Computational or data element representing a *Physical Entity*. Virtual Entities can be either Active or Passive *Digital Entities*. (IoT-A).

VLAN
Virtual LAN. A group of devices on one or more LANs that are configured (using management software) so that they can communicate as if they were attached to the same physical network cable, when in fact they are located on a number of different LAN segments. Because VLANs are based on logical instead of physical connections, they are extremely flexible (Cisco).

VM
Virtual Machine. A virtualization construct where multiple virtual devices each with its own independent operating system can run on the same physical computer, typically a server (Authors).

VoIP
Voice over IP. VoIP carries normal voice traffic, such as telephone calls and faxes, over an IP-based network. DSP segments the voice signal into frames, which then are coupled in groups of two and stored in voice packets. These voice packets are transported using IP in compliance with ITU-T specification H.323 (Cisco).

VPN
Virtual Private Network. A network connection between two peers over the public network that is made private by strict authentication of users and the encryption of all data traffic. You can establish VPNs between clients, such as PCs, or a headend, such as the security appliance. (Cisco).

VSA
Vendor-specific attribute. An attribute in a RADIUS packet that is defined by a vendor rather than by RADIUS RFCs. The RADIUS protocol uses IANA-assigned vendor numbers to help identify VSAs. This lets different vendors have VSAs of the same number. The combination of a vendor number and a VSA number makes a VSA unique. For example, the cisco-av-pair VSA is attribute 1 in the set of VSAs related to vendor number 9. Each vendor can define up to 256 VSAs. A RADIUS packet contains any VSAs attribute 26, named Vendor-specific. VSAs are sometimes referred to as subattributes (Cisco).

WAN

wide-area network. Data communications network that serves users across a broad geographic area and often uses transmission devices provided by common carriers (Cisco).

WCCP

Web Cache Communication Protocol. Transparently redirects selected types of traffic to a group of web cache engines to optimize resource usage and lower response times (Cisco).

Websense

A content filtering solution that manages employee access to the Internet. Websense uses a policy engine and a URL database to control user access to websites (Cisco).

WEP

Wired Equivalent Privacy. A security protocol for wireless LANs, defined in the IEEE 802.11b standard (Cisco).

Wi-Fi

Wireless Fidelity, Wireless Internet (Various).

WINS

Windows Internet Naming Service. A Windows system that determines the IP address associated with a particular network device, also known as "name resolution." WINS uses a distributed database that is automatically updated with the NetBIOS names of network devices currently available and the IP address assigned to each one. WINS provides a distributed database for registering and querying dynamic NetBIOS names to IP address mapping in a routed network environment. It is the best choice for NetBIOS name resolution in such a routed network because it is designed to solve the problems that occur with name resolution in complex networks (Cisco).

Wireless communication technologies

"Wireless communication is the transfer of information over a distance without the use of enhanced electrical conductors or 'wires'. The distances involved may be short (a few meters as in television remote control) or long (thousands or millions of kilometers for radio communications). When the context is clear, the term is often shortened to 'wireless'. Wireless communication is generally considered to be a branch of telecommunications" ([Wikipedia WI]).

Wireless Sensors and Actuators Network

"Wireless sensor and actuator networks (WSANs) are networks of nodes that sense and, potentially, control their environment. They communicate the information through wireless links enabling interaction between people or computers and the surrounding environment" ([OECD2009]).

Wireline communication technologies

"A term associated with a network or terminal that uses metallic wire conductors (and/or optical fibres) for telecommunications" ([setzer-messtechnik2010]).

WoT
The Web of Things (WoT) is a term used to describe approaches, software architectural styles, and programming patterns that allow real-world objects to be part of the World Wide Web. Similarly to what the Web (Application Layer) is to the Internet (Network Layer), the Web of Things provides an Application Layer that simplifies the creation of Internet of Things applications (Wikipedia).

WSN
Wireless Sensor Network. A network of typically low powered sensors connected over a wireless network often employing mesh technology (Various).

X.509
In cryptography, X.509 is a standard for a public key infrastructure (PKI) to manage digital certificates and public-key encryption and a key part of the Transport Layer Security protocol used to secure web and email communication.

An ITU-T standard, X.509 specifies formats for public key certificates, certificate revocation lists, attribute certificates, and a certification path validation algorithm (Wikipedia).

xauth
See IKE Extended Authentication (Cisco).

xlate
An xlate, also referred to as a translation entry, represents the mapping of one IP address to another, or the mapping of one IP address/port pair to another (Cisco).

XML
Extensible Markup Language (W3C).

XMPP
Extensible Messaging and Presence Protocol. Standardized in IETF RFC 6120 and 6121 (IETF).

ZigBee
Short-range wireless protocol promoted by the ZigBee Alliance (ZigBee Alliance).

Z-Wave
Short-range wireless protocol, initially developed by a small Danish company called Zensys. Focuses on home automation applications (Z-Wave Alliance).

References

1. Internet page on Wikipedia, online at: https://en.wikipedia.org/wiki/Actuator
2. Cisco Glossary, online at: http://www.cisco.com/c/en/us/td/docs/security/asa/asa80/configuration/guide/conf_gd/glossary.html
3. Association for Automatic Identification and Mobility, online at: http://www.aimglobal.org/
4. Information Model, Deliverable D3.1, Autonomic Internet (AutoI) Project. Online at: http://ist-autoi.eu/autoi/d/AutoI_Deliverable_D3.1_-_Information_Model.pdf
5. Information architecture reference model. Online at: http://cwe.ccsds.org/sea/docs/SEA-IA/Draft%20Documents/IA%20Reference%20Model/ccsds_rasim_20060308.pdf
6. Computer Dictionary Definition, online at: http://www.yourdictionary.com/computer/m2-m
7. E-FRAME project, available online at: http://www.frame-online.net/top-menu/the-architecture-2/faqs/stakeholder-aspiration.html
8. EPC Global glossary (GS1), online at: http://www.epcglobalinc.org/home/GS1_EPCglobal_Glossary_V35_KS_June_09_2009.pdf
9. ETSI Technical report ETR 173, Terminal Equipment (TE); Functional model for multimedia applications. Available online: http://www.etsi.org/deliver/etsi_etr/100_199/173/01_60/etr_173e01p.pdf
10. ETSI Corporate telecommunication Networks (CN); Mobility for enterprise communication, online at: http://www.etsi.org/deliver/etsi_tr/102400_102499/102477/01.01.01_60/tr_102477v010101p.pdf
11. IEEE 1471-2000, "IEEE Recommended Practice for Architectural Description of Software-Intensive Systems"
12. The Internet of Things summary at ITU, online at: http://www.itu.int/osg/spu/publications/internetofthings/InternetofThings_summary.pdf
13. Information technology—Vocabulary—Part 1: Fundamental terms, online at: http://www.iso.org/iso/iso_catalogue/catalogue_tc/catalogue_detail.htm?csnumber=7229
14. ISO 27001: An Introduction To Information, Network and Internet Security
15. Open GeoSpatial portal, the OpenGIS abstract specification Topic 12: the OpenGIS Service architecture. Online at: http://portal.opengeospatial.org/files/?artifact_id=1221
16. Reference Model for Service Oriented Architecture 1.0 http://docs.oasis-open.org/soa-rm/v1.0/soa-rm.pdf
17. "Smart Sensor Networks: Technologies and Applications for Green Growth", December 2009, online at: http://www.oecd.org/dataoecd/39/62/44379113.pdf
18. [Sclater2007] Sclater, N., Mechanisms and Mechanical Devices Sourcebook, 4th Edition (2007), 25, McGraw-Hill
19. setzer-messtechnik glossary, July 2010, online at: http://www.setzer-messtechnik.at/grundlagen/rf-glossary.php?lang=en
20. Open Group, TOGAF 9, 2009
21. Energy harvesting page on Wikipedia, online at: http://en.wikipedia.org/wiki/Energy_harvesting
22. Internet page on Wikipedia, online at: http://en.wikipedia.org/wiki/Internet
23. Microcontroller page at Wikipedia, online at: http://en.wikipedia.org/wiki/Microcontroller
24. Software Architecture with Viewpoints and Perspectives, online at: http://www.viewpoints-and-perspectives.info/doc/spa191-viewpoints-and-perspectives.pdf
25. Wireless page on Wikipedia, online at: http://en.wikipedia.org/wiki/Wireless
26. Internet of Things – Architecture (IoT-A): http://www.iot-a.eu/public/terminology/copy_of_term
27. OneM2M: http://www.onem2m.org/images/files/deliverables/TS-0001-Functional_Architecture-V1_6_1.pdf

28. D. Willis, A. Dasgupta, S. Banerjee, "Paradrop: a multi-tenant platform for dynamically installed third party services on home gateways," In: SIGCOMM workshop on Distributed cloud computing. ACM (2014)

29. Xu, Wenyuan, et al. "Jamming sensor networks: attack and defense strategies." *Network, IEEE* 20.3 (2006): 41-47.

30. Ye, Wei, John Heidemann, and Deborah Estrin. "Medium access control with coordinated adaptive sleeping for wireless sensor networks." *Networking, IEEE/ACM Transactions on* 12.3 (2004): 493-506.

31. Van Dam, Tijs, and Koen Langendoen. "An adaptive energy-efficient MAC protocol for wireless sensor networks." *Proceedings of the 1st international conference on Embedded networked sensor systems.* ACM, 2003.

32. Dyer, Kevin P., et al. "Peek-a-boo, i still see you: Why efficient traffic analysis countermeasures fail." *Security and Privacy (SP), 2012 IEEE Symposium on.* IEEE, 2012.

33. Park, Junho, et al. "An Energy-Efficient Selective Forwarding Attack Detection Scheme Using Lazy Detection in Wireless Sensor Networks." *Ubiquitous Information Technologies and Applications.* Springer Netherlands, 2013. 157-164.

34. Bysani, Leela Krishna, and Ashok Kumar Turuk. "A survey on selective forwarding attack in wireless sensor networks." *Devices and Communications (ICDeCom), 2011 International Conference on.* IEEE, 2011.

35. Xiao, Bin, Bo Yu, and Chuanshan Gao. "CHEMAS: Identify suspect nodes in selective forwarding attacks." *Journal of Parallel and Distributed Computing* 67.11 (2007): 1218-1230.

36. Thulasiraman, Preetha, Srinivasan Ramasubramanian, and Marwan Krunz. "Disjoint multipath routing to two distinct drains in a multi-drain sensor network." *INFOCOM 2007. 26th IEEE International Conference on Computer Communications. IEEE.* IEEE, 2007.

37. Sun, Hung-Min, Chien-Ming Chen, and Ying-Chu Hsiao. "An efficient countermeasure to the selective forwarding attack in wireless sensor networks." *TENCON 2007-2007 IEEE Region 10 Conference.* IEEE, 2007.

38. Grau, Alan. "Can you trust your fridge?." *Spectrum, IEEE* 52.3 (2015): 50-56.

39. Li, Chunxiao, Anand Raghunathan, and Niraj K. Jha. "Hijacking an insulin pump: Security attacks and defenses for a diabetes therapy system." *e-Health Networking Applications and Services (Healthcom), 2011 13th IEEE International Conference on.* IEEE, 2011.

40. D. Evans, "The internet of things how the next evolution of the internet is changing everything. Technical report", CISCO IBSG, April 2011.

41. R. Thomas, et al. "Hey, you, get off of my cloud: exploring information leakage in third-party compute clouds." *Proceedings of the 16th ACM conference on Computer and communications security.* ACM, 2009.

42. M. Dabbagh, B. Hamdaoui, M. Guizai and Ammar Rayes, "Release-time aware VM placement," in Globecom Workshops (GC Wkshps), pp.122-126, 8-12 Dec. 2014.

43. M. Dabbagh, B. Hamdaoui, M. Guizani, A. Rayes, "Toward energy-efficient cloud computing: Prediction, consolidation, and overcommitment," in Network, IEEE, vol. 29, no.2, pp.56-61, March-April 2015.

44. M. Dabbagh, B. Hamdaoui, M. Guizani, A. Rayes, "Efficient datacenter resource utilization through cloud resource overcommitment," in IEEE Conference on Computer Communications Workshops (INFOCOM WKSHPS), pp. 330-335, 2015.

45. R. Boutaba, Q. Zhang, and M. Zhani. "Virtual Machine Migration in Cloud Computing Environments: Benefits, Challenges, and Approaches." *Communication Infrastructures for Cloud Computing. H. Mouftah and B. Kantarci (Eds.). IGI-Global, USA* (2013): 383-408.

46. D. Perez-Botero, "A Brief Tutorial on Live Virtual Machine Migration From a Security Perspective." *University of Princeton, USA* (2011).

47. W. Zhang, et al. "Performance degradation-aware virtual machine live migration in virtualized servers." *International Conference on Parallel and Distributed Computing, Applications and Technologies (PDCAT),* 2012.

48. V. Venkatanathan, T. Ristenpart, and M. Swift. "Scheduler-based defenses against cross-VM side-channels." *Usenix Security*. 2014.
49. T. Kim, M. Peinado, and G. Mainar-Ruiz "Stealthmem: System-level protection against cache-based side channel attacks in the cloud," *In Proceedings of USENIX Conference on Security Symposium, Security'12. USENIX Association*, 2012.
50. H. Raj, R. Nathuji, A. Singh, and P. England. "Resource management for isolation enhanced cloud services," *In Proceedings of the 2009 ACM workshop on Cloud computing security*, pages 77–84. ACM, 2009.
51. Y. Zhang and M. K. Reiter, "Duppel: Retrofitting commodity operating systems to mitigate cache side channels in the cloud" *In Proceedings of the 2013 ACM SIGSAC Conference on Computer; Communications Security*, CCS '13. ACM, 2013.
52. P. Li, D. Gao, and M. K. Reiter, "Mitigating access driven timing channels in clouds using stopwatch," *IEEE/IFIP International Conference on Dependable Systems and Networks (DSN)*, 0:1–12, 2013.
53. R. Martin, J. Demme, and S. Sethumadhavan, "Timewarp: Rethinking timekeeping and performance monitoring mechanisms to mitigate sidechannel attacks," *In Proceedings of the 39th Annual International Symposium on Computer Architecture*, 2012.
54. Fangfei Zhou et al. "Scheduler vulnerabilities and coordinated attacks in cloud computing." *10th IEEE International Symposium on Network Computing and Applications (NCA), 2011.*
55. K. Panagiotis, and M. Bora. "Cloud security tactics: Virtualization and the VMM." *Application of Information and Communication Technologies (AICT), 2012 6th International Conference on*. IEEE, 2012.
56. F. Zhang et al. "CloudVisor: retrofitting protection of virtual machines in multi-tenant cloud with nested virtualization." *Proceedings of the Twenty-Third ACM Symposium on Operating Systems Principles*. ACM, 2011.
57. T. Taleb, and A. Ksentini, "Follow me cloud: interworking federated clouds and distributed mobile networks," *IEEE Network*, 2013.
58. E. Damiani et al. "A reputation-based approach for choosing reliable resources in peer-to-peer networks." *Proceedings of the 9th ACM conference on Computer and communications security*. ACM, 2002.
59. W. Itani et al. "Reputation as a Service: A System for Ranking Service Providers in Cloud Systems." *Security, Privacy and Trust in Cloud Systems*. Springer Berlin Heidelberg, 2014. 375-406.
60. J. Sahoo, M. Subasish, and L. Radha, "Virtualization: A survey on concepts, taxonomy and associated security issues." *Second International Conference on Computer and Network Technology (ICCNT)*, 2010.
61. S.Yi, Q. Zhengrui, and L. Qun, "Security and Privacy Issues of Fog Computing: A Survey." In *Wireless Algorithms, Systems, and Applications*, pp. 685-695. Springer International Publishing, 2015.
62. E. Oriwoh, J. David, E. Gregory, and S. Paul, "Internet of Things Forensics: Challenges and approaches." In *9th International Conference Conference on Collaborative Computing: Networking, Applications and Worksharing (Collaboratecom)*, pp. 608-615. IEEE, 2013.
63. Z. Brakerski, and V. Vinod, "Efficient fully homomorphic encryption from (standard) LWE." *SIAM Journal on Computing* 43.2 (2014): 831-871.
64. E. Lauter, "Practical applications of homomorphic encryption." In *Proceedings of the 2012 ACM Workshop on Cloud computing security workshop*. ACM, 2012.
65. C. Hennebert, and D. Jessye "Security protocols and privacy issues into 6lowpan stack: A synthesis." *Internet of Things Journal, IEEE* 1.5 (2014): 384-398.
66. Daily Tech Blogs On Line: http://www.dailytech.com/Five+Charged+in+Largest+Financial+Hacking+Case+in+US+History/article32050.htm
67. M. Miller, "Car hacking' just got real: In experiment, hackers disable SUV on busy highway", *the Washington Post*, 2015, online: http://www.washingtonpost.com/news/morning-mix/wp/2015/07/22/car-hacking-just-got-real-hackers-disable-suv-on-busy-highway/

68. "2015 Data Breach Investigation Report", Verizon Incorporation, 2015.
69. M. Dabbagh et al. "Fast dynamic internet mapping." *Future Generation Computer Systems*, pp 55-66, 2014.
70. Forrester, "Security: The Vital Element of the Internet of Things," 2015, online: http://www. cisco.com/web/solutions/trends/iot/vital-element.pdf
71. F. Adib and D. Katabi. "See through walls with WiFi!," volume 43. ACM, 2013.
72. S. Kumar, S. Gil, D. Katabi, and D. Rus, "Accurate indoor localization with zero start-up cost," In *Proceedings of the 20th Annual International Conference on Mobile Computing and Networking*, pages 483–
73. 494. ACM, 2014.
74. G. Wang, Y. Zou, Z. Zhou, K. Wu, and L. Ni, "We can hear you with Wi-Fi!" In *Proceedings of the 20th Annual International Conference on Mobile Computing and Networking*, pages 593–604. ACM, 2014.
75. Y. Qiao, O. Zhang, W. Zhou, K. Srinivasan, and A. Arora, "PhyCloak: Obfuscating Sensing from Communication Signals," in *Proceedings of the 13th USENIX Symposium on Networked Systems Design and Implementation (NSDI)*, 2016.
76. T. Yu, et al. "Handling a trillion (unfixable) flaws on a billion devices: Rethinking network security for the Internet-of-Things." *Proceedings of the 14th ACM Workshop on Hot Topics in Networks*, 2015.
77. M. Dabbagh, B. Hamdaoui, M. Guizani and A. Rayes, "Software-defined networking security: pros and cons," IEEE Communications Magazine, 2015.
78. "What is PKI? - A Complete overview, January 23, 2015". Retrieved 2015-02-24, Online: https://www.comodo.com/resources/small-business/digital-certificates1.php
79. Jump up ^ "What is a Public Key Infrastructure - A Simple Overview, April 17, 2015", Onlie: http://www.net-security-training.co.uk/what-is-a-public-key-infrastructure/
80. XML 1.0, Fifth Edition, W3C Recommendation, Nov 26, 2008, Editors: T. Bray, J. Paoli, C. M. Sperberg-McQueen and E. Maler, Online: http://www.w3.org/TR/REC-xml/
81. Z-Wave Alliance, http://z-wavealliance.org/
82. ZigBee Alliance, http://www.zigbee.org/
83. OASIS, https://www.oasis-open.org/committees/tc_cat.php?cat=iot
84. IEEE Standard 802.15.4-2011, September 2011.
85. LoRa Alliance: https://www.lora-alliance.org/What-Is-LoRa/Technology
86. D. Guinard, V. Trifa1, F. Mattern, E. Wilde, "5 From the Internet of Things to the Web of Things: Resource Oriented Architecture and Best Practices1", Springer. pp. 97–129. ISBN 978-3-642-19156-5., Online: http://www.vs.inf.ethz.ch/publ/papers/dguinard-fromth-2010.pdf

Appendix B: IoT Projects for Engineering Students

To pass this course, engineering students are required to build an IoT solution, demonstrate their results, and then write detailed reports describing their findings. This Appendix lists the main elements of standard IoT projects and then provides examples of such projects with supplement information, i.e., sensor types and the expected outcomes. Appendixes C–F show examples of IoT Reports.

One of two well-known educational platforms is often used by the students: Arduino IoT and Raspberry Pi. Arduino is an open-source platform that was designed by hobbyists. Arduino IoT Cloud is an application helps students to connect IoT devices and allows them to exchange data with basic mentoring capability. Raspberry Pi platform was developed by Raspberry Pi Foundation to promote computer science education. Arduino is microcontroller board, while Raspberry Pi is a microprocessor based mini-computer. The Microcontroller on the Arduino board contains the CPU, RAM, and ROM. Raspberry Pi typically requires an Operating System to run.

Typical Elements of IoT Projects

In general, IoT projects will include the following main components:

- **Sensors**: to detect and capture data.
- **Switch** (e.g., Raspberry Pi, Arduino Uno): to receive, process, and analyze data from sensors and other sources. Results may be sent to other devices for notification.
- **Electrical Board (optional)**: to connect sensors, switches, LEDs (if needed), and other devices in a consistent and secure way. The board allows students to connect multiple devices to each other and allow them to exchange real-time

data. Common boards include 3.2″ × 2.1″ solderless breadboard with four bus lines spanning the length of the board and 30 rows of pins, enough for up to four 14-pin DIP ICs or three 16-pin DIP ICs.

- **Wireless Module** (e.g., Wi-Fi) to integrate the system onto cloud and send updates to specified devices.
- **Cloud Platform/Application** (e.g., Microsoft Azure, Amazon AWS, IBM Watson): for detailed data storage, services monitoring, analysis with advanced capabilities such as artificial intelligent, machine learning, object/face recognition, data trending, predictions, and forecasting.

It should be noted that some of the above elements may be already integrated, e.g., Raspberry Pi 4 with integrated USBs, Giga Ethernet port HDMI ports.[1]

Examples of IoT Projects

Hundreds of IoT Projects are available on various Internet IoT training sites. Table B.1 lists over a dozen of typical IoT projects for students. Additional projects may be found in https://create.arduino.cc/projecthub/products/arduino-iot-cloud and https://create.arduino.cc/projecthub/projects/tags/iot.

Table B.1 Examples of IoT Projects

IoT Projects	Sensors used	Expected outcome
Home Security system Motion-Sensing Alarm using IR Sensors	Infrared (IR) Sensor	The solution includes IR sensors to detect unusual movement, Wi-Fi Module to integrate the system onto cloud and send updates to user, and Arduino Uno/Raspberry Pi to capture and process sensor's data and then notify homeowner(s) when a harmful activity (e.g., front door is opened) takes place in the home. The system can also store collected data in a cloud platform for further interpretation. Alarm should be sounded in the home when a major issue is detected.
Touch Dimmer Switched Circuit Project	Touch Sensor	LED/Light is turned on when a sensor is touched.
Weather Monitoring System: Thermometer and Humidity	Temperature sensor (e.g., Arduino LM35) and Humidity sensor	Temperature (in Fahrenheit of Celsius) and Humidity readings are displayed, with two decimal digit accuracy, on digital thermometer.

(continued)

---

[1] https://www.raspberrypi.org/products/raspberry-pi-4-model-b/

Table B.1 (continued)

IoT Projects	Sensors used	Expected outcome
Automatic Lighting System: LDR Controlled Bulb	LDR (Light Dependent Resistor) Sensor	LDR sensors detect the changes in the sunlight intensity and send the data to the Arduino Uno/Raspberry Pi for interpretation. LED/Light is turned on when the intensity is low (or sensor is covered). Arduino Uno/Raspberry Pi receives data from sensors and switch the light on or off. Relay drivers can be used to convert the voltage to operate the light.
Sun tracker Using LDR	LDR (Light Dependent Resistor) Sensor	Sensor/Machine is turned in 3-dimension following the sun/light source.
Smart Irrigation System	Moisture Sensor	Soil moisture level is measured from a sample of dry soil first. The result may be displayed on Laptop/Smart-Phone/etc. Water is added to the sample soil and the moisture level is measured again.
Smart Water and Flood Monitoring System	Rain and Water Sensor	Water is placed on the rain and water sensor. Alarm is sounded, LED is illuminated, and an email/text message is issued. Alarm should stop once the sensor is dried out.
Accelerometer Based Hand-Gesture-Controlled Robot	Accelerometer Sensor	Accelerometer sensor based machine (e.g., small vehicle) moves and turns according to a sensor-enabled hand. LED may be taped into a student's hand.
Line Following Robot	IR Sensor	Small vehicle follows a specific trajectory (e.g., based on a line on the street).
Fix Distance Alarm	Ultrasonic Sensor	Alarm is sounded once an object approaches the sensor. Distance of the object is measured and reported.
Smart Blind Stick	Ultrasonic Sensor	Alarm is sounded once an objective is detected by the blind stick.
Motion-Sensor Lamp	PIR (Passive Infrared) Sensor	When a hand is waved in-front of the sensor, the lamp/light is turned on/off. Two modes may be tested: repeatable and non-repeatable triggered modes.
Home Automation System	DHT (digital temperature and humidity) Sensor	The idea is building a single system to control electrical appliances in the home. It can be integrated with a Raspberry Pi board to make it an IoT device and then can be controlled from a remote location via Internet.
Smart Trash Can Smart Mailbox	RFID Reader	The idea is designing a system that notifies waste truck driver when the bin is nearly full (or to notify a homeowner when a mailbox has mail). In addition to sensors, the solution may include: RFID Reader (to scan the code of the trash can integrated with a RFID Tag), RFID Tags and Raspberry Pi to process the data and send notifications to the truck driver.

References

1. Top 10 Sensors and Projects, July 31, 2021, online: https://www.etechnophiles. com/top-10-arduino-sensors-projects-beginners
2. Top 10 Arduino-Sensors with Projects for Beginners, July 5, 2021, online: https://www.youtube.com/watch?v=cAKnTSJb-SE
3. 100+ Ultimate List of IOT Projects For Engineering Students, July 5, 2018, online: https://www.electronicshub.org/iot-project-ideas/
4. List of Latest IOT Projects for Engineering Students, V. Vidyakar, July 5, 2018, online: http://www.skyfilabs.com/blog/ list-of-latest-iot-projects-for-engineering-students
5. IoT Projects, July 5, 2018, online: http://nevonprojects.com/iot-projects/
6. Top 5 IoT Projects (Best Internet of Things Home Automation Projects), Maker Pro, Nov 29, 2017, online: https://www.youtube.com/watch?v=MREnJ7a3BV0

Appendix C: IoT Project 1—Parking Availability App Using IoT

Miguel Covarrubias Martinez
Under the supervision of Professor Ammar Rayes

Abstract Finding parking in densely populated areas has become an issue due to the high volume of cars in the road. People often struggle to choose between the many parking garages in the area because they do not know if parking spaces will be available. In this paper we explore a solution to this problem that involves using IoT devices and the cloud. The idea is to make all the parking garages in the area "smart" by attaching ultrasonic sensors to each of the parking spaces in order to detect if the parking spaces are occupied. Micro-controllers will collect this information and upload it to a web server in the cloud where it can be accessed by users via a web application.

Keywords IoT, Ultrasonic, Sensor, Arduino, Webapp

Introduction

People tend to use their cars over public transportation because it can be more convenient. This has led to many parking garages running out of space. People currently do not have access to information regarding how busy parking garages are in terms of spaces available to park. Currently, there are parking garages that display parking availability before you actually enter the garage but they are not connected to the cloud. We can solve this problem by taking advantage of the popularity of IoT devices and the cloud. For this project, I will monitor the availability of parking spaces from different parking garages by attaching object detection sensors to

M. C. Martinez
Computer Engineering Department, San Jose State University, San Jose, CA, USA
e-mail: miguel.covarrubias@sjsu.edu

© The Author(s), under exclusive license to Springer Nature Switzerland AG 2022
A. Rayes, S. Salam, *Internet of Things from Hype to Reality*,
https://doi.org/10.1007/978-3-030-90158-5

parking spaces. How does sensor work? "Sensors typically collect data using physical interfaces (inputs) that sense the environment and then convert input signals into electrical signals (outputs) that are understood by the communication and computing devices" [1, p. 70]. There are different sensors such as IR (Infrared) or ultrasonic that are used for object detection. In this project I used ultrasonic sensors because they can be calibrated to detect objects within specific distances. Using physical sensors will provide the users with most accurate information about parking availability.

Problem Statement/Project Architecture

In this paper I will present a solution for finding available parking spaces in parking garages. The solution will use IoT devices located in the physical parking garages and the statuses about the parking spaces will be sent to a web server in the cloud where it can be accessed by users via a web application. This architecture covers the sensor layer, network layer, data processing layer, and application layer as shown in Fig. C.1. In the sensor layer I am using an Arduino microcontroller that receives the signals from the ultrasonic sensors. The microcontroller then processes the signals and evaluates if any of the parking spaces is occupied or open to use. This information is then sent over the Internet as a post request to the web server (IoT Data Processing Hub) for storage. The Arduino will also turn on the local green LED light if a parking space is open to use or it turn on the red LED light if a parking space occupied. Once the web server receives the payload from the Arduino it will then store the status of the parking garages in a memory data structure. The structured live data can then be accessed by users via a web application. Users will then be able to make decisions on where to go park based on the live data coming from the parking garages IoT devices.

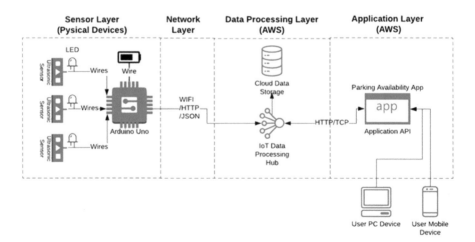

Fig. C.1 Architecture

Method(s) System Design

1.1.1 Sensor Layer (Physical Devices)

First, the main physical components of this IoT project are the Arduino Uno Wi-Fi Rev2, ultrasonic sensors, LED's wires, breadboard, and the power source as shown in Fig. C.2.

In this model there are three ultrasonic sensors with two LED lights (green and red) and each of the sensors represents a parking space in the parking garage. The sensors are placed in front of the parking spaces in order for the sensors to detect objects in front of them. The way the ultrasonic sensors work is by using sound waves above 20 kHz range to detect the proximity of objects [2, p. 1]. The waves (pulses) that the sensor emits are reflected back towards the sensor by objects within the field of view of the sensor [2, p. 2] as shown in Fig. C.3. By calculating the time it takes for the pulse to get back to the sensor and by using the speed of sound (29 μs/cm) the distance of the object in centimeters can be calculated as in (C.1).

$$d_{\text{OneWay}} = \frac{t_{\text{RoundTrip}} \times v_{\text{Sound}}}{2}$$

C.1

In this project I have defined a distance of 7 cm for the ultrasonic sensors to detect objects. This is an ideal value for the small parking garage model that I have created. For use in a real parking space, we would only need to update the distance value. The Arduino takes readings from the censors every 2 s and if an object is detected within 7 cm then the Arduino sends an "on" signal to the appropriate red

Fig. C.2 Arduino circuit diagram

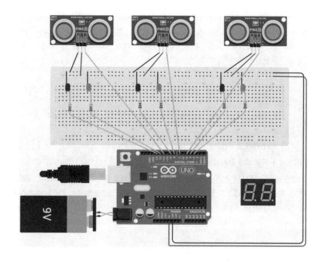

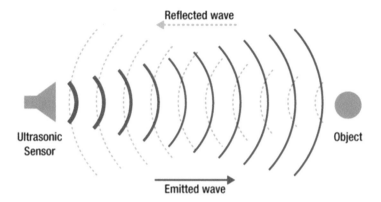

Fig. C.3 Ultrasonic time-of-flight measurement [2]

LED in the parking garage. In addition, the green LEDs in the parking garage will turn on if the sensors do not report objects within the defined distance.

Network Layer

The Arduino that I am using has a Wi-Fi module that allows it to connect to the Internet. This is very useful because it can send information about the statuses of the different parking spaces to a web server for storage. In the network, the Arduino has its own assigned IP address, and this can be used as a unique identifier. However, for this project I have decided to pre-program the only Arduino with a unique defined name. When it comes to sending data to the Internet from IoT devices, the data serialization format is important. This is because you want the applications to interpret the data from IoT devices with minimal format translations [1, p. 132]. Furthermore, it is good practice to only send the minimal required information to save bandwidth. Popular data serialization formats include XML, JSON, and EXI. For this project I decided to use the JSON format for the data payload which is sent as a HTTP post request from the Arduino every 2 s to the web server as in (C.2):

$$\{ \text{"parkingId"} : \text{"Garage1"}, \text{"spotId"} : 1,$$
$$\text{"isAvailable"} : true, \text{"totalAvailableSpots"} : 3\} \qquad \text{C.2}$$

In average, the payload size for each post request is 100 bytes and this payload is sent every 2 s for each parking space. Using this information, we can determine how busy our network can get depending on the number of parking spaces. Figure C.4 shows the code that the microcontroller executes to send the post request to the web server API.

```
// report back via serial if we get a connection
if (client.connect(server, 80)) {
  // Make a HTTP request:
  Serial.println("connected to server");
  client.println("POST /api/updatespot HTTP/1.1");
  client.println("Host: 192.168.1.36:80"); // AWS web server
  //client.println("Host: 192.168.1.36:80"); // local network
  client.println("Content-Type: application/json");
  client.print("Content-Length: ");
  client.println(measureJsonPretty(payload));
  client.println();
  serializeJsonPretty(payload, client);
} else {
    Serial.println("NOT connected to server");
}
```

Fig. C.4 Arduino code post request

Fig. C.5 Actual Web App
from a mobile device

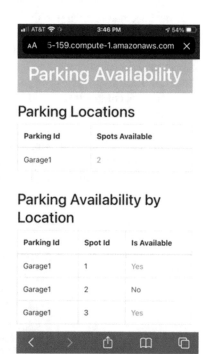

Data Processing and Application Layer

The web server (IoT Data Processing Hub) as shown in Fig. C.1 collects and processes the data sent by the Arduino. Then the data is stored in a memory data structure that can be accessed by REST API. For this implementation I decided to only keep the most recent status of the parking spaces in the garages, but in an actual production implementation it would be a good idea to keep the historic data for analytics. Both the Web UI and Web server were created using the React and Node. js frameworks. Figure C.5 shows how the web application looks like from a mobile device.

Deployment

The Web UI and Web server were built on top of Docker images to make the deployment to the cloud easy. For this project I decided to use AWS (Amazon Web Services) as my cloud platform because I was able to create a free student account. To deploy the applications, I created an EC2 (Elastic Compute Cloud) instance in AWS and transferred the Docker image from my personal computer to the instance. Once the Docker image was transferred in the EC2 instance it only took one Docker command to deploy the application. I had to make sure that the proper ports on the instance were exposed to the public in order to access the applications from anywhere. The EC2 instance had its own public IP that I used as the API endpoint when I deployed the Arduino code. It was trivial to verify if the application was working since I was able to see the parking spaces statuses change after I placed/removed cars into my garage model. Deploying the web application is a straightforward process but when it comes to deploying new code into the microcontrollers it becomes more difficult. In order to accomplish this and more there needs to be in place an IoT Service Platform. "The functions of the IoT Services Platform include the ability to deploy, configure, troubleshoot, secure, manage, and monitor IoT devices. They also include the ability to manage applications in terms of software/firmware installation, patching, starting/stopping, debugging, and monitoring" [1, p. 181]. However, due to time constrains I was not able to implement this key component of the IoT workflow.

Fig. C.6 Parking garage model

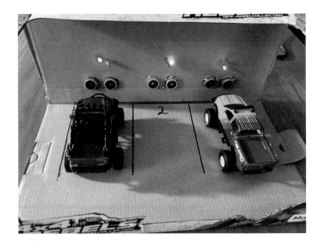

Fig. C.7 Actual Arduino circuit attached to garage model

Evaluation Methodology/Results

The parking garage model that I created is shown in Fig. C.6. In this model you can see the red LED lights turned on when the ultrasonic sensors detect an object in front of them. You can also see that the parking space in the middle has the green LED turned on because there is no object in front of the sensor. Figure C.5 shows what a user would see from the web application according to Fig. C.6. I also tested this model by pacing the car about eight centimeters away from the sensor and as I expected the red LED did not turn on. In Fig. C.7 we can see the actual implemented Arduino circuit that is attached to parking garage model.

I wanted to add an LCD display to show the number of parking spaces available but LCD display requires several more pins and I did not have a breadboard extender. Also, it is hard to see in the image but the LED lights are using a resistor in order to limit the current going through it and prevent that from burns.

Conclusion and Future Work

I was very satisfied with what I was able to accomplish in this project. I applied in real life the concepts that I learned in class. This project was a good proof of concept that demonstrated that it is possible to use IoT and the cloud to inform users about parking availability within parking garages. There are many things that can be improved on this prototype such as the website UI/UX, security, and fault management with respect to the sensors and microcontrollers. I believe that this project would be very useful if it was implemented in real life and the cost would be low because the IoT devices are not expensive. In addition to using ultrasonic sensors, it would be a good idea to use a type of pressure sensor to be able to differentiate between cars that are actually parked or people that are just standing on the parking space. Furthermore, developing an analytics dashboard with information for each parking garage such as the most busy hours, average parking space usage, and the number of current users looking for parking would be very useful. I believe IoT has a great future ahead and I believe this is just the start.

Acknowledgment I would like to thank professor Ammar Rayes for sharing his knowledge and lecturing the IoT class at SJSU.

References

1. Rayes and S. Salam, Internet of Things From Hype to Reality: The Road to Digitization. Cham: Springer, 2019.
2. Toa, M. and Whitehead, A., 2020. Ultrasonic Sensing Basics. [ebook] TexasInstruments. Available at: https://www.ti.com/lit/an/slaa907c/slaa907c.pdf [Accessed 16 November 2020]

Appendix D: IoT Project 2—Sensor Activated Lights with Cloud Data

Emma Peatfield
Under the supervision of Professor Ammar Rayes

Abstract Thoughtful design is not always something that successfully makes its way into the infrastructure of a living space. Sometimes it is the priority, but other times, it completely lacks. One of the areas where thoughtful design is sometimes lacking is good lighting, either by natural lighting or light fixtures. A lot of households and work spaces end up with rooms or closets that have little to no natural light and the light source in that room is either weak or nonexistent. This makes it hard for people to utilize these spaces. Luckily, this is an issue that can be easily solved if we have access to a portable light source that can be mounted in some of these places. This project looks at creating that type of light source while also incorporating modern cloud technologies to collect and display metrics. It experiments with using an Arduino Uno and a Node MCU Wi-Fi Module to create a light source which then sends data to the cloud. This data is stored and displayed for testing and analytical purposes.

Keywords IoT, Internet of Things, Arduino, Sensor, Grafana, AWS, EC2, InfluxDB, Docker

Introduction

How many times have you lived in a place that did not have any kind of light within a dark corner or closet? In many apartments and living spaces, there are plenty of areas that are not reached by natural light and that do not have their own light

E. Peatfield
Computer Engineering Department, San Jose State University, San Jose, CA, USA
e-mail: emma.peatfield@sjsu.edu

source. Thanks to this, most people likely have an area of their home where they wish they had more light, especially in a closet-like space. In my apartment alone, I can name a handful of places where a small light would make the world of a difference. This light does not have to be something that is complicated or expensive, just a simple fixture that could illuminate the space. If these spaces were more illuminated, I feel as if I could get so much more use out of a currently dark space. This project looks to solve that problem with a light fixture for one of my closets that has little to no natural light. To do this, I will utilize Arduino accessories, sensors, and modern cloud technologies.

Problem Statement and Architecture

Having a space with no access to natural light becomes a real pain when you would like to realistically use it. Without the light, the space gets wasted and the rest of the space in the apartment has to compensate for that. With this project, I hope to solve that problem by creating a small light fixture that turns on when the door opens and off when the door closes. In creating this, I would also like to have the device send metrics to the cloud, so that I can track usage statistics of the light source. This will be very useful for testing and tracking the actual usage of the light in the future. With this project, I also hope to inspire others to build on my idea to solve problems in their own homes with IoT Devices.

For this proposed solution, I will need to have a light source and a door sensor connected to an Arduino Uno, which will then connect to the Internet so that data can be sent to the cloud. To do this, I will need an extra Wi-Fi module and to create an instance on the cloud where I can store and display data. This will be done using AWS services and Docker containers. In Fig. D.1, you can see the basic architecture of the project that I created for my proposed solution. I have incorporated various aspects of IoT, so that I could gain knowledge in a multitude of ways while working on this solution.

As pictured, I will have a door sensor that connects to an Arduino Uno and a Wi-Fi Module circuit. This circuit will then connect both to a light source and to the cloud. The cloud will be moderated with an EC2 Instance, a database hosted with InfluxDB, and a front-end dashboard display using Grafana. I will go more into detail on this in the next section.

Equipment and System Design

When working on the design for this project, a few things were taken into account. First, I knew this fixture would need to have some sort of power source since the closet does not have any kind of source within it. Also, I knew that I would need to find a way to connect my project to the Internet since I wanted to send and store metrics within the cloud. This was something that I knew I could not do with the

Arduino module that I currently had. Now that those two problems were noted, I ordered necessary parts and started designing a circuit for the project.

In this section I will discuss the hardware parts that I used, such as sensors, light sources, and modules. Also, I will describe a few of the main software tools that I used for connecting everything to the cloud, and I will go into detail about the final circuit design for the project once I had everything in place.

Hardware Used

One of the first major pieces of equipment that I used was an Arduino Uno. This is a small micro-controller that I bought a few years back, but I had never had the chance to work with it until now. It was something that was pretty simple to work with and gave me exactly what I needed to connect my sensor and light source together. Once I connected everything on here, I could use the Arduino IDE to write a short program to control my circuit and basically tell the Arduino what to do and when (Fig. D.2).

The next major piece that I used in my project was a Node MCU 12E Wi-Fi Module. This piece was a tool that I did not have, so I had to order it if I wanted to connect my project to the Internet. I had never worked with one of these before, so it was a little challenging to get started. Without this, I never would have been able to collect metrics for the project. Thus, it was instrumental in getting a lot of things working and certainly gave me more freedom with the project (Fig. D.3).

The final major piece of hardware that I used was a door sensor. The one that I used for this project was a magnetic door sensor, which can be found on any electronics website or on Amazon. It connects to the circuit on the one side and the other side is used on the opening door. When the circuit is closed, it sends a signal on the connected side, letting the light know to turn off. And when the sensors are not connected together, the signal is sent to turn the light on. This sensor is pretty basic, but it worked nicely for what I needed (Fig. D.4).

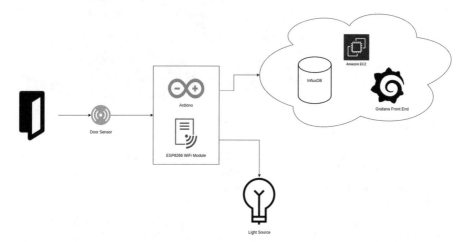

Fig. D.1 Project architecture

Fig. D.2 Arduino Uno and Breadboard

Fig. D.3 Node MCU Wi-Fi module

Software Used

For this project, I did not want to just have a basic door light built with my Arduino, so I decided that I wanted to track various metrics and store them on the cloud. Doing this, would be very helpful when testing the circuit, and also would keep track of usage statistics for the light source. Again, without purchasing the Wi-Fi module, this would not have been possible. Thus, once I had this module and completed the set up, I was able to store information by using a few more tools.

Fig. D.4 Door sensor

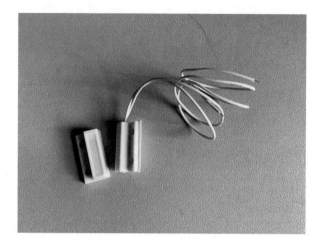

The first tool I used was an Amazon EC2 Instance. This was my first time setting one of these up on my own, so it took some research, but luckily Amazon has great tutorials. I used this tool to create a Linux instance that I knew I could install the necessary packages on without too many roadblocks. Working with an EC2 Instance was flexible for every need that I had, and it was free for the tiers that I chose.

Next tool that I used was Docker. Docker is a container platform that basically packages environments for development and makes it easier to develop and deploy products [3]. I installed Docker onto the EC2 Instance, as well as Docker Compose, which is "a tool for defining and running multi-container Docker applications" [2]. By using Docker compose, it was very easy to start up each container at once and each kept external storage in case they got shut down by accident, or the system crashes. All that was needed was a YAML file with information for each container that I used.

Another tool that was very important for this project was InfluxDB. InfluxDB is a database that I used to store all of the data coming in from the Arduino and Node MCU. I did this by creating an InfluxDB container using Docker and Docker Compose. Storing data with this tool was simple and it connected seamlessly to my final software tool, Grafana.

Grafana is a tool that I used to create dashboards for displaying and observing data [1]. I am using this tool to display my data on a dashboard that could be customized for my needs. As I mentioned, it connects directly to InfluxDB, so that I can retrieve whatever data I need and then Grafana displays what I specify. It always looks very beautiful and clean, especially if you use one of their many built-in plugins. As long as you have data coming in, you can create tables, graphs, meters, and so much more.

Using these tools for this project really helped me take the data collection to the next level and show what you could do with even just simple metrics. All of this, of course, would not have been possible without the hardware tools that I mentioned earlier and the circuit design that I will discuss next.

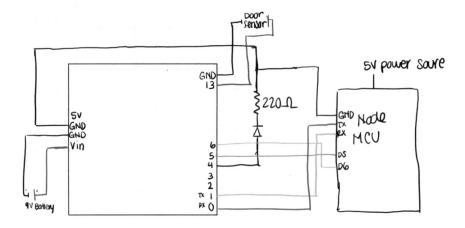

Fig. D.5 Project circuit

Circuit

Designing the circuit was important for having a successful project. I knew it needed to be compact and also connect so that everything worked as intended. I first started with just the Arduino circuit that used the lightsource and the door sensor. Once I had that working together, I worked on connecting the Node MCU Wi-Fi Module to those, and tested that out. Both of these were connected to my laptop for their source of power, but after testing each component out, I was able to connect the circuit to power sources that were detached from my laptop (Fig. D.5).

Methodology

This project gave me a chance to work with not only sensors and computers, but also with cloud technologies and Docker containers. Having this wide range was important for me so that I could get the most out of this experience. The project was completed in quite a few steps, which I will lay out in this section.

First, I was tasked to design a circuit for how I wanted this to work. The first thing I did was connect the door sensor, resistor, and the light source to the Arduino Uno. That part was not too challenging and moved pretty quickly. The next task was connecting the Node MCU and connecting the circuit to the Internet so that I could collect metrics from the project to store in the cloud.

Connecting the Node MCU to the Internet was not as hard as I thought it would be. Thus, I was able to start working on setting up my EC2 Instance, so that I could begin sending data. This part was also rather straightforward since Amazon has a lot of great tutorials in their documentation for AWS services. Once this was all done,

I installed Docker and Docker Compose onto the instance so that I could utilize containers in this project. I worked on setting up these containers that would be home to InfluxDB and Grafana. These required only a little customization at the start so that I could ensure things were set up properly.

Now that the containers were ready to go, I had to work on sending data to them for collection. This task was one of the most challenging parts of the project. It was quite simple to get the Node MCU connected to the Internet, but it was a bigger challenge to get the data sent to the EC2 instance. Figuring out how to properly format the data for the POST request was difficult. Each way I tried, I seemed to get an invalid response. Eventually, I figured out how to format the data by concatenating strings. This works very nicely, however, I would like to find a more efficient way of doing this in the future.

From this point on, the main parts of the project were complete. I now had to figure out how to power the Uno and Node MCU without my laptop so that I could place the project in my closet. I had a 9 V battery power adapter that I was able to use for the Arduino, but I had to end up using a portable charger for the Node MCU since I did not have any other adapters, and could not purchase one at this point in time. This worked fine to power both of the modules, however, I did have an issue with sending data between the two. I realized that I had not connected the two for serial communication correctly. It worked while plugged into the laptop, but not separately. To fix this, I quickly connected each Tx connector to the other Rx, and vice versa. This allowed the two devices to transmit and receive data to each other. After this, the project was complete and ready to be placed in the closet.

Evaluation

As far as goals went, I met every goal with this project. The project met the requirements necessary to be a light fixture within a small space that turns on when the door opens. It also sends metrics to the cloud where I can view them with ease. This project was able to be placed inside my closet and it worked as intended. I filmed a few videos of it working inside of the closet that are available to watch in my demo video [4]. I am very satisfied with how this project turned out and I cannot wait to add more and build off of it. I worked really hard to get the final circuit out and making it portable was an added bonus. Figure D.6 is a picture of the final circuit detached from the closet.

Once the final circuit was complete and powered on, adding it into the closet was simple since the door sensors that I bought came with an adhesive backing. I just placed the project on a shelf near the door and attached the sensors. Once in place, I was able to use Grafana to view the metrics that I was collecting. Grafana displayed the metrics beautifully, and I was able to customize what I wanted it to show and how.

As of now, I have a table displaying the time a metric came in, the status of the door, the time the door has been open in seconds, and the total amount of open time in seconds. Figure D.7 is a screenshot of the dashboard where you can view this table.

Fig. D.6 Final project

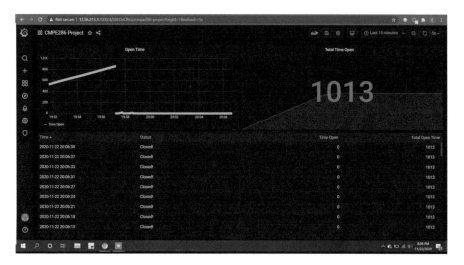

Fig. D.7 Grafana screenshot of the dashboard

Also in this dashboard is a big number metric that is showing the total amount of time that the door has been open with the light on. This is useful for tracking the amount of usage that a light source may get. It may mean you will expect the light bulb to go out since you will have an estimate of its usage. There is also a graph on this dashboard that graphs the seconds that the door has been open. As you can see in Fig. D.7, the time increases and then flattens out to zero when the door was closed. What is great about Grafana is that you can choose a time frame of data to look at, so you can see the history of the metrics.

Having the ability to see these metrics was very helpful while I was testing out my project. It helped me see if there were any errors in my code logic or within my circuit. Overall, I am very happy with how this project turned out and how well the metrics were able to be sent to the cloud from my Arduino and Node MCU Wi-Fi Module.

Future Work

Although this project was very successful, and I met all of my initial goals, I would love to build on it. A few things I would like to change are the lightsource, Amazon Alexa integration, and a fixture to display the project.

For this phase, I was not able to add a real lightbulb fixture, I could only add a basic LED light. That is something that I would love to change, once I have the funds to do so. This would be an improvement for how much light I could get within the dark space and I could also buy a bulb that has some sort of dimming ability. That way I could adjust the light for my needs based on the time of day and how dark the different spaces may be.

After changing the lightbulb, I would also love to incorporate Amazon Alexa into this project. If I was able to do this, I could control the light without the need for a door sensor at all times. I would also be able to change the brightness of the light with my voice.

Finally, if possible, I would love to create some sort of holder for the light, so that I can mount it on the wall or ceiling. This might require a 3D printer, but it could be possible without one. This would be a project much farther in the future.

Conclusion

The finished project met all of my goals and I am very happy with the end result. Adding cloud services worked very well for testing and enabled me to see how everything was being recorded. I am also very impressed with how well I was able to display data using InfluxDB and Grafana. This was all possible with only a few snags while creating it. In the future, I hope that I could build upon this project and create something even more impressive. This project just shows that even some of our simplest problems can be solved with IoT.

References

1. "Grafana: The Open Observability Platform." *Grafana Labs*, Grafana Labs 2020, http://grafana.com/
2. "Overview of Docker Compose." *Docker Documentation*, Docker Inc., Nov. 2020, https://docs.docker.com/compose/
3. "What is a Container?" *Docker,* Docker Inc., 2020, www.docker.com/resources/what-container
4. Peatfield, Emma, director. *Cmpe 286 Demo,* 2020, https://youtu.be/_F9VbPc0xVk

Appendix

Code: https://github.com/epeatfield/cmpe286Project

Appendix E: Warehouse Inventory Management System

Ayesha Siddiqua and Vidushi Jain
Under the Supervision of Professor Ammar Rayes

Abstract One of the reasons behind every successful package shipment is proper communication from warehouse to the point of destination. The real-time developing information provided during the shipment allows the team to manage the multiple shipments efficiently. Proper management also helps in maintaining per shipment requirements for better service to the customer. In the transportation of medicines it is a must requirement to maintain the temperature of the shipment between a certain range. For example, the developing COVID vaccines are required to be maintained between 2 and 8 °C during transportation. Ineffectiveness in maintaining the temperature and other environmental metrics will result in the loss of complete vaccination shipment, in the worse-case scenario it can cost the life of patient. Proper network coverage are among several factors that impede proper communication during shipment. Identifying, customizing, and reporting vital environmental metrics are also a significant challenge. Conventionally tracking of a shipment was done manually, which was a tedious task involving a lot of cost and manual labor. In addition to manual tracking, RFID tags are also used for tracking the packages inside the warehouse. Though these tags provided wireless communication, the tags effectiveness is affected with the amount of metal and liquid present around the package. Our proposed solution to the above problem is to develop a complete-intelligent Warehouse Inventory Management system for the better tracking. It deals with the convergence of various services for customizing requirements for each shipment and timely management of it. NRF52832 Bluetooth low energy (BLE) module is used to design the shipment beacons. The BLE tags are ideal to track location or movement of inventory, individual packages, pallets, or equipment, regardless of whether they are indoors, outdoors, or in transit. This BLE-based solu-

A. Siddiqua · V. Jain
Computer Engineering Department, San Jose State University (SJSU), San Jose, CA, USA
e-mail: ayesha.siddiqua@sjsu.edu; vidushi.jain@sjsu.edu

© The Author(s), under exclusive license to Springer Nature Switzerland AG 2022
A. Rayes, S. Salam, *Internet of Things from Hype to Reality*,
https://doi.org/10.1007/978-3-030-90158-5

tion offers real-time condition monitoring for COVID and CANCER vaccines including their temperature, humidity, altitude(from pressure), and in some cases ambient light for the closed packaging of the sterilizing kits. The Raspberry Pi device is used to design a Warehouse tracker. The device transmits real-time sensor data collected from multiple shipment beacons over BLE along with the alerts for any temperature excursions due to unfavorable conditions of the vaccines to the Azure Internet of Things (IoT) Hub. Thus, warehouse tracker IoT system presents a complete tracking system to the warehouse industry. The Azure platform offers real-time data visualization and device tracking. It provides visualization of sensor data from multiple trackers with the help of charts and text from the Azure explorer.

Keywords Internet of Things, BLE, High reliability, Live-tracking, Real-time visibility, IoT

Introduction

In the early 1990s, the demand for expansion of interaction between the devices beyond just the computers connected to the web led to the development of several smart objects which can be controlled in real time over the Internet such as WearCam, toaster, etc. [1]. A global identification system based on RFID was developed to provide the devices the capability to "observe, identify and understand the world" with the help of sensors by the late 1990s. RFID technology was extensively utilized in commercial products to decrease the dependency of the inputs from humans [2]. To boost Internet Protocols (IP) in the network of the Internet of Things (IoT) devices, the IPSO Alliance was launched by the companies. In 2006 the Federal Communications Commission (FCC) approved the "white space spectrum" for the IoT devices, which helped to reduce the gap between the rural digital connectivity with the help of the broadband services. By 2016, Narrow band IoT was developed by 3GPP to provide cellular services over a broader area with the help of Low-Power Wide Area Network radio technology [3, 4]. The 3G and 4G technology are not sufficient today to meet the requirements of fifth-generation (5G) wireless devices, and this was the motivation for the usage of an unlicensed spectrum for the IoT devices [5]. This technology improvement and faster connectivity can be utilized in wide IoT applications. They could be used to track shipments of vaccination in healthcare industry to large distance with lesser utilization of power. The future concentration of the IoT systems is towards autonomous vehicle communication which can be used for easy transport and shipment tracking without further manual labor and guarantees reliable delivery [6].

The basic building blocks of smart products are hardware, software, sensors, and communication networks. In smart surveillance applications, the maturity of installed sensors plays a vital role in tracking a shipment and also in maintaining customized parameters required for proper maintenance of the shipment. For quick detection of any unusual situations in the surrounding helps to prevent the damage in advance. Shipment management mainly relies on temperature and pressure monitoring; using relevant sensors facilitates detecting change in temperature and humidity content in the air [7]. The main reasons for change in preset temperature are because of human mistakes or system faults which harms patients' lives if it was during the medicine transport. After successful detection of change in parameters the proposed warehouse tracker network sends an alert to the Azure IoT Hub using MQTT communication protocol. Traditional methods fail to detect change in the customized environmental parameters.

Problems in the Existing Tracking and Monitoring Systems

In the fast moving world, the time is money and the business efficiency is critical to any competitive edge. Making an essential to nowhere your products, equipment and assets are, at all times. Unfortunately, packets in transit often go missing or delivered to the wrong location as shown in Fig. E.1. An assets or inventory in a storage or in use can be difficult to locate in a warehouse or a port. The current tracking and the monitoring solution such as RFID tags are limited in functionality and expensive to set up. The lack of the visibility to the location, the condition of the individual packages and assets can be result in wasted time, missing inventory, customer dissatisfaction, loss of profits, and big headaches to the firms involved in the shipments.

Solutions to the Limitations of Existing Tracking Systems

In the healthcare industries, the collection of data points performs a significant role that includes gathering, analyzing, and data processing from various sources. One of the most important use case of data collection in the healthcare is needed to monitor the vaccines conditions during the shipment [8]. The shipment beacon solution can work with the warehouse trackers or mobile phones to provide package or item level monitoring for all the important goods such as vaccines. It will be the

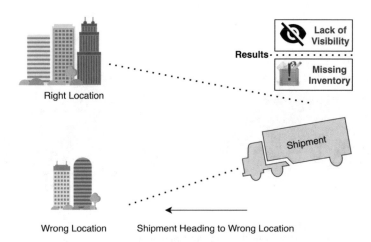

Fig. E.1 Problems in current monitoring systems

industries most affordable, in-transit indoor or the outdoor battery operated solution. The shipment beacon is Bluetooth low energy (BLE) based hot-spot beacon solution which provides end-to-end supply chain and asset monitoring solution across the enterprise, whereas the warehouse tracker is the GSM or Wi-Fi based hotspot solution, which provides reliable device-cloud communication for a fleet of devices.

This project report aims at describing the integration of the shipment beacon and the warehouse tracker for monitoring the healthcare shipments. It includes the design implementation and development of the warehouse inventory management system. This report is divided into seven parts. The following Section describes the architecture of the overall design implementation. However, the section "Method(s)/ System Design" describes the methods and system designs components of our solution. It additionally includes the software components of the system are emphasized to present the update of data collected from the shipment beacon to the warehouse tracker. The design ease outs the tracking of the devices and provides their real-time status. Additionally, in the section "Project Implementation" that includes system implementation of the proposed solution for shipment beacon and for the warehouse tracker. Then we have system integration in the section "System Integration" and its basic unit testing section "Testing and Verification" and finally testing ends with the evaluation tests section "Evaluation Methodology and Results" results of our solution. At last, the section "Summary, Conclusions and Recommendations" concludes the paper (Fig. E.2).

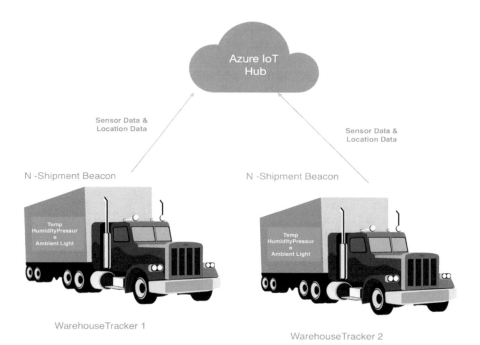

Fig. E.2 Warehouse inventory management system design

Project Architecture

This section contains the design and development of BLE beacon device and the Internet based tracker for monitoring shipments and the conditions for the health-care medicines. The High-level design architecture consists of a BLE based wireless MCU that will transmit BLE Beacons after every 10 s (configurable from App) in a connection-less mode that will be collected by BLE enabled central devices (Raspberry Pi). This device will be used for monitoring physical status of the shipments using temperature, humidity and the motion sensors. Multiple users can receive data whoever will be present in the range of the device (Fig. E.3).

This project includes a shipment beacon for real-time monitoring of the vaccines using multiple sensors which collects temperature, pressure, humidity, and ambient light from the surroundings. The shipment beacon transmits the sensor data to the warehouse tracker, which is actually the Raspberry Pi processor. The Raspberry Pi acts as a hotspot to get the data from BLE and connects to the Azure IoT Hub via MQTT protocol and sends the data to Azure platform. The azure IoT Hub ensures the security and reliability of communication. This section details the BLE beacon device architecture design. It is designed with nRF52832 wireless MCU targeting for BLE applications from Nordic Semiconductors. This device will transmit the beacons along with the sensor data and these beacons will be collected by central devices like mobile phones and the Raspberry Pi devices. Figure E.4 shows the overall architecture of the warehouse inventory management system.

The architecture consists of three layers, the sensor layer, communication layer, and the data processing layer. The first layer is the sensor layer which contains the physical sensors. It deals with reading and writing of the sensor register values directly from the physical devices. The type of data collected from this layer depends on the type of application. The data can be collected from environmental parameters such as temperature, pressure, and humidity or human physical activities [9]. In our project application, the data is collected from environmental sensors. Later, the

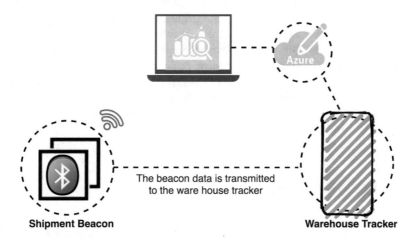

The beacon data is transmitted to the ware house tracker

Shipment Beacon   **Warehouse Tracker**

Fig. E.3 High-level design architecture

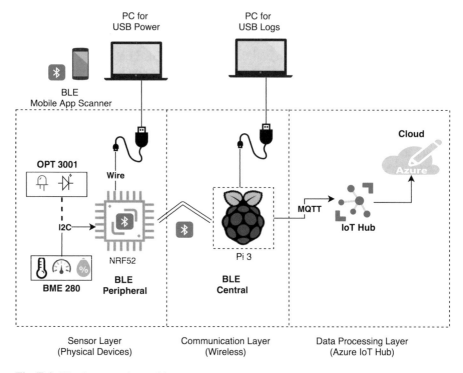

Fig. E.4 Warehouse tracker architecture

collected data from the sensors is concatenated with the BLE beacon payload and transferred to the second layer. The intermediate layer represents the communication layer. This layer will act as a BLE central hot-spot for the first layer. This layer can perform tasks, run algorithms required for temperature excursions, and contain enough memory to collect the BLE data from the beacons. The Last layer in the architecture is the cloud layer. This layer is essential for providing remote access to the data and for online monitoring of the shipments and for vaccines diagnostics [10]. The second layer is used to establish the connection with the azure server using Internet services. This layer performs a major role when it comes to long-term storage of data and to minimize congestion at the core network by performing an in-depth analysis of mega data at the cloud layer [11]. The warehouse inventory management system architecture from Fig. E.4 used to collect real-time condition monitoring of the vaccines and transfer the data to the mobile phone or to the azure server for its computation and analysis.

Method(s)/System Design

This section of the report presents the hardware and the software components used in the warehouse inventory management system design.

Shipment Beacon

The shipment beacon consists of two components nRF52832 BLE module and the environmental sensors (Fig. E.5)

1. **BLE Module nRF52832**: The shipment beacon's firmware built on nRF52832. The nRF52832 is a wireless MCU targeting Bluetooth Smart application. The device is a member of the nRF52xx family of cost-effective, ultra-low power, 2.4-GHz RF devices. Very low active RF and MCU current and low-power mode current consumption provide excellent battery lifetime.

The nRF52832 contains a 32-bit ARM Cortex-M4 running at 64-MHz as the main processor and a rich Peripheral feature set, including a unique ultra-low-power sensor controller, ideal for interfacing external sensors and/or collecting analog and digital data autonomously while the rest of the system is in sleep mode. The key features of the nRF52832 Micro-Controller are as follows:

(a) Powerful ARMCortex-M4.
(b) Up to 64-MHz Clock Speed.
(c) 512 KB of In-System Programmable Flash.
(d) 64-KB SRAM.
(e) Supports Over-The-Air Upgrade (OTA).

2. **Humidity, Temperature Pressure Sensor: BME280**: These sensors are packed with embedded functions with flexible user-programmable options, configurable interrupt pins. Embedded interrupt functions allow for overall power savings relieving the host processor from continuously polling data. There is access to both low-pass filtered data and high-pass filtered data, which minimizes the data analysis required for tilt detection/faster transitions, temperature detection, pressure detection, humidity detection, and light intensity.

The device can be configured to generate inertial wake-up interrupt signals from any combination of the configurable embedded functions allowing the temperature, pressure, humidity, and light sensor to monitor events and remain in a low-power mode during periods of inactivity. The sensor interface is communicating with controller via Serial Bus Interface.

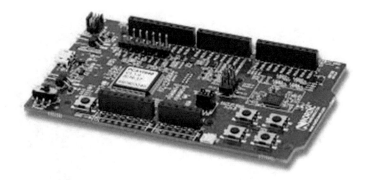

Fig. E.5 BLE MCU nRF52832

(a) SDA is for data receiving and transmitting to and from controller.
(b) SCL is for controlling the data on sensor interface that is transmitting and receiving to and from controller.

3. ***Working of Shipment beacon****:* The shipment beacon not only records the sensor data of the vaccines conditions but also send alerts in real time when an exception occurs. It also reduces the need of download the data manually via USB cable at the end of every shipment. It provides the cold chain real-time visibility for the vaccines which needs to maintain the stringent requirement of the cold chain temperature range, i.e., 2–8 °C.

In the shipment beacon design, the nrf52832 BLE microcontroller will perform the role of communication as master and sensors act as either as a slave receiver or transmitter. The master must generate the Start(S)/Stop(P) condition for the inter-integrated circuit(I2C) interface and provide the serial clock on SCL pin. BME280 Sensor will be used to measure temperature, humidity, and pressure of the containers/boxes in which it will be placed. BME280 is a 8 Pin, I2C/SPI based sensor IC from Bosch Sensortec. It will be powered up directly via coin cell battery/USB and kept in low-power mode or shutdown mode as per requirement. It will measure the temperature, relative humidity, and pressure of a shipment container and send to BLE module through I2C. The BLE module will send it in form of packets to a central device like smart phone or the Raspberry Pi (Fig. E.6).

As shown in Fig. E.7 nrf52832 BLE module will advertise the BLE packet which contains the payload as mentioned in the Table E.1 which includes the sensor and the battery information in the advertisement.

Warehouse Tracker

The real-time sensor data sent out by multiple shipment beacons is received by the warehouse tracker. The warehouse tracker is implemented using Raspberry Pi as shown in Fig. E.8 the Raspberry Pi after receiving the data from multiple nRF52 boards is sent in real time to the Azure IoT Hub for further processing.

Fig. E.6 Environmental
sensor BME280

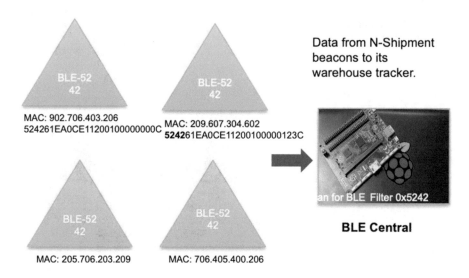

MAC: 902.706.403.206
524261EA0CE11200100000000C

MAC: 209.607.304.602
524261EA0CE11200100000123C

MAC: 205.706.203.209   MAC: 706.405.400.206

Data from N-Shipment beacons to its warehouse tracker.

an for BLE Filter 0x5242

BLE Central

BLE Peripheral - Will broadcast the sensor data with mac address name of the BLE chip.

Fig. E.7 Working of BLE tag

Table E.1 BLE broadcast data format

Info	0x2F (Temp)
	0x04 (Temp)
	0x64 (Battery)
	0x01 (Light)
	0x48 (Humidity)
	0x64 (Humidity)
	0x03 (Pressure MSB)
	0x4A (Pressure LSB)
	:
	:
	:
	0x00

Fig. E.8 Architecture of
warehouse tracker

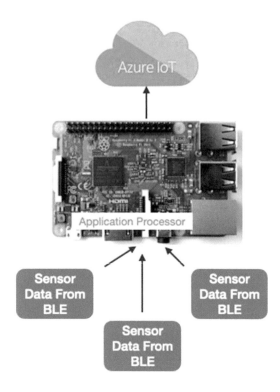

Project Implementation

1.1.1 BLE Advertisement and BME280 Sensor Interfacing

Interfacing of the BME280 sensor with controller is done by the Serial Bus Interface. It is a three step process to get the values from the sensor. The first step is the sensor initialization. During initialization the driver registers the slave device and basic initialization is done for the temperature, pressure and for the humidity sensors. In the second step, slave device, i.e., sensor will read the values of the respective sensors from the registers and will broadcast the sensor readings to the central device, i.e., Raspberry Pi. The software flow diagram for the master (nrf52832) and slave interaction (BME280) is shown in Fig. E.9.

The code snippet of the BLE advertisement payload which includes the sensors information from array index 5 to 12 is shown below:

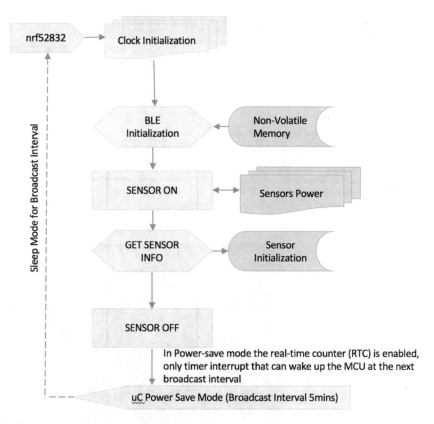

Fig. E.9 Software Flow diagram for BLE and Sensor interfacing

BLE Advertisement on Mobile App

As described in the BLE advertisement payload code snippet we can verify the advertisement payload by connecting the shipment beacon with the NRF connect open source mobile application which can be easily downloadable from the App Store or the Play Store depending on the users smartphone. We can connect the shipment beacon with the NRF Connect mobile app (android or iOS) by following the steps shown in Fig. E.10.

After connecting with the mobile application the payload of the shipment beacon will look like as shown in Fig. E.11. The payload includes the temperature, pressure, humidity, and the light sensor readings as marked in Fig. E.11. The payload advertised by the beacon is in 8.8 Fixed-Point Format (FPF).

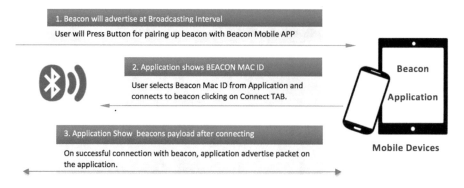

Fig. E.10 Shipment Beacon and NRF Connect App interfacing

Fig. E.11 Shipment
Beacon Advertisement on
Mobile APP

The FPF can be converted at the Raspberry Pi side by using the following code:

```
float mTemp = Measured Temperature;
uint8_t firstByte = (uint8_t) mTemp;
Uint8_t second Byte = (uint8_t) (mTemp x 256);
```

```
//! Example:
//! Temperature of 30.81 is sent as 0x1E and 0xCF in
      8.8 Fixed-Point Format

mTemp = 30.81
firstByte = (uint8_t)30.81 = 30 = 0x1E
secondByte = 30.81 * 256 = 7887.36 = (uint8_t) 0
      x1ECF = 0xCF
```

Raspberry Pi and Microsoft Azure IoT Hub Connection

Implementation of Microsoft Azure IoT Hub was a simple process to setup. A Microsoft account is required to utilize any Azure services including Azure IoT Hub. Upon making an account and setting a service subscription, several IoT hub resource groups were made to accommodate the devices needed to connect to the IoT server. Each device is provided a private connection string which can be used to authenticate secure communication to the server. Shared Access Signature tokens were also generated using the Azure IoT Explorer tool (on Windows) to authenticate communication between device and server. Additionally the Azure IoT tool included telemetry to monitor device-to-cloud messages.

Configuration of Raspberry Pi for connection with Azure IoT Hub

- Get the connection string from the Azure IoT Hub page after adding the device to Azure platform.
- Add the connection string mentioned on the Azure Portal to establish connection with the Azure platform to the python script.
- Connect the device client (Fig. E.12).

- `conn_str = "HostName=5G-IoT-System-For-Emergency-`
- `     Responders.azure-devices.net;DeviceId=`
- `     application_device1;SharedAccessKey=`
- `     x84oYfc8Wm41L7nfMzNm87X7YmFbC+TtHX4ny+bV8ck="`
- `     device_client = IoTHubDeviceClient.`
- `     create_from_connection_string(conn_str)`
- `      # Connect the device client.`
- `     await device_client.connect()`

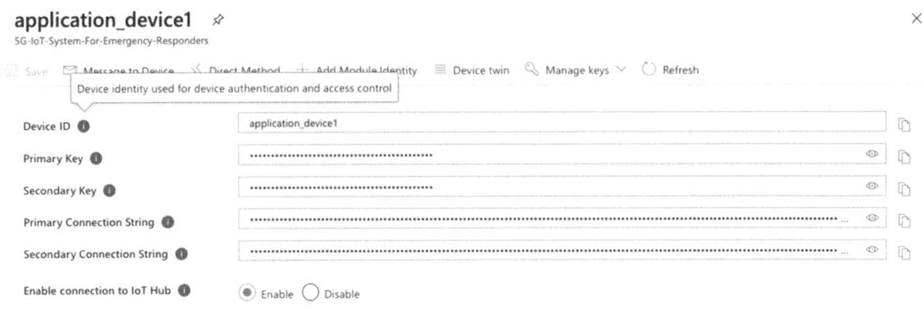

application_device1 ✗
5G-IoT-System-For-Emergency-Responders

×

⬚ Save   ✉ Message to Device   ✗ Direct Method   + Add Module Identity   ☰ Device twin   ✎ Manage keys ∨   ↻ Refresh

Device identity used for device authentication and access control

Device ID ⓘ	application_device1
Primary Key ⓘ	••
Secondary Key ⓘ	••••••••••••••••••••••••••••••••••••
Primary Connection String ⓘ	••
Secondary Connection String ⓘ	•••
Enable connection to IoT Hub ⓘ	⦿ Enable ◯ Disable

Fig. E.12 Application Device creation on Azure platform

Real-Time Sensor Data Transfer

The Raspberry Pi collects the data from multiple shipment beacons via the BLE communication and parses the data to send the temperature, humidity, pressure, and ambient light parameters to the Azure IoT Hub.

Configuration of Raspberry Pi for Environmental Data Transmission to Azure Hub

- Get the data from the BLE beacons.
- Use the python script to parse the sensor data.
- Add the connection string mentioned on the Azure Portal to establish connection with the Azure platform.
- The script periodically captures the data in real time.
- The captured data is transmitted to the Azure IoT Hub for every 2 s.

```
while count < 45:
        raw_bytes_array = sim_data_raw_data[idx]
        pressure_lsb = raw_bytes_array[12]
        pressure_msb = raw_bytes_array[11]
        curr_pressure = (pressure_msb * 256 +
    pressure_lsb) #Pascal

        humidity_lsb = raw_bytes_array[9]
        humidity_msb = raw_bytes_array[10]
        curr_humidity = ((humidity_msb*256) + (
    humidity_lsb)) #g/kg

        temp_lsb = raw_bytes_array[5]
        temp_msb = raw_bytes_array[6]
        curr_temp = (temp_msb * 256 + temp_lsb)/100
```

```
#Celcius

        curr_light = False
        light_msb = raw_bytes_array[8] if light_msb >= 1:
            curr_light = True

        message_properties = {}
message_properties["deviceId"] = "Shipment
    -05"
        message_properties["pressure"] =
curr_pressure
        message_properties["ambientLight"] =
curr_light
        message_properties["temperature"] =
curr_temp
        message_properties["humidity"] =
curr_humidity

        if curr_temp > threshold_temp :
            print ("WARNING: Vaccine shipment temp
    exceeded set threshold!!!")

            message_properties['Warning'] = "Temp
Exceeded"

        if curr_pressure > threshold_pressure :
            print ("WARNING: Vaccine shipment temp
    exceeded set threshold!!!")
          message_properties['Warning'] = "
    Pressure Exceeded"

        if curr_humidity > threshold_humidity:
            print ("WARNING: Vaccine shipment temp
    exceeded set threshold!!!")
            message_properties['Warning'] = "
    Humidity Exceeded"

      msg = Message(json.dumps(message_properties)
    )
      msg.message_id = uuid.uuid4()
      msg.content_type = 'application/json'
```

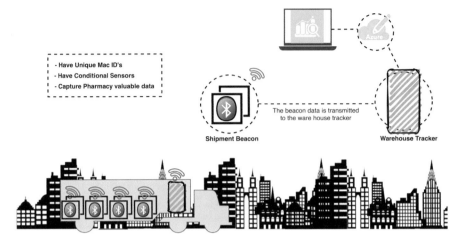

Fig. E.13 Real-time vaccine monitoring

```
    await device_client.send_message(msg)
    print("Message successfully sent:" + str(msg
))
    print()

    count = count + 1
    time.sleep(2)
```

System Integration

The shipment beacons each have unique MAC address and fitted with condition sensors which collects valuable data for the vaccines. The data is transmitted to the warehouse trackers which contains the Raspberry Pi. The warehouse tracker accesses the hot-spot via Internet service. This can be used in transit, as a part of shipment or in warehouses. The warehouse trackers transmit the real-time data from the shipment beacon to the azure IoT hub via MQTT protocol. It improves the inventory forecasting and keeping projects on track by monitoring vaccines in use and vaccines that are idle sitting in the cold chain environment.

As shown in Fig. E.13, the shipment beacons can be stick on the surface of the containers or multi-modal simple affix the beacon to the cases and the pallets that you are shipping and place the portable wire-free warehouse tracker in the shipment truck. The sensor data is uploaded as frequently as every 5 min to the azure cloud. So, that user can take the action when it counts to protect your vaccines with real-time fore-sights and insight about temperature. This project mainly concerns the temperature excursions and the information about the hot-spot inside the shipment truck.

Fig. E.14 Passive vaccine monitoring mode

In the current scenarios, the vaccine shipment for the COVID. It can be used with great advantages. Typically medicines, including COVID and cancer vaccines are shipped in a cold chain temperature range with stringent requirement of maintaining 2–8 °C. Any anomaly in the temperature during the cold chain shipment will lead to spoilage of the product and can adversely affect the patients' life. Hence, real-time condition monitoring and the passive monitoring mode are useful for various vaccines condition monitoring.

Although active monitoring allows us to track the shipment/condition in real time it is bound by a dependency on a hot-spot/back-haul device. The passive monitoring mode enables device to store data over the duration of the shipment to be retrieved later directly through the APP. If the warehouse tracker is unable to communicate in real-time due to the overseas shipment of vaccines. The condition data stored in the shipment beacons and automatically transmitted or pulled by the user by a mobile app. This provides the zero touch data upload feature from the datalogger at any point in transit or upon the arrival. This guarantees no loss of data over the shipment which helps make a decision regarding the state of the shipment (Fig. E.14).

Message successfully sent:{"deviceId": "Shipment-05", "pressure": 1016, "ambientLight": false, "temperature": 20.13, "humidity": 74}
Message successfully sent:{"deviceId": "Shipment-05", "pressure": 1015, "ambientLight": false, "temperature": 20.15, "humidity": 71}
Message successfully sent:{"deviceId": "Shipment-05", "pressure": 1017, "ambientLight": true, "temperature": 20.0, "humidity": 80}
Message successfully sent:{"deviceId": "Shipment-05", "pressure": 1018, "ambientLight": true, "temperature": 20.11, "humidity": 68}
Message successfully sent:{"deviceId": "Shipment-05", "pressure": 1007, "ambientLight": true, "temperature": 20.09, "humidity": 63}
Message successfully sent:{"deviceId": "Shipment-05", "pressure": 1006, "ambientLight": true, "temperature": 20.04, "humidity": 64}
Message successfully sent:{"deviceId": "Shipment-05", "pressure": 1012, "ambientLight": false, "temperature": 20.14, "humidity": 72}
Message successfully sent:{"deviceId": "Shipment-05", "pressure": 1016, "ambientLight": false, "temperature": 20.13, "humidity": 74}
Message successfully sent:{"deviceId": "Shipment-05", "pressure": 1015, "ambientLight": false, "temperature": 20.15, "humidity": 71}
Message successfully sent:{"deviceId": "Shipment-05", "pressure": 1017, "ambientLight": true, "temperature": 20.0, "humidity": 80}
Message successfully sent:{"deviceId": "Shipment-05", "pressure": 1018, "ambientLight": true, "temperature": 20.11, "humidity": 68}
Message successfully sent:{"deviceId": "Shipment-05", "pressure": 1007, "ambientLight": true, "temperature": 20.09, "humidity": 63}

Fig. E.15 Application Device Terminal data

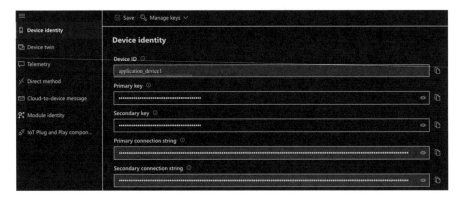

Fig. E.16 Azure explorer connection and services

Testing and Verification

1.1.2 Real-Time Environmental Sensor Testing

The python script parses the temperature, humidity, pressure, and ambient light and constructs a message before sending it to the Azure IoT Hub. The message is printed out in the terminal and verifies w.r.t. to the logs received from the BLE beacons. As shown in Fig. E.15 we are packaging the message along with Device ID for easy processing at later stage.

Real-Time Azure IoT Hub Integration Testing

The connection is established with the Azure Explorer using the connection string for the Application device to verify the messages sent to the platform.

Figure E.16 shows the application device connection with the explorer and the services offered by the explorer.

Fig. E.17 Telemetry connection

The telemetry services offered by the explorer are utilized to visualize the data in real time as shown in Fig. E.17.

Click the start button highlighted in the mentioned Fig. E.17 to start receiving the messages. The data in shown in Fig. E.18.

Evaluation Methodology and Results

1.1.3   Real-Time Sensor Evaluation

The sensor data to the Azure IoT Hub is verified by sending data to the Hub for a long period of time. Periodical verification is done on the explorer to verify the messages being sent. As shown in Fig. E.19 we have sent more than 2000 messages to evaluate the real-time data transfer.

Bluetooth Low Energy Range Evaluation Testing

In BLE-based projects range testing is very important. Range testing depends on the RSSI value of the broadcasting beacons from the device. RSSI stands for Received Signal Strength Indicator. It is the strength of the beacon's signal as seen on the receiving device, e.g., a smartphone. The signal strength depends on distance and broadcasting power value. At maximum broadcasting power (+4 dBm) the RSSI ranges from -26 (a few inches) to -100 (40–50 m distance). Figure E.20 shows the graph of distance vs. RSSI for one shipment beacon. The data for the range testing is plotted in Table E.2 which shows the two RSSI readings at the same distance to make sure the consistency of the results.

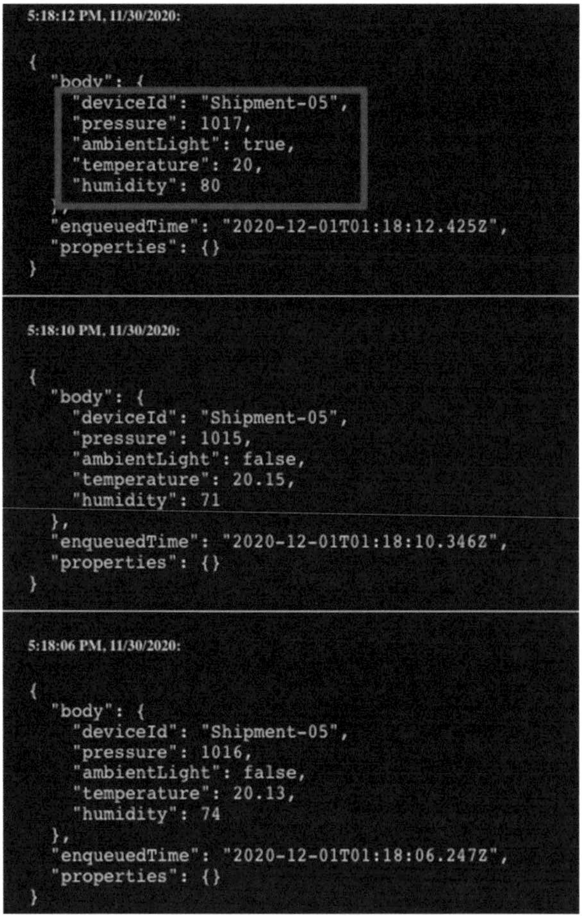

Fig. E.18 Real-time data
transfer to Azure IoT

5:18:12 PM, 11/30/2020:

{
  "body": {
    "deviceId": "Shipment-05",
    "pressure": 1017,
    "ambientLight": true,
    "temperature": 20,
    "humidity": 80
  },
  "enqueuedTime": "2020-12-01T01:18:12.425Z",
  "properties": {}
}

5:18:10 PM, 11/30/2020:

{
  "body": {
    "deviceId": "Shipment-05",
    "pressure": 1015,
    "ambientLight": false,
    "temperature": 20.15,
    "humidity": 71
  },
  "enqueuedTime": "2020-12-01T01:18:10.346Z",
  "properties": {}
}

5:18:06 PM, 11/30/2020:

{
  "body": {
    "deviceId": "Shipment-05",
    "pressure": 1016,
    "ambientLight": false,
    "temperature": 20.13,
    "humidity": 74
  },
  "enqueuedTime": "2020-12-01T01:18:06.247Z",
  "properties": {}
}

However, there is no standard application available for testing the BLE range. For testing range in meters, we have connected the BLE beacon with the NRF application and they started walking away from the beacon until it lost the connection. Apart from that we have done the different casings testing, by keeping the 2 BLE devices, i.e., beacon with the BLE-based speakers at 2 m apart, and then checked the RSSI values in the NRF application.

Summary, Conclusions, and Recommendations

The proposed Warehouse Inventory Management presented several implementations toward improving the efficiency of the communication networks currently employed in warehouse management. The key implementations that the system

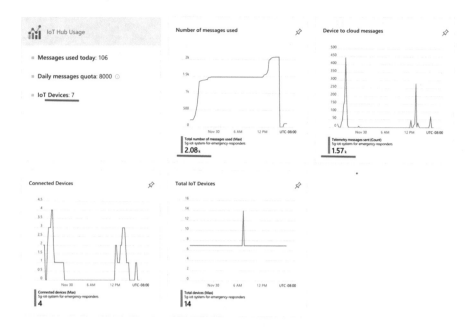

Fig. E.19 Sending data to the Azure IoT Hub for a long time

introduces are device-to-cloud communication, real-time monitoring of environmental (temperature, humidity, light, pressure) sensors and remote management of application device is done by Azure IoT platform. All data from each IoT shipment beacon was simultaneously sent to the Azure IoT Hub in real time. All features were successfully implemented, tested, and evaluated by recording numerous trials of operating each device individually and as a complete integrated system. These features would not only improve timely communication between the shipment beacon and warehouse tracker in real time for better inventory management, but it would also reduce the manual effort required in maintaining the inventory and tracking the lost package.

In future live-video streaming can be implemented on the Application device for better real-time monitoring. BLE beacons have shorter connectivity range, one can look out for long-connectivity network technology.

Table E.2 RSSI Readings

Distance	RSSI (BLE)
0 m	−35
0 m	−38
2 m	−55
2 m	−64
3.2 m	−55
3.2 m	−56
8 m	−64
8 m	−66
12 m	−81
12 m	−74
17 m	−81
17 m	−74
23 m	−76
23 m	−78
32 m	−75
32 m	−79
40 m	−92
40 m	−92

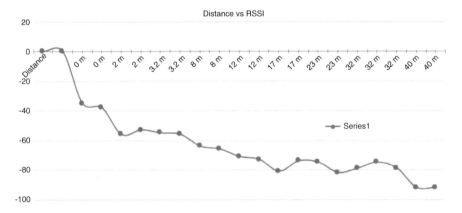

Fig. E.20 BLE range evaluation testing

References

1. L. Chettri and R. Bera, "A Comprehensive Survey on Internet of Things (IoT) Toward 5G Wireless Systems," in IEEE Internet of Things Journal, vol. 7, no. 1, pp. 16-32, Jan. 2020.
2. Whitmore, A., Agarwal, A. Da Xu, L. The Internet of Things—A survey of topics and trends. Inf Syst Front 17, 261–274 (2015). https://doi.org/10.1007/s10796-014-9489-2.

3. Grant, Svetlana (September 1, 2016). "3GPP Low Power Wide Area Technologies - GSMA White Paper" (PDF). gsma.com. GSMA. p. 49. Retrieved October 17, 2016.
4. J. M. Khurpade, D. Rao and P. D. Sanghavi, "A Survey on IoT and 5G Network," 2018 International Conference on Smart City and Emerging Technology (ICSCET), Mumbai, 2018, pp. 1-3.
5. F. Khan, Z. Pi, and S. Rajagopal, "Millimeter-wave mobile broadband with large scale spatial processing for 5G mobile communication," inProc. 50th Annu. Allerton Conf. Commun. Control Comput. (Allerton), 2012, pp. 1517–1523.
6. H. S. Ma, E. Zhang, S. Li, Z. Lv, and J. Hu, "A V2X Design for 5G Network Based on Requirements of Autonomous Driving," SAE Technical Paper Series, Sep. 2016.
7. E. K. Markakis et al., "Efficient Next Generation Emergency Communications over Multi-Access Edge Computing," in IEEE Communications Magazine, vol. 55, no. 11, pp. 92-97, Nov. 2017.
8. Natallia Sakovich. "The Importance of Data Collection in Healthcare," Sam Solutions, April 9, 2019. [Online].
9. S. Salkic, B.C. Ustundag, T. Uzunovic, and E. Golubovic, "Edge Computing Framework for Wearable Sensor-Based Human Activity Recognition," Lecture Notes in Networks and Systems, pp. 376–387, Jul. 2019.
10. G. Manogaran, P. Shakeel, H. Fouad, Y. Nam, S. Baskar, N. Chilamkurti, and R. Sundarasekar, "Wearable IoT Smart-Log Patch: An Edge Computing-Based Bayesian Deep Learning Network System for Multi Access Physical Monitoring System," Sensors, vol. 19, no. 13, p. 3030, Jul. 2019.
11. M.M. Shurman and M.K. Aljarah, "Collaborative execution of distributed mobile and IoT applications running at the edge," 2017 International Conference on Electrical and Computing Technologies and Applications (ICECTA), Nov. 2017.

Appendix F: IoT Fumigation Robot

Jesus De Haro and Nicholas Kaiser
Under the Supervision of Professor Ammar Rayes

Abstract The goal of this project was to build a robot capable of carrying a fumigation apparatus and dispensing its chemicals, an Android application to control the robot's movement and dispensation of the fumigation chemicals, and a mock fumigation apparatus to imitate a real fumigation apparatus. The motivation for this project was to improve the safety of fumigation workers by building a robot to perform this task. This allows fumigation workers to perform their job at a safe distance by controlling the movement of the robot and dispensation of the fumigation chemicals via an Android application, while the robot carries the fumigation apparatus. The overall goal was accomplished successfully, resulting in a mock fumigation apparatus and a robot chassis capable of carrying a fumigation apparatus and dispensing its chemicals with the robot's movement and chemical dispensation capability controlled via an Android application.

Keywords IoT, Fumigation, Raspberry Pi, Robot

Introduction: Written by Nicholas K

The purpose of this project was to build a robot capable of carrying a fumigation apparatus and dispensing its chemicals via an Android application in order to reduce the exposure of fumigation workers to harsh chemicals. Additionally, a mock fumigation apparatus was built in order to emulate the functionality of a real fumigation

J. De Haro · N. Kaiser
Computer Engineering Department, College of Engineering, San Jose State University, San Jose, CA, USA
e-mail: jesus.deharodereza@sjsu.edu; nicholas.kaiser@sjsu.edu

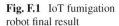

Fig. F.1 IoT fumigation
robot final result

apparatus, which is a costly piece of equipment that we could not obtain for this
project (Fig. F.1).

Fumigation is a method of pest control that completely fills an area with gaseous
pesticides to suffocate or poison the pests within [1]. The pesticides are also
extremely harmful to the human workers who are exposed to them during the dispensation process [2]. This project aims to improve the health and safety of these
fumigation workers while still allowing them to perform fumigation services.

The first technical goal of this project was to interface an L298n motor driver, a
Raspberry Pi 3B+, a servo motor, and two 12 V DC motors, and to actuate the
motors by issuing commands from an Android application connected to the
Raspberry Pi over a Wi-Fi interface. This technical goal was accomplished by completing several objectives. The first objective was to design and implement the interface between all electrical components in order to actuate the motors. The second
objective was to design and implement an Android application capable of sending
movement direction commands to control the DC motors, and rotation direction
commands to control the servo motors. The third objective was to establish communication between the Android application and Raspberry Pi in order to actuate
the motors using the Android application.

The second technical goal of this project was to interface a 5 V 2 Channel relay,
a 12 V submersible water pump, and a Raspberry Pi 3B+ in order to be able to pump
water by turning the water pump on and off by issuing commands from an Android
application connected to the Raspberry Pi over a Wi-Fi interface. This technical
goal was accomplished by completing the following objective. The fourth objective
was to design and implement the interface between all electrical components in
order to turn the water pump on and off. The interface between the Raspberry Pi and
Android app will already be set up in the previous technical goal and so the water
pump circuitry can simply be added to the existing system.

The functional goal of this project was to design a robot chassis capable of carrying a fumigation apparatus and pulling or releasing its trigger in order to control
the dispensation of fumigation chemicals. This functional goal was accomplished

by completing several objectives. The fifth objective was to design and build a chassis to carry a fumigation apparatus, and hold the servo motor in place so that it can pull or release the trigger on the fumigation apparatus. The sixth objective was to integrate the circuitry onto the chassis in order to drive the wheels of the robot, therefore controlling the robot's movement direction.

Methodology: Written by Nicholas K

This section discusses the objectives, challenges, problem formulation, and design of the entire project in more detail.

Objectives

- The first objective was to design and implement the interface between the L298n motor driver, two 12 V DC motors, servo motor, and Raspberry Pi in order to actuate all motors in a specific way. The two DC motors were required to do the following; both rotate clockwise, both rotate counter-clockwise, and both rotate in opposite directions of each other. The servo motor was required to rotate clockwise 180°, and rotate counter-clockwise 180°.
- The second objective was to design and implement an Android application with buttons in order to move the chassis left, right, forward, and reverse, and buttons in order to turn the servo motor 180° clockwise or counter-clockwise.
- The third objective was to establish communication between the Android application and Raspberry Pi so that the DC and servo motors could be actuated based on the button pressed in the Android application.
- The fourth objective was to design and implement the interface between the 12 V submersible water pump, 5 V 2 channel relay, and Raspberry Pi in order to turn the water pump on and off at the touch of a button on our Android application.
- The fifth objective was to design and build a sturdy but lightweight chassis that was capable of carrying a fumigation apparatus, and pulling or releasing its trigger in order to dispense the chemicals stored inside the fumigation apparatus.
- The sixth objective was to integrate the Raspberry Pi, motor driver, and motors onto the chassis in order to drive the robot left, right, forward, reverse, and to pull or release the fumigation apparatus's trigger with the servo motor mounted onto the chassis.

Challenges

- The first challenge encountered was finding the right size motors that would support our weight requirements. Typical 3.3 V or 5 V DC hobby motors were not capable of supporting the weight of our relatively heavy wooden chassis, let alone a fumigation apparatus in addition to the weight of the chassis. Larger motors have greater power requirements and so our options were limited since the motors had to be able to run on a portable battery due to the nature of our project. Fortunately, we were able to find some lower RPM rated 12 V DC motors that would fit our power requirement limitations, but could still support the weight of our chassis and fumigation apparatus.

• The second challenge encountered was finding a suitable, but inexpensive battery and battery charger for our two DC motors. Due to the power requirements of our motors, a power source that could supply enough current for a reasonable amount of time was on the expensive side. We ended up sacrificing battery life for affordability since our project was a prototype and a smaller mAh power supply would get the job done for demonstration purposes. We ended up going with a smaller mAh battery and battery charger which helped keep the cost of this project down.

• The third challenge encountered was mounting the two DC motors onto the wheels. The types of wheels that small DC motors (12 V or less) are designed to "plug into" were not suitable for our project. We needed taller wheels to suspend our chassis a sufficient height off of the ground, and we needed stronger wheels that could support the weight of a fumigation apparatus in addition to the chassis weight. Due to these requirements, we had to use 7-in. plastic wheels which cannot be directly mounted onto the shaft of our DC motors. We had to design our own hub that would attach to the DC motor shaft on the one end, and attach to the wheel on the other end. The wheels have thick bits of plastic placed in inconvenient places, which made it hard to center the hub onto the wheel.

Problem Formulation and Design

The design of the entire system consists of a Raspberry Pi 3B+ and its power supply, L298n motor driver, a servo motor, a 5 V 2 channel relay, two 12 V DC motors, and a LiPo battery to power the DC motors and 12 V submersible water pump, all mounted inside a wooden box-shaped chassis and a mock fumigation apparatus made out of PVC pipe, plastic tubing, and cardboard. Additionally, an Android phone is connected to the Raspberry Pi over a Wi-Fi interface. The overall system design was broken down into several smaller modules that were tested individually in order to ensure correct functionality upon integration into the full system.

The first module is the Raspberry Pi—Android application module, which requires the Raspberry Pi to produce the specific outputs shown in Table F.1 based on the commands it receives from the Android application over a Wi-Fi interface.

The second module is the Raspberry Pi—motor driver—DC motors module, which requires both DC motors to produce the outputs shown in Table F.2 based on the input that the motor driver receives from the Raspberry Pi, and requires the servo motor to produce the outputs also shown in Table F.2 based on the input it receives from the Raspberry Pi over a PWM interface.

The full system design was created by merging the two separate modules. The system design requires the motors to produce the outputs show in Table F.3 based on the commands that the Raspberry Pi receives from the Android application over the Wi-Fi interface.

Table F.1 Raspberry Pi output with Android App input

Android App Button pressed	Raspberry Pi output
Left	Rotate chassis left
Right	Rotate chassis right
Forward	Move chassis forward straight
Reverse	Move chassis reverse straight
Stop	Halt all movement
Open	Pull fumigation apparatus trigger and turn water pump ON
Close	Release fumigation apparatus trigger and turn water pump OFF

Table F.2 Motor output with Raspberry Pi input

Raspberry Pi command	Motor output
Rotate chassis left	L DC motor: turn clockwise R DC motor: turn clockwise
Rotate chassis right	L DC motor: turn counter-clockwise R DC motor: turn counter-clockwise
Move chassis forward straight	L DC motor: turn counter-clockwise R DC motor: turn clockwise
Move chassis reverse straight	L DC motor: turn clockwise R DC motor: turn counter-clockwise
Halt all movement	L DC motor: don't turn R DC motor: don't turn
Pull fumigation apparatus trigger	Servo motor: rotate 180° clockwise
Release fumigation apparatus trigger	Servo motor: rotate 180° counter-clockwise

Table F.3 Motor output with Android App input

Android App Button pressed	Motor output
Left	L DC motor: turn clockwise R DC motor: turn clockwise
Right	L DC motor: turn counter-clockwise R DC motor: turn counter-clockwise
Forward	L DC motor: turn counter-clockwise R DC motor: turn clockwise
Reverse	L DC motor: turn clockwise R DC motor: turn counter-clockwise
Stop	L DC motor: don't turn R DC motor: don't turn
Open	Servo motor: rotate 180° clockwise Water pump: turn ON
Close	Servo motor: rotate 180° counter-clockwise Water pump: turn OFF

Implementation

This section covers the hardware and software design of the entire project in more detail.

Chassis Design: Written by Nicholas K

The fumigation robot's chassis consists of a larger wooden box and a smaller wooden box stacked on top of each other. The smaller wooden box is where the robot's circuitry is placed, and the larger wooden box is where the fumigation apparatus is placed. A cutout was made in the bottom of the larger wooden box so that the smaller wooden box could be nailed to the bottom of the cutout in order to secure it in place. A swivel wheel was attached to the rear of the chassis in order to prevent the back end of the larger wooden box from dragging on the floor (Figs. F.2, F.3, and F.4).

In order to mount the DC motors on the chassis, two mounting brackets were screwed onto the bottom of the larger wooden box. The DC motors were fastened to the mounting brackets by screwing them into the brackets using the screw holes present on the faceplate of the DC motors (Fig. F.5).

In order to mount the wheels onto the DC motor shafts, a mounting hub was fastened to the one side of a bottle cap, and a wheel was fastened to the other side of the bottle cap. The drive shafts of each DC motor were inserted into a mounting hub, and then screws were used in order to keep the motor shafts from slipping out of the mounting hubs (Fig. F.6).

Mock Fumigation Apparatus Design: Written by Nicholas K

The mock fumigation apparatus consists of a cardboard box with cutouts to insert a PVC pipe, and a cutout for the water pump's power wires. The PVC pipe is secured to the cardboard box by screwing long screws into the pipe onto each side of the cardboard box in order to prevent the pipe from slipping out. The PVC pipe also has a hole drilled into it about half way in order to route the tubing that is connected to the water pump through the pipe (Figs. F.7 and F.8).

A plastic container was placed inside the cardboard box in order to hold the pump and the water that the pump is submerged in. The plastic container has a hole drilled into the center of its lid in order to route the tubing through the PVC pipe, and a hole drilled in the corner of its lid in order to route the water pump's power wires through the container.

The servo motor was adhered to the front of the mock fumigation apparatus right below the PVC pipe. On a real fumigation apparatus, this is where the trigger would be located. Our mock fumigation apparatus does not have a trigger and so the servo motor was mounted here for proof-of-concept purposes (Fig. F.9).

Fig. F.2 Chassis
front view

Fig. F.3 Chassis top view

Hardware Design: Written by Jesus D

The processor used in the robot is a Raspberry Pi 3B. This hardware was chosen for its capabilities and the quantity of resources and support that can be found on the Internet. This board contains more GPIO pins than required to complete this project,

Fig. F.4 Chassis
bottom view

Fig. F.5 Chassis with
wheels bottom view

Fig. F.6 Wheel mounted
onto DC motor

Fig. F.7 Mock fumigation
apparatus side view

with pulse-width modulation (PWM), and has Wi-Fi capabilities for remote communication for the IoT project. Of the 27 accessible GPIO pins, only 7 are used to control the other components.

One PWM pin goes to the servo motor to control the direction and speed of rotation, which could be considered the amount of pressure applied to the fumigation apparatus, controlling the quantity of spray that is expelled. The other four pins are connected to the motor driver's four "In" pins. Pins In1 and In2 are for controlling DC motor 1 with the driver's Out1A and B output pins. Pins In3 and In4 control DC motor 2 with pins Out2A and B.

For this architecture, two separate power sources are required, a 5-V power source to power the Raspberry Pi, servo motor, and the relay-board, and a 12-V power supply for the motor driver, the two DC motors, and the water pump. The smaller supply's voltage would not be sufficient to drive the motors and water pump,

Fig. F.8 Mock fumigation
apparatus back view

Fig. F.9 Mock fumigation
apparatus inside view

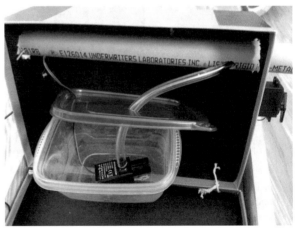

hence a larger 12-V supply was required. The 12 V are also supplied at the normally open pin of relay 1 (NO1), and ground is placed at the normally closed pin (NC2) of relay 2. Both common pins of the relay board, COM1 and COM2, are connected to the pump's power and ground wires, respectively. A diagram of the described architecture is shown in Fig. F.10.

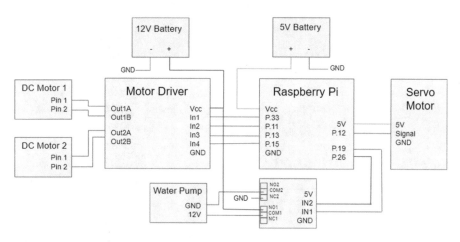

Fig. F.10 Hardware architecture of the fumigation robot

With this hardware design, we are able to control the robot by moving forward, in reverse, or rotating the body in clockwise or counterclockwise fashion. The servo motor is also controlled by making it rotate clockwise and counterclockwise. Finally, the water pump used to dispense the "pesticide" is activated by setting both relays in the closed configuration.

Software Design: Written by Jesus D

Software flowchart for this project is shown in Fig. F.11. The robot's programming was done in Python, and is only a few steps. The code for this project was developed referencing RootSaid [3] and Instructables Circuits [4]. When the code is executed, it starts by initializing the socket for communicating over a Wi-Fi network, then five GPIO pins (pins 11, 12, 13, 15, 33, 35, and 37) are initialized as outputs. PWM is enabled on pin 12 for use in controlling the rotational speed and direction of the servo motor.

After initializing Wi-Fi communication and GPIO pins, the Raspberry Pi now waits for input from the user that is sending commands via the Android app "RootSaid—WiFi Command Center," which can be downloaded from the Google Store. The app itself contains three tabs. In this project we only use two of them. One is for setting up the IP address and port for communication between the app and Raspberry Pi. The second tab is used for sending commands. The third tab is for powering on/off smart appliances, which is not relevant to this project. Once the user has entered the IP address and port number, the link symbol to the left must be pressed for the app to know where it will be sending data. With this app, we send the Raspberry Pi seven different commands: "forward," "backward," "left," "right," "stop," "action 1," or "action2." The two tabs that are used in RootSaid app are shown in Fig. F.12.

Fig. F.11 Software
flowchart

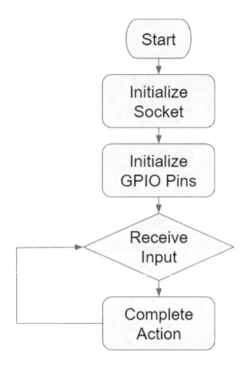

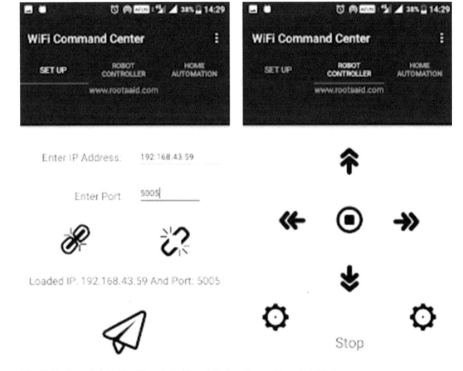

Fig. F.12 RootSaid's Set-Up tab (left) and Robot Controller tab (right)

To be able to execute the code, we needed to import Python libraries into our code: RPi.GPIO and Socket. RPi.GPIO is a standard library used in Raspberry Pi to set up the GPIO pins and its peripherals. The pins are configured as an output and they are controlled with APIs. With this library, the PWM output is also controlled. With the socket library, we established communication over Wi-Fi for the Raspberry Pi to receive commands from the Android application. The source code can be viewed in Appendix A.

Functions were created to make the code easier to read with what the pin output results in. For example, if the input to In2 is high and In1 is low, this will result in the left motor to push the robot forward. Therefore, functions were named left_motor_forward(), right_motor_forward(), left_motor_reverse(), right_motor_reverse(), spray_on() and spray_off(), and motors_stop() to make the code easier to read.

Testing and Verification: Written by Jesus D

The first test was establishing and verifying communication between the mobile phone via RootSaid—Wi-Fi Command center app and the Raspberry Pi over the Wi-Fi network. Connecting the Pi to a monitor and using the "ipconfig" command, we can acquire the IP address to enter in the mobile app's "Setup" tab. Running the same code shown in step 7 of RootSaid [3], print statements for each command can be used to verify Wi-Fi communication between the app and robot, which will be essential for the following tests.

Next, to test the DC motors, the motor driver (L298N), 12-V battery, Wi-Fi network, RootSaid android app, and the Raspberry Pi were required. Making the connections from the controller to the motor driver and DC motors shown in Fig. F.6, we used the code from the previous test-verification. At each directional command, the behavior was observed for comparison with the described behavior in Table F.3 for verification.

Testing the water pump required the relays, 12-V battery, Raspberry Pi, water for the pump, and code allowing control of the relays to better understand the configuration needed to activate the pump. To test the parts, they were connected as shown in Fig. F.6. The pump was placed in a container with enough water to be above the impeller. Once everything is connected and enough water is supplied, relays 1 and 2 are opened and closed to find the combination that activates the pump. The testing python script, shown in Appendix B, accepts the relay number to be controlled and configuration to be set. After sending the configuration command, the current state of each relay would be displayed to note each relay's state and the water pump's action, as shown in Fig. F.13. With this, the output values of the pins required to set each relay as closed are noted and properly implemented in the fumigation robot's code.

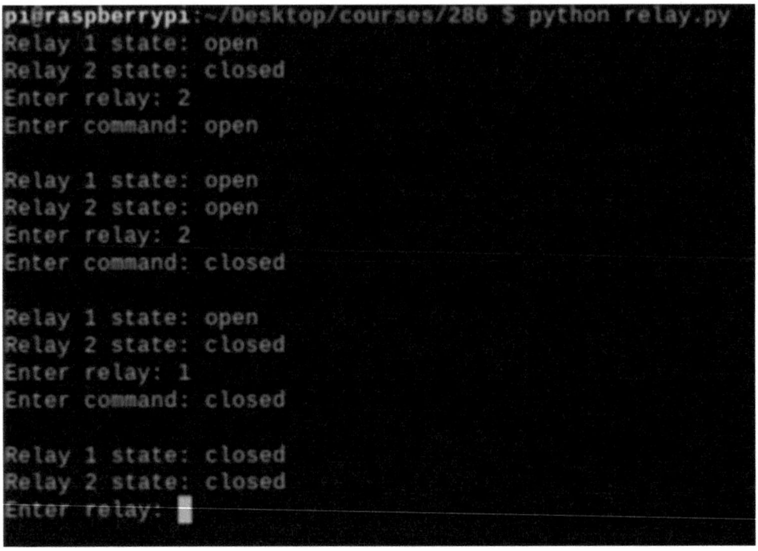

Fig. F.13 Relay testing script output

Conclusion: Written by Nicholas K

The first technical goal of this project was to interface an L298n motor driver, a Raspberry Pi 3B+, a servo motor, and two 12 V DC motors, and to actuate the motors by issuing commands from an Android application connected to the Raspberry Pi over a Wi-Fi interface. The second technical goal of this project was to interface a 12 V submersible water pump, a Raspberry Pi 3B+, and a 5 V 2 channel relay in order to turn the water pump on and off by issuing commands from an Android application. The purpose of this project was to integrate the aforementioned circuitry onto a robot chassis in order to build a robot capable of carrying a fumigation apparatus and dispensing its chemicals via a user operated Android application in order to reduce the exposure of fumigation workers to harsh chemicals. The primary objectives of this project were to first design and implement the interface between all electrical components in order to actuate the motors. The second objective was to design and implement an Android application capable of sending movement direction commands to control the DC motors, and rotation direction commands to control the servo motors. The third objective was to establish communication between the Android application and Raspberry Pi in order to actuate the motors using the Android application. The fourth objective was to design and implement the interface between all electrical components in order to turn the water pump on and off. The fifth objective was to design and build a chassis to carry a fumigation apparatus, and hold the servo motor in place so that it can pull or release the trigger on the fumigation apparatus. The sixth and final objective was to integrate the circuitry onto the chassis in order to drive the wheels of the robot, therefore controlling the robot's movement direction. The overall goals and individual objectives were all completed successfully, rendering an IoT fumigation robot that improves the health and safety of fumigation workers while still allowing them to perform fumigation services. Lessons learned throughout the entire duration of this project consisted of gaining a better understanding of the Wi-Fi IoT protocol, becoming more familiar with the Raspberry Pi embedded hardware platform as well as the Python programming language, and learning the basics of DC and servo motors. Further improvements can be made to this project, such as designing a better hub so that the wheels are mounted to the DC motors more securely, and designing and building a larger chassis such that the size and weight of a full-size fumigation apparatus can be supported.

Acknowledgment We would like to thank San Jose State University for providing lots of space for us to conduct drive tests with our robot.

References

1. Bessin, R. B. (2018, November 30). *Fumigation Safety*. Kentucky Pesticide Safety Education. https://www.uky.edu/Ag/Entomology/PSEP/fumsafety.html
2. Department of Consumer Affairs Structural Pest Control Board. (2019). *Questions & Answers About Fumigation*. https://www.pestboard.ca.gov/forms/fumigate.pdf
3. "WiFi Controlled Robot using Raspberry Pi – Android Controlled ...". [Online]. Available: https://rootsaid.com/robot-control-over-wifi/. [Accessed: 02-May-2021].
4. Williamwaw and Instructables, "IOT Water Pistol/plant Waterer," Instructables, 04-Jun-2019. [Online]. Available: https://www.instructables.com/OK-Google-Water-PistolPlant-Waterer/. [Accessed: 23-Apr-2021].

Appendix A: Source Code

```
import RPi.GPIO as GPIO
import socket
import time

#UDP_IP = "192.168.0.14"
UDP_IP = "192.168.43.40"
UDP_PORT = 5050


pin_IN1 = 33
pin_IN2 = 11
pin_IN3 = 13
pin_IN4 = 15
pin_SERVO = 12
```

```python
relay1 = 35
relay2 = 37

GPIO.setwarnings(False)
GPIO.setmode(GPIO.BOARD)
GPIO.setup(pin_IN1, GPIO.OUT) # Pins 33 and 11 for left motor (IN1
and IN2, respectively)
GPIO.setup(pin_IN2, GPIO.OUT)
GPIO.setup(pin_IN3, GPIO.OUT) # Pins 13 and 15 for right motor (IN3
and IN4, respectively)
GPIO.setup(pin_IN4, GPIO.OUT)

GPIO.setup(relay1, GPIO.OUT)
GPIO.setup(relay2, GPIO.OUT)

GPIO.output(relay1, True)
GPIO.output(relay2, True)

def left_motor_forward():
    GPIO.output(pin_IN2, True)
    GPIO.output(pin_IN1, False)

def left_motor_reverse():
    GPIO.output(pin_IN2, False)
    GPIO.output(pin_IN1, True)

def right_motor_forward():
    GPIO.output(pin_IN4, True)
    GPIO.output(pin_IN3, False)

def right_motor_reverse():
    GPIO.output(pin_IN4, False)
    GPIO.output(pin_IN3, True)

def motors_stop():
    GPIO.output(pin_IN1, False)
    GPIO.output(pin_IN2, False)
    GPIO.output(pin_IN3, False)
    GPIO.output(pin_IN4, False)

def servo_stop():
    global servo
    servo.ChangeDutyCycle(0)

def spray_on():
```

```python
        GPIO.output(relay2, True)
        GPIO.output(relay1, False)

def spray_off():
    GPIO.output(relay2, False)
    GPIO.output(relay1, True)

def main():
    global UDP_IP
    global UDP_PORT

    sock = socket.socket(socket.AF_INET, socket.SOCK_DGRAM)
    sock.bind((UDP_IP, UDP_PORT))

    GPIO.setup(pin_SERVO, GPIO.OUT)
    servo = GPIO.PWM(pin_SERVO, 50)
    servo.start(0)

    t = 0.15


    while True:
        data, addr = sock.recvfrom(1024)
        print("data: " + str(data))
        print("addr: " + str(addr))
        raw = data

        if raw == "forward":
            left_motor_forward()
            right_motor_forward()
            print("Robot Move Forward")
        elif raw == "stop":
            motors_stop()
            print("Robot Stop")
        elif raw == "backward":
            left_motor_reverse()
            right_motor_reverse()
            print("Robot Move Backward")
        elif raw == "left":
            left_motor_reverse()
            right_motor_forward()
            print("Robot Move Left")
        elif raw == "right":
            left_motor_forward()
            right_motor_reverse()
            print("Robot Move Right")
```

```
        elif raw == "action 1":
            spray_on()
            servo.ChangeDutyCycle(2.5)
            time.sleep(t)
            servo.ChangeDutyCycle(0)
            print("Robot Action 1: Servo Open")
        elif raw == "action2":
            spray_off()
            servo.ChangeDutyCycle(12.5)
            time.sleep(t)
            servo.ChangeDutyCycle(0)
            print("Robot Action 2: Servo Close")
        else:
            spray_off()
            motors_stop()
            print("STOP")
        print("")


if __name__ == "__main__":
    try:
        print("Starting controller.py\n")
        main()
    except KeyboardInterrupt:
        print("Exiting controller.py")
```

Appendix B: Relay Test Code

```
import RPi.GPIO as GPIO
import time

GPIO.setwarnings(False)
GPIO.setmode(GPIO.BOARD)

relay1 = 35
relay2 = 37

GPIO.setup(relay1, GPIO.OUT)
GPIO.setup(relay2, GPIO.OUT)

r1_state = "closed"
r2_state = "open"
```

```python
print("Relay 1 state: " + r1_state)
print("Relay 2 state: " + r2_state)

while(True):
    relay_num = input("Enter relay: ")
    relay_cmd = raw_input("Enter command: ")

    if relay_num == 1:
        if relay_cmd == "closed":
            GPIO.output(relay1, False)
            r1_state = "closed"
        elif relay_cmd == "open":
            GPIO.output(relay1, True)
            r1_state = "open"
        else:
            print("Invalid command: " + relay_cmd)
    elif relay_num == 2:
        if relay_cmd == "open":
            GPIO.output(relay2, False)
            r2_state = "open"
        elif relay_cmd == "closed":
            GPIO.output(relay2, True)
            r2_state = "closed"
        else:
            print("Invalid command: " + relay_cmd)
    else:
        print("Not a relay: " + str(relay_num))

    print("\nRelay 1 state: " + r1_state)
    print("Relay 2 state: " + r2_state)
```

Index

© The Author(s), under exclusive license to Springer Nature Switzerland AG 2022   441
A. Rayes, S. Salam, *Internet of Things from Hype to Reality*,
https://doi.org/10.1007/978-3-030-90158-5

Printed in the United States
by Baker & Taylor Publisher Services